REDISCOVERING THE FORGOTTEN VIENNA CIRCLE

BOSTON STUDIES IN THE PHILOSOPHY OF SCIENCE

VOLUME 133

REDISCOVERING THE FORGOTTEN VIENNA CIRCLE

Austrian Studies on Otto Neurath and the Vienna Circle

Edited by

THOMAS E. UEBEL
Northwestern University, Illinois, U.S.A.

SPRINGER SCIENCE+BUSINESS MEDIA, B.V.

Library of Congress Cataloging-in-Publication Data

Rediscovering the forgotten Vienna Circle : Austrian studies on Otto Neurath and the Vienna Circle / edited by Thomas E. Uebel.
p. cm. -- (Boston studies in the philosophy of science ; v. 133)
Includes translations from the German.
Includes bibliographical references and index.
ISBN 978-94-010-5404-1 ISBN 978-94-011-3182-7 (eBook)
DOI 10.1007/978-94-011-3182-7
1. Science--Philosophy--History. 2. Science--Methodology--History. 3. Vienna circle--History. 4. Neurath, Otto, 1882-1945--Knowledge--Science. I. Uebel, Thomas E. (Thomas Ernst), 1952- . II. Series.
Q174.B67 vol. 133
[Q174.8]
001'.01 s--dc20
[501] 91-13982

ISBN 978-94-010-5404-1

Printed on acid-free paper

For Robert S. Cohen

Otto Neurath on December 21, 1945.

TABLE OF CONTENTS

NEURATH AND THE VIENNA CIRCLE

APPLICATIONS

PART III. NEURATH AS POLITICIAN OF KNOWLEDGE: THE PARTISANSHIP OF ENLIGHTENMENT

ACKNOWLEDGEMENTS

I am indebted to many people for their help, both as editor and translator. Sincere thanks:

To Rudolf Haller, who actively supported the project from the beginning and made its realisation possible by securing a grant from the *Fonds zur Förderung der wissenschaftlichen Forschung*, Wien; and to the *Präsidium* of the *Fonds*; to Elisabeth Nemeth and Friedrich Stadler for their suggestions and encouragement and to Karl H. Müller and Paul Neurath for contributing unpublished work; to all the authors and the copyright holders for their permission to translate and/or reproduce the material here.

To Hugh Aitken, Marina Blümel, Mike Brown, A. X. Cambridge, Ian Dovell, Jenny Garland, Philip Schulte, Gabriel Segal, Dan Seymour and Jo Simpson for their help in London where this work was begun; to Renate and Oskar Bley and Doris Uebel for support from back home and to Phyllis Jordan from close to it.

To Arthur Fine and David Hull for the award of a postdoctoral fellowship which allowed me to complete this work in the stimulating environment of the Department of Philosophy at Northwestern University.

To Nancy Cartwright, Barbara Fultner, William Regh and Thomas A. Ryckman for advice on the translations.

To Wendy Ward for her expertise in transferring the manuscript files into camera-ready copy and care in entering proof corrections.

And finally, to Robert S. Cohen, the general editor of this series, for his encouragement, suggestions and help.

Evanston, Ill., February 1991 T. E. U.

TRANSLATOR'S NOTE

In view of the difference between the concepts *sentence*, *statement* and *proposition*, on the one hand, and, on the other, the not uncommon usage of the former two as interchangeable, the German terms *Satz* and *Aussage* have been translated throughout as *sentence* and *statement* respectively.

Where available, existing English translations of cited materials have been used (and their spelling retained); unless specifically indicated, page reference in the text and the footnotes are to the editions last mentioned under the relevant entry in the bibliography. Translations of texts unavailable in English from the German are by the present translator.

The name index lists the specific references in the texts and footnotes only.

INTRODUCTORY ESSAY

THOMAS E. UEBEL

OTTO NEURATH AND THE NEURATH RECEPTION: PUZZLE AND PROMISE

> It will be stimulating when we new critics of our language will be criticized by means of the procedures we proposed. (Neurath 1941a, p. 217)

Otto Neurath is being rediscovered. The present collection of essays on his philosophical thought, the first in the English language, makes accessible the important results—so far available only in German—of a distinctive group of scholars who have been in the forefront of this rediscovery. The essays translated in this volume treat what has been a forgotten dimension of the history of Anglo-American analytical philosophy in the Vienna Circle; as such, they provide a complement to the recently reawakened interest in the theories of Rudolf Carnap and Moritz Schlick. At the same time, the essays collected here document a recent chapter of Continental analytical philosophy—what could be called 'the Austrian school of Neurath criticism'—and so testify to a particularly interesting aspect of the 'internationalisation' of the analytical idiom. It is the merit of these essays to show that, when taken seriously, Neurath reveals himself to be a serious thinker, one whose belated recognition is bound to transform the picture of the past of analytical philosophy.

What makes the rediscovery of Neurath important is the breadth and the topicality of his thought. As a philosopher, Neurath commands the attention of historians of the analytic tradition and of epistemologists and philosophers of science; as the inventor of picture languages, he engages students of communication processes; as a utopian economist, he challenges economists and political theorists; as an activist of the late enlightenment, Neurath interests social historians of pre-World War II Central Europe; as all of these together, finally, Neurath questions the self-understanding of each of the 'separate' disciplines whose practitioners his work engages. (While the present collection focusses on Neurath's philosophical thought, it is only fitting that its connection to his other activities is noted throughout.) For present-day readers, moreover, there is also a particularly timely resonance in the rediscovery of Neurath: his criticism of competing philosophies of science in the Vienna Circle anticipated much of the recent critiques of the

T. E. Uebel (ed.), Rediscovering the Forgotten Vienna Circle, 3–22.

Circle's later orthodox version, the 'received view' of scientific theories. His attempts to replace this view (still before it was 'received') therefore cannot but strike a strangely familiar chord in contemporary readers.

The 'case' of Neurath thus presents both a puzzle and a promise. There is the question, for example, of how Neurath came to anticipate these post-positivist criticisms at such an early time, and whether he did so on conclusive grounds. Then there is the question of what Neurath wanted to put in place of the dogmas he discarded—what were his criticisms in aid of? Finally, of course, there is the question why his work was neglected for so long. Once these puzzles are answered, however, and Neurath's project is reconstituted, a stunning possibility opens up: we still stand to learn from Neurath's explorations. Therein lies its promise.

The present collection contains introductions to Neurath's philosophical thought from various angles; investigations of various stages of the development of his metascientific thought and of the philosophical background against which his contributions must be read; analyses of the application of his theories to the social sciences; and explorations of the socio-political context which Neurath saw himself responding to in his work. Several of these essays place Neurath's philosophical and educational efforts in a biographical context (chs. 2, 16, 17, 19)[1] and there is also readily available in English a fascinating biographical collage of documents and memoirs by associates and friends spanning his entire life (Neurath 1973, ch. 1). My own introductory remarks here are intended to tease out the puzzle and promise of Neurath's work somewhat further so as to place the essays that follow in the larger picture towards which they contribute. In section 1, I summarise somewhat pointedly the importance of Otto Neurath for the development of analytical philosophy; in section 2, I characterise the point of Neurath's work as a whole; in section 3, I review the reception which his philosophy has found so far; in section 4, I introduce the essays of the present volume; in the concluding section 5, I venture some remarks on outstanding problems in the rediscovery and interpretation of Neurath's philosophy.

1. The Importance of Otto Neurath to Analytical Philosophy

That there is something of value to be rediscovered in Neurath is a rather surprising finding, given his philosophical reputation. Thinkers as acute and from as diverse traditions as Lakatos and Blumenberg thought it safe to write off Neurath completely. The former called Neurath's resistance to Popper's criticisms "typically blockheaded" (Lakatos 1970, p. 156 fn.), the latter presented Neurath's *index verborum prohibitorum* as a denouement of pettymindedness that needs only to be pointed out in order to be dismissed (Blumenberg 1987, p.127). Add to this the in-house reputation of sloppiness in the formulation of theses (Carnap 1934/7, p. 321; 1963, p. 52; Feigl 1963, p. 229 fn.; Hempel 1969, pp. 166/7)—and the widespread belief that he held to an untenable coherence theory of truth—and his stature is clearly that of the original—confused—neo-positivist caveman. But the facts are otherwise. Thus let me assert somewhat polemically that analytical philosophy would not be what it is today, had there not been Otto Neurath. This is likely to strike most readers as simply preposterous. Yet such a claim is not mere hagiography. Neurath's work did influence the course of early analytical philosophy quite profoundly and continues to cast its shadow over contemporary efforts to come to terms with the legacy of logical empiricism.

Consider, to begin with, the manner in which the Vienna Circle broke onto the philosophical scene in the 1930's. Had the Vienna Circle not created such a hoopla with its seemingly outrageous claims, it is questionable whether the forum for its kind of philosophy would have been as wide as it turned out to be (and so propelled what for the longest time seemed a self-consciously formed myth into its own tradition). It was Neurath who gave the *Vienna Circle* and its successor tradition of *logical empiricism* its names and many of the sloganeering terms that became its trademarks ('unified science', 'physicalism'). Moreover, much of the rich body of work left behind by the Circle simply would not exist had it not been for the occasions which Neurath's incessant organisational activity provided for his colleagues from the late 1920's in Vienna to the late 1930's in the dispersion of exile—and for the discussions spawned thereby.

If, on these grounds, we are prepared to grant Neurath an important role, it may nevertheless appear, by this same reasoning, that Neurath's low standing as a philosopher is, after all, commensurate with the role he played. He was the propagandist and organiser—as Carnap put it to him, the "big locomotive" of the Circle[2]—but not a theoretician of importance. This impression is misleading. Neurath provided the Vienna Circle with one of its three distinctive poles of philosophical orientation, alongside Carnap and Schlick. Though his own aims remained ultimately frustrated in the Vienna Circle, Neurath still managed to set its agenda in decisive terms (Carnap 1932b, p. 74 fn.; 1934/7, pp. 320/21): the notorious protocol sentence debate was provoked in large part by his anti-phenomenalist militancy. The conceptions of physicalism and unified science which the later 'received view' of scientific theories came to adopt diverged radically from Neurath's own versions, to be sure; but that these programs were in one or another form inscribed on the banners of the Vienna Circle and logical empiricist philosophy was due to Neurath. (For better or worse, Neurath's ideas deeply influenced Carnap, the most lastingly influential of the Vienna Circle philosophers in turn, until he was 'rescued' by Tarski's semantic conception of truth.)

By crediting Neurath thus, we already acknowledge him as an important philosopher, albeit one whose own theoretical contributions were not of lasting significance. But this would be to miss his importance still. His importance is not only limited to the history of the tradition he helped found. Note that some of the very arguments which in the 1960's led to the emancipation of analytical philosophy from the 'dogmas' of logical empiricism generally and the 'received view' in particular were in fact anticipated by Neurath in his critique of the emerging Vienna Circle orthodoxy in the 1930's. This remarkable fact surely secures Neurath at least the status of 'cult-hero' for 'anti-positivists'.

Having noted his anticipation of the overcoming of neo-positivist orthodoxy, however, we are led still further, for we must ask what his early criticisms were in aid of. When we do so we find that one of the more important contemporary problems discussed in analytical philosophy turns out to have been very much his already. Neurath's anticipated the pursuit of historically informed theorising about science (associated with Kuhn); the denial of an 'iron law' of scientific method (associated with Feyerabend); the naturalisation of epistemology (associated with Quine).[3] How anyone could have anticipated all of these more or less at once is puzzling already; ultimately at issue, however, is the fact that all these anticipations leave Neurath with a problem. Neurath may rightly claim that an inquiry informed by these orientations is descriptively far more adequate to the practice of science than any rational reconstruction; but can he still muster for them the normative authority which epistemology and the theory of science need, if they are not to sink into cultural relativism—or, closer to Neurath's heart, if they may still serve the purposes of critique? Neurath's philosophy of science is of interest still today, if only because we are still confronted by this problem which his view already faced in the 1930's and 40's: can naturalism sustain normativity?[4] At a minimum, we stand to learn from his mistakes.

This is but one example of how Neurath's anticipations lead into contemporary discussions. In Otto Neurath we have a philosopher who left his mark on a tradition apparently only as founding member and chief organiser and propagandist, but who in fact set important themes for this tradition, even if his own views did not carry the day. Indeed, it seems that what were central problems for Neurath can only now find the audience which he vainly sought for in the tradition he helped found, partly because he already anticipated its later emancipation from its dogmatic beginnings.

2. *Neurath's Philosophical Project*

The contemporary problem just raised for Neurath's anticipation of 'post-dogmatic' empiricism suggests that Neurath's work is in fact located at what amounts to a point of potential contact between analytic philosophy of science and Continental critical theory.[5] If it is true to say that classical critical theorists tended to take the 'received view' of scientific theories as gospel and so were reinforced in their insistance on the legitimacy of a separate domain of philosophical discourse, while analytic theorists of science tended to dismiss as irrelevant any concern with the socio-political context in which science was practiced, then Neurath may well serve as the patron saint for all those who are concerned to overcome the limitations of either approach. Neurath's scientific metatheory was to be a critical theory of knowledge. To see this we have to place his philosophical efforts in the context of his other scientific and practical educational work

Born on December 10, 1882 in Vienna, Otto Neurath studied physics and mathematics, then history, political economy and philosophy at the universities of Vienna and Berlin from 1902 to 1906. After many years of teaching economics at a Viennese business school and several research trips to the Southeast of the Austro-Hungarian empire, he attained his *Habilitation* in Max Weber's department in Heidelberg in 1917. As a result of turning from economic theory to efforts at its practical realisation in post-war Germany, however, he was deported in 1919 and was never able to take up his academic appointment. Back in Vienna he worked in various capacities for the workers movement and in 1925 founded the Social and Economic Museum; with the artist Gerd Arntz he there developed and employed the Vienna Method of pictorial statistics which was intended to popularise sociological knowledge and thereby to educate and empower the wider population.

Unable to return from abroad to Vienna after the suppression of all social-democratic activity in February 1934, he set up shop in The Hague, continuing to organise international congresses for the Circle and pursuing his educational activities through publications and exhibitions. In 1941 he had to flee the advancing Germans to England; he continued his work in Oxford until his sudden death on December 22, 1945—reportedly preparing for his return to Vienna.

Absent from the previous paragraph is any mention of his work in philosophy—Neurath was never an academic philosopher. Yet it would be false to think of Neurath as only a part-time philosopher; rather, philosophy constituted the very centre of his diverse activities. This remains the case even though Neurath's numerous writings are by no means exclusively about philosophical or general methodological issues in the sciences; most, in fact concern topics in economics, the development of pictorial statistics and various educational and political issues.[6] Despite their great variety, all of Neurath's publications are united by the common theme of improving the social and economic conditions of broad humanity. What then, the reader will ask, makes this central concern a philosophical one? It is Neurath's view of the role which knowledge plays in the project of emancipation and the nature of his contribution to the understanding of the concept of knowledge itself.

Essential to the recognition of the depth of the philosophical commitments which Neurath brought to the Circle around Schlick is the recognition of his involvement in the first Vienna Circle of 1907-12 with Hans Hahn and Philipp Frank.[7] Neurath's familiarity with the theses of Duhem dates from this time; in his case it complemented their common upbringing in the "Machian tradition" (Neurath 1946a, p. 230) and meshed with his own expertise in history, economics, early German social science, and Schröder's logical algebra.[8] By early 1919, when he decided to "conclude [his] life of contemplation and to begin one of action" (1919a, p. 124), Neurath had essayed a program for a unified scientific metatheory (1910a), explored and dismissed utilitarian calculi as a substitute for 'decision' (1912), embraced antifoundationalist fallibilist holism (1913a, 1913b), provided directions for historical-critical conceptual history of science (1915a, 1916a), and offered his own comprehensive reconceptualisation of economics (1917a). During his detention which followed the suppression of the Bavarian soviet republic only soon after, he wrote his devastating critique of Spengler's unscientific historiography and reaffirmed his overall position (1921a). Adaptions and elaborations of his socialisation plans and practical organisational work for the workers movement left no time for metatheoretical work[9], however, until his opposition to Carnap's epistemological position of 'methodological solipsism' prompted him to develop 'physicalism' as in important respects a non-reductive naturalism and his pluralist 'encyclopedic' conception of 'unified science'.

The doctrines Neurath developed in the Circle around Schlick clearly built on the antifoundationalist holism adopted earlier, but there is more here than a mere continuation of work which, once begun, had been temporally set aside to engage in practical work. Somewhat simplified, while the early Neurath sought a general theory of science to orient his pursuit of social science (Neurath rejected the distinction of *Geistes-* and *Naturwissenschaften* already in 1910), the later Neurath found himself forced to defend the antifoundationalist-holist conclusions he had reached and, in turn, sought to show them applicable to other sciences as well. If finding a general theory of science was instrumental for the early Neurath's reconceptualisation of economics, then this reconceptualisation was instrumental for the later Neurath for developing and refining the general theory of science.

The central theme throughout was the absence of epistemic foundations and the irreducible contextuality of knowledge and justification. This continuity is illustrated by his frequent employment of the metaphor to which most of his latter day fame is owed: the seafarers who have to repair their boat at sea without ever being able to pull into dry dock (Neurath 1913a, p. 457; 1921a, p. 199; 1932d, p. 92; 1937a, p. 181; 1944, p. 47). First used in the year after the first Vienna Circle was concluded, it also expresses the positive moral he drew from his exposure to conventionalism: scientific knowledge is formed and is to be *reformed* by collective decisions.

Neurath's guiding idea was the provision of a conception of 'controllable rationality'. Neurath took the enlightenment idea of scientific knowledge as liberator from the reign of dogma and sought to import it from the domain of the natural sciences, the natural world, to that of the social sciences, the social world. What allowed science to serve as a liberator was its empirical method, its reliance on intersubjective evidence and the adjudication of theory acceptance in its light. Whenever Neurath equated the growth and spread of scientific knowledge with the growth and spread of enlightenment, he had this conception of controllable rationality in mind. Neurath's philosophy was dedicated to its development, and so continued the line of the 19th century Viennese enlightenment thinkers like Ernst Mach and Josef Popper-Lynkeus (and their friend, his father Wilhelm Neurath).[10]

Like his colleagues in the Vienna Circle (and like some of the 'school philosophers' he opposed), Neurath sought to comprehend the upheavals in the scientific understanding of the world which the preceding turn-of-the-century had brought. Neurath explored ways of overcoming the dilemma of foundationalism and relativism which only grew more intense as the 20th century grew older, but whose roots—and whose pseudo-solutions which waylaid progress all along—he already clearly discerned early on. The question for a critical theory of knowledge was how to conceive of knowledge such that it can play a role not only in reproducing the world, but in changing it. Neurath's distinctive answer consisted in exploring his guiding idea of controllable rationality which also may be expressed as *conceptual responsibility by collective management.* For Neurath, scientific knowledge of nature or society did not simply "flow from its subject matter" (Neurath 1928b, p. 296), nor was it to be understood as a clean logical "system" (Neurath 1935a, 116) freely created by solitary geniuses (Neurath 1931c, p. 369). Neurath's sketches of a non-dogmatic scientific 'encyclopedism' stressed, indeed exalted the creative aspect of science, but it was a creation that was negotiated in the collective of scientists so as to answer to criteria of acceptance both 'internal' and 'external' to science itself, criteria which in turn were not pre-given but arrived at in collective work and reflection about the historical situation.[11]

It is, I believe, this theme and strategy of problem solution which animates Neurath's work from early on and which, besides his anticipation of 'post-dogmatic' analytic empiricism and its problems, makes for its lasting interest. To recover and defend this conception constitutes the deepest challenge of the Neurath interpretation.

3. *The Reception of Neurath's Philosophy*

How was it possible that for nearly half a century Neurath has been forgotten in the philosophical tradition of Anglo-American analytical philosophy that owes so much to the Vienna Circle? Given what I said so far, it is hard to resist the platitude that he was 'ahead of his time'. Neurath's characteristically elliptical mode of expression did not help, of course, but he is hardly the only philosopher to be accused of obscurity (even among those valued in the analytic tradition). But it is possible to be more specific. The constellation of influences leading to his demise from the philosophical scene (and his rediscovery) rather reflects the content and context-relatedness of his philosophy.

Unlike his ISOTYPE work on picture languages and their application in statistics and education which was continued by Marie Neurath, Otto Neurath's philosophy certainly suffered an "eclipse" after his sudden death in 1945 (Johnston 1972, p. 195.). Several circumstances conspired to bring this about. Back home in Austria, Vienna Circle philosophy had been all but rooted out under national socialism, and those of its former members who returned to university teaching (Kraft, Juhos) had never belonged to its radical wing with which Neurath was identified, and so were unlikely to rekindle interest in it in the climate of (what Neurath would have called) reconstituted 'school philosophy'. In Germany meanwhile, the returning Frankfurt School soon began their *Positivismusstreit* with the critical rationalists of Popperian persuasion, for the most part neglecting the differences between Popper and the Vienna Circle (to say nothing of the differences in the Circle itself). In Britain, Vienna Circle philosophy was represented and gradually weakened by Ayer who early on had taken on its radical rhetoric, but not the positions of its radical wing. In America finally, where most of the Circle's former members and associates could be found after the war (e.g. Carnap, Feigl, Frank, Hempel, Reichenbach), the emigrés pursued their own versions of the Vienna Circle program—becoming in turn a philosophical orthodoxy from whose formalist trend Neurath had already felt alienated in the late 1930's. (By comparison, Neurath's own writings in English could not but appear as those of a somewhat cranky outsider.) Add to this 'internal' theoretical dynamic the 'external' facts about the American political culture in the 1940's and '50's, and it is hardly surprising that Neurath, who never cared to dissociate his philosophy from his politics, all but disappeared from view.

Yet the impetus of Neurath's philosophical work was not totally lost after 1945. Work in Neurath's spirit was carried on by his old comrade-in-arms Philipp Frank. His American publications often address the very question that lies at the centre of Neurath's thought: the importance of scientific knowledge to everyday life not only in providing technological progress, but in transforming man's self-conception as an epistemic agent (e.g. Frank 1946; cf. Kemble 1965, Geiringer-Mises 1965). And while his former Vienna Circle collegues continued to sidestep 'external' influences on scientific thought as 'pragmatical aspects' (to be dealt with 'later') in their increasingly sophisticated rational reconstructions of theories, Frank made it a point to stress the role of "extra-scientific reasons" in theory acceptance (Frank 1956b; 1957, pp. 354-360).

Otto Neurath's important role in the Vienna Circle, finally, was forcefully recalled by Rudolf Carnap in his autobiography (Carnap 1963, pp. 22-4, 51/2, 57), who so cleared what he may have felt an outstanding debt to a philosopher who then was becoming all but forgotten. But given the best intentions, Carnap's recollections could at best tell half the story of Neurath's importance. Carnap may well have saved Neurath from oblivion, but he preserved him only from the perspective of the 'received view'.

It was in the work of Robert S. Cohen that the direction of Frank's later work and Carnap's hints came together—with the help of Marie Neurath.[12] Cohen's article on Neurath for the *Encyclopedia of Philosophy* (Cohen 1967) gave much needed body to Carnap's remarks, as did the portrait-by-documents-and-recollections of Neurath in his and Marie Neurath's first collection of translations from Otto Neurath's work, *Empiricism and Sociology* (Neurath 1973), followed by *Philosophical Writings 1912-1945* (Neurath 1983)—both part of the valuable series *The Vienna Circle Collection.*[13]

A major part of Neurath's work has been available in English for some time now, but while the first of these two collections may well have been the catalyst for the rediscovery of Neurath in Europe, no such interest has so far stirred Anglo-American circles. Apart from some standard treatments of the Circle's protocol sentence debate, a handful of contributions to the Schlick-Neurath Centenary, and a series of papers by C. G. Hempel, concern with Neurath in the English-speaking world has been virtually non-existent.[14] Unlike Carnap whose reassessment by students of the *a priori* across history and of philosophers of logic is well under way,[15] and also unlike Schlick's philosophy which continues to attract interest,[16]Neurath's own views have remained shrouded in the 'received view' until very recently[17]—while the metaphor of Neurath's boat has found the widest coinage, and naturalistic epistemology and socio-historical approaches to scientific knowledge are flourishing as never before.

Fortunately, the judgement of undue neglect holds only for Anglo-American analytic philosophy. By contrast, a substantial body of Neurath scholarship has by now accumulated on the European continent, especially in Austria and Germany, but also Italy and elsewhere.[18] (I should add, however, that the historical interest of Continental analytic philosophers of science is by no means restricted to Neurath.[19])

The rediscovery of Neurath in German language analytical philosophy has been particularly lively[20]; in fact, it seems to have preceded the reawakening of interest in the Vienna Circle in general (note the recently instituted series *Wiener Kreis - Schriften zum Logischen Empirismus*[21]). It seems fair to say that the German interest in the Vienna Circle (Mohn, Hegselmann) was initially prompted by the rediscovery of the political dimension of Vienna Circle philosophy which flatly contradicted the common lore of it as a more or less quietist technocratism, however well intentioned.[22] In other words, this early work attempted to regain the historical Vienna Circle after the confusions of the *Positivismusstreit*. Later German contributions to the Neurath rediscovery, at the1982 Bielefeld Colloquium on the History of Logical Empiricism and elsewhere are already less marked by this concern and instead focus on the internal dynamic of the Circle (esp. Heidelberger); two more recent monographs and related papers (Koppelberg, Hofmann-Grüneberg) concern the place of Neurath in the continuing tradition of analytic philosophy and the rehabilitation of Neurath's 'theory of truth.'[23]

The Austrian Neurath rediscovery has exhibited from the start a different dynamic and concern. I noted at the beginning of this introduction that this volume also documents the 'internationalisation' of the analytical idiom. What gives this particular 'internationalisation' its poignancy is that for its practitioners it involved the rediscovery of an indigenous tradition whose very loss 'at home' was decisive for it becoming, to all intents and purposes, an Anglo-American tradition. Pointedly formulated as the thesis of an 'Austrian philosophy' (Haller), this concern is not a parochial one, however; it is to be understood rather as the rediscovery and continuation of a tradition which, embodied in the Vienna Circle, places Austria, on the eve of its darkest period, in the context of European enlightenment. It is, I believe, this particular location of the Austrian authors relative to their subject matter that may be held responsible for the tenacity of their efforts to recover what standard histories of logical empiricism discarded. (Thus something of the distinctive nature of Neurath's thought is reflected in its rediscovery: to wit, its concern with the wider context in which science and its metatheory are pursued.)

Rudolf Haller's opening address at the 1977 exhibition of Neurath's picture statistical work in Graz represents the fortuitous coming together of burgeoning Austrian analytical philosophy, R. S. Cohen's philosophical efforts and Marie Neurath's educational ISOTYPE work and its continuation at the University of Reading.[24] The next five years saw the publication of Haller and Heiner Rutte's edition of Neurath's *Gesammelte philosophische und methodologische Schriften* (Neurath 1981) and of numerous studies on Neurath: contributions to the Third International Wittgenstein Symposium of 1978 (Berghel, Hübner, E. Köhler 1979)[25]; a valuable, still unpublished dissertation on lesser known aspects of Neurath's early work (K. Fleck 1979); two monographs dealing with Neurath's conception of the place of Vienna Circle philosophy in the intellectual climate of its day (Nemeth 1981) and the documentation of the Viennese tradition of 'late enlightenment' from Mach to the Vienna Circle (Stadler 1982b); contributions to the 1982 Schlick-Neurath Centenary (Haller 1982e)[26], the Bielefeld Colloquium (Dahms 1985a)[27] and the 1982 exhibition of the graphical work of Neurath and his collaborators, celebrated by a catalogue with research papers providing an unparalleled prospectus of Neurath's broad field of activity (Stadler 1982a).[28] Since then, Haller and Kinross's edition of Neurath's *Bildpädagogische Schriften* has been published (Neurath 1991) and work on Neurath and related topics continues unabated.[29]

4. *About the Studies in this Volume*

The present collection of translations of important papers of the Austrian school of Neurath criticism aims to make Neurath's work accessible.[30] Such work, it must be admitted, is required. Neurath is not an easy writer and thinker to comprehend. Not only was his output prodigious, spanning a very wide, seemingly disparate domain of interests, but his style was contrapuntal in the extreme, requiring of the reader extensive and not always obvious scene-setting for his arguments. Moreover, the very context against the background of which his writings have to be read—which was almost completely lost by the Americanisation of the Vienna Circle philosophy—must be reconstructed. The papers collected here explore Neurath's philosophy with just this methodological desideratum clearly in mind.

The present volume reflects both facilitating and documentary intentions: both the course and breadth of Neurath's philosophical thought and the course and breadth of the Austrian school of Neurath criticism are reflected here. This volume can therefore be read in a number of ways—sequentially, by individual authors, or by focus on specific topics. If the contributions are read sequentially, a reader previously unfamiliar with Neurath's works will, we hope, be in a position to approach them with the sophistication their study requires. If this volume is read by authors, readers will find the fertility of

Neurath's explorations reflected in the characteristically different perspectives with which the authors assembled here treat their various topics. My remarks are limited to introducing the contributions in their sequential order.

Part I—'The Case of Neurath and the History and Historiography of Science and Philosophy'—sets out to confront the reader right away with Neurath's powerful intellectual personality, his anticipation of 'post-dogmatic' empiricism, and the manifold consequences of taking Neurath's insights seriously. This part consists of four essays. Rudolf Haller's occasional lecture 'On Otto Neurath' (ch. 2) recaptures for us the beginning of the Austrian Neurath renaissance: who was this heretofore nearly forgotten founding member of the Vienna Circle? The answer: a *renaissance man* to whom philosophy was as much integral as it was incidental. Neurath's nevertheless revolutionary naturalistic philosophical impetus is unpacked further when Haller returns to the topic in the following year with 'History and the System of Science in Otto Neurath' (ch. 3). Neurath's anticipation of the socio-historical turn of analytical philosophy of science is squarely placed before the reader. This is already a striking fact in itself, but moreover has momentous consequences. What if we were to take Neurath's theses and suggestions seriously—and consider his own historiographical efforts and apply his 'externalist' insight to the Vienna Circle itself? In 'On the Historiography of Austrian Philosophy' (ch. 4), Haller's investigation confirms the acuity of the historian Neurath in discerning a distinct Austrian philosophical tradition. In 'Aspects of the Social Background and Position of the Vienna Circle at the University of Vienna' (ch. 5), Friedrich Stadler follows Neurath's suggestion that the 'external' influences on science and philosophy must be integrated into its historiography—and so of the Vienna Circle itself. His report on the institutional practices and the political climate at the University of Vienna (in which the Vienna Circle was located as a minority voice) makes vivid not only its adversarial relation to classical 'school' philosophy, but also the urgency of its opposition. The importance of the recognition of this socio-political context for a proper understanding of Neurath's work can hardly be overestimated.

Part II—'Neurath as a Metatheoretician: Epistemology and Methodology'—presents detailed investigations of the background and various aspects and stages of his philosophical thought and its application, prior to, in and, as it were, after the Vienna Circle. To complement the Austrian re-introduction by Haller set forth previously, this part begins with an introduction to Neurath the *philosopher*. In 'The Philosopher Otto

Neurath' (ch. 6), Heiner Rutte gives a perspicuous overview of the salient aspects of Neurath's view of scientific theories and their ultimate combination in Neurath's later 'encyclopedic' conception of unified science.

As I stressed, it is very important to note that Neurath's engagement in the Vienna Circle marks but the second half of his philosophical career. In 'The First Vienna Circle' (ch. 7), Haller lays out the philosophical themes of the main authors read in the years before 1912 in the discussion group composed of Hans Hahn, Philipp Frank and Neurath. As Haller's magisterial account makes clear, this discussion group's confrontation of the Machian tradition with French conventionalism had no less effect on Neurath than the far later publication of Duhem in English on the 'anti-positivists' of the 1950's and '60's. In 'On Neurath's Writings on Logic, Ethics, and Physics' (ch. 8), Eckehard Köhler reviews Neurath's writings in these areas during and shortly after the years of this first Vienna Circle. Far from uncritical of Neurath's efforts[31], Köhler notes several revolutionary aspects of Neurath's early thought. Haller returns with 'The Neurath Principle: Its Grounds and Consequences' (ch. 9) to describe and analyse Neurath's early embrace of antifoundationalist fallibilist holism in 1913 and to trace it through later writings. In essence, this was the very position which was to prove so controversial in the Vienna Circle roughly 20 years later: here lies the connection between the Neurath of the first and the second Circles.

The first two of the papers in the section 'Neurath and the Vienna Circle' deal with general aspects of the philosophical outlook which Neurath did so much to promote (they provide an 'internalist' complement to the 'externalist' background provided in ch. 5). In 'Metaphysics in the Vienna Circle' (ch. 10), Köhler returns for a critical examination of one of the two most controversial doctrines of the philosophy of the Vienna Circle, the meaninglessness of metaphysics. He chronicles what he deems its ultimate failure, but does not fail to note the more than understandable reasons which led to its initial embrace. In 'Ethics and the Problem of Value in the Vienna Circle' (ch. 11), Rutte sets out to correct the widespread misperception about the Vienna Circle as a group of simple-minded noncognitivists and shows, by contrast, the varied ways in which value-theoretical questions were treated by various members of the Circle. Both papers set out important background for the evaluation of Neurath's philosophical project.

The following four papers deal with the divisions in the Vienna Circle. In 'Otto Neurath—Moritz Schlick: On the Philosophical and Political Antagonisms in the Vienna Circle' (ch. 12), Stadler presents the intellectual 'physiognomies' of the two great opponents within the Circle, Neurath and Schlick, and brings out clearly how their views on numerous issues were internally related, but mutually contradictory. In 'Neurath contra Schlick. On the Discussion of Truth in the Vienna Circle' (ch. 13) and 'On Neurath's Empiricism and his Critique of Empiricism' (ch. 14), Rutte investigates the debate between Neurath and Schlick from two perspectives: once as concerning the nature of the basis of the truth claims which science makes and once as concerning the grounds of the acceptance of scientific theories. One way to read these two contributions is as an investigation of the debate from the perspective of each participant, first from that of Schlick, next from that of Neurath; this does not exhaust the interest of these pieces, however, for in each the disputants also pressed to justify their approach more clearly than is immediately apparent from their original contributions. In 'Two Ways of Experiential Justification' (ch. 15), Haller turns his attention to the Schlick-Neurath debate. He concludes that ultimately, both disputants' positions are unsatisfactory.[32] These four papers are included here despite some inevitable overlap among them, for the contrasting perspectives and foci which they bring to bear on the Circle's vexing protocol sentence debate: more was involved in it than meets the eye and more searching criticism than is usual in traditional treatments of the matter is required before Neurath's views are rejected—or accepted in the form of one or another reconstruction.

The next three papers discuss the applications of Neurath's conception of scientific knowledge. In 'Otto Neurath's Contribution to the Theory of the Social Sciences' (ch. 16), Lola Fleck connects Neurath's early metascientific investigation with his own late contribution to the Encyclopedia he founded, exhibits the continuity of approach and spells out Neurath's often misunderstood intention of 'unified science' (and so continues Rutte's description of his encyclopedism in ch. 6). In 'Sociological Thought with Otto Neurath' (ch. 17), Otto Neurath's son, the sociologist Paul Neurath, provides a reconsideration of his father's critique of the sociologies prevalent during his time and his alternative approach (thus specifying Neurath's metatheoretic reaction to the currents of thought whose proponents Stadler identified as political opponents of the Circle in ch. 5). In 'Neurath's Theory of Pictorial-Statistical Representation' (ch. 18), Karl H. Müller attends to Neurath's work in visual pedagogy and investigates the relation between Neurath's metatheory of science and his efforts to develop a system of pictorial language and representation (continuing an inquiry begun by Haller in ch. 2); specifying the relation of his underlying conception to Carnap's and

Wittgenstein's, he presents some striking parallels between the views of Neurath and the later Wittgenstein.

Part III—'Neurath as Politician of Knowledge: the Partisanship of Enlightenment'—is dedicated to investigations of the context in which Neurath saw his philosophical efforts located. In 'Otto Neurath: Encyclopedist, Adult Educationalist and School Reformer' (ch. 19), Stadler provides an introduction to Neurath the *Aufklärer*, the man of enlightenment. He approaches Neurath's project from the perspective of its practical socio-political dimension and provides important details of his work in 'Red Vienna'. In 'Otto Neurath and Adult Education: Unity of Science, Materialism and Comprehensive Enlightenment' (ch. 20), Johann Dvorak specifies Neurath's approach towards people's or adult education and usefully distinguishes it both from both unpolitical 'do-goodism' and party-political indoctrination, common then as now. In 'The Unity of Planned Economy and the Unity of Science' (ch. 21) and 'Neurath's Utopias: the Will to Hope' (ch. 22), finally, Elisabeth Nemeth investigates two of the most controversial aspects of Neurath's thinking, his stress on the need for economic planning and the unabashed utopianism of his economic models. What her sensitive discussion brings out is not only the connection between Neurath's reconceptualisation of economics and his early metatheoretical conclusions and the conscious and willful character of his utopianism as the projection of an alternative aim in already darkening times; what is particularly interesting is her result that, ultimately, Neurath required the reconceptualisation of our concept of scientific knowledge itself.

It is this result that unites the contributions of part III, but it also connects with Haller's initial portrait of Neurath and so suggests a guiding thread through Neurath's work from the early embrace of his distinctive anti-foundationalist position all the way through the protocol sentence debate.

5. *Outstanding Problems of Neurath Interpretation*

The reader will discover that while there exists a certain amount of cumulative agreement about Neurath amongst the contributors, no particular party line of interpretation is pursued in my selection. What unites these papers is their appreciation and rendition of the distinctive nature of Neurath's thinking which received short shrift in the standard histories of the Vienna Circle and even still in Carnap's portrait: the theoretical-philosophical arguments which fortified what is often falsely held to be an only unreflective practical-political attitude to science, philosophy and the Vienna Circle itself. But while there are few open interpretive disputes in evidence here, the careful reader will doubtlessly find further instances where the implications of the view expressed by one contributor run against those of another. This, I would urge, is as it should be: the present volume does not only seek to assemble the necessary building blocks for a systematic interpretation of Neurath's philosophy, but also to continue to probe what Neurath himself would have called the 'blurred margins' of his own theoretical efforts and to highlight them from different angles. As I noted, the interpretation of Neurath's work is an ongoing concern. In the essays collected here, Neurath's historical importance has been rediscovered and much of his own program has been recovered, to be sure; the contributors are likely to agree, however, that questions both of the genesis of Neurath's project and of its adequacy to its goal remain open.

The papers assembled here are clear on various open problems and tensions in Neurath's work. Consider Rutte's reservations and Haller's clear doubts concerning the tenability of Neurath's epistemological views; the ultimate reemergence of metaphysics and ontology as applied logic, noted by Köhler; the theoretical lacuna left by Neurath's ambiguous abandonment of truth talk, leading some critics to attribute to him the coherence theory of truth, others to attribute exclusive concern with theory acceptance; and finally, the often perceived tension, alluded to by Stadler, between the Circle's denial of unconditional normative ethics and Neurath's view of the political partisanship of the 'scientific world conception'. Considering these difficulties, we may of course still applaud Neurath's attempts at what Stadler calls the 'democratisation' of knowledge, what Nemeth calls making science 'manageable', or what I called 'controllable rationality'; but we must wonder whether such a project can be brought to fruition at all, especially under the constraints imposed on metatheoretic reasoning by Neurath's self-consciously 'anti-philosophical' stance.

Each of the noted problems deserves urgent attention if it were our purpose to fully rehabilitate Neurath as a philosopher who 'got it right', but I doubt whether *all* of his views would withstand close scrutiny, however sympathetically they may be reconstructed in the light of more recent philosophical developments (i.e. 'deflationism'). The promise which the rediscovery of his work holds out is, I believe, not dependent on this. Let me therefore suggest in conclusion that each of the separate problems noted may be linked also to the as yet uncompleted task of giving body to his idea of 'controllable rationality' and thus making Neurath's overall project of a critical theory of knowledge philosophically respectable.

It is at this juncture, of course, that the topic which occupied the German critics early on once again assumes centre stage: the relation of Neurath's socialism to his scientific metatheory. This important question of Neurath interpretation is particularly complex, for it must be noted that it is still an open question just what kind of a Marxist Neurath was and what aspect, if any, of his Marxism is essential to his project.[33] It is also at this juncture that the question of the genesis of his program is profitably raised: What prompted Neurath to challenge the supposed context-independence of the concept of knowledge? This question asks us to reach still beyond the formative influence of the reading in the first Vienna Circle, for neither Hahn nor Frank adopted a position quite as radical as Neurath's.[34]

Fortunately, continuing progress on the question of the viability of Neurath's philosophical project is not wholly dependent on answering these specific historical questions. Progress may be achieved by further studies which—building on the results presented here and pertinent discussions in the Neurath literature at large (e.g. Nemeth 1981, Zolo 1987/9)—exhibit Neurath's guiding idea not only as a desideratum for philosophical theories imposed from without, or as a practical aim which 'rationalises' Neurath's sometimes strange theoretical proposals, but which show whether, and if so how, Neurath's theoretical contributions themselves actually served his pragmatic aim. Required, in other words, are further studies which bring the appreciation of Neurath's emancipatory orientation to bear on the epistemological views he expressed, for example, in the Vienna Circle's protocol sentence debate or in later expositions of his encyclopedic view of the unity of science. It is only in this way that it will be possible to decide which of Neurath's proposals can be dropped without jeopardising his overall project and whether Neurath's project has, after all, a chance of success. Only if his metascientific proposals successfully mediate his practical intentions can his whole project be said to succeed.

Already at the present time, however, the merit of the Austrian contributions to such an 'International Encyclopedia of Unified Neurath' is clear: enough of the initial puzzles about Neurath's work have been solved to be able to discern the promise of his project. If the present volume should help to stimulate work on its 'cash-value', it will have reached its aim.

Northwestern University

NOTES

1 Haller 1979a, Lola Fleck 1982a, Paul Neurath 1991, Stadler 1989a.

2 Carnap to Neurath, 23/8/1945, quoted in Hegselmann 1987.

3 One or another (or several) of these anticipations are noted in Haller 1979a [ch. 2],1979c [ch. 3], 1982a [ch. 9], Rutte 1982a [ch. 6], Hegselmann 1979a and Koppelberg 1987, 1990 and discussed in Uebel 1991a. Further anticipations to contemplate are Wartofsky's 'historical epistemology' (1973, 1987) and several aspects of 'social constructivism.'

4 See the debate about normative naturalistic epistemology (e.g. Putnam 1982, Kim 1988, Laudan 1987, 1990).

5 Neurath once attempted such a meeting himself with his contribution to the *Zeitschrift für Sozialforschung* (1937c), remaining surprisingly unpolemical despite Horkheimer's previous hatchet-job on the Vienna Circle (1937). For an investigation of the parallels between the research programs of Frankfurt and Vienna, see Müller 1985 and 1989, pp. 85-113.

6 Neurath's output is is voluminous by any standards and still being discovered. A first bibliography in Neurath 1973 lists 278 original articles and monographs to which a further 61 must be added from the supplementary bibliography in Neurath 1983 and 14 from the bibliography in Stadler 1982a (as numbered there: 1, 153d, 155c, 164c, 164e, 171a, 171b, 171c, 177b, 185b, 207a, 207b, 207c, 207h). Further additions are forthcoming: Koppelberg noted Neurath 1946c (an important find), Uebel 1989 suggested adding Neurath 1906b (alternative imprint of part I of his doctoral dissertation), 1930d, 1939/46, Neurath & Schumann 1919, Neurath & J. Frank 1930, and Neurath 1932e-h (four unsigned abstracts of lectures summarised in Stadler 1982b). In addition, note must be taken of the work of Neurath's Social and Economic Museum; see Stadler 1982a, pp. 246-250 for a list of its exhibitions and publications.

7 See Frank 1949a and Haller 1985a [ch. 7].

8 See Köhler 1982 [ch. 8], Lola Fleck 1982a [ch. 16], Stadler 1989a [ch. 19].

9 With the exception of his debate with Helene Bauer on the possibility of economic planning without recourse to the unit of money in 1923 (cf. Nemeth 1982a [ch. 21]), there is a distinct hiatus in his metascientific writings between 1919 (when the *Anti-Spengler* (1921a) was written) and 1928 (when he started his campaign against 'methodological solipsism' in 1928b).

10 Fellow first Vienna Circle member Philipp Frank's 1917 and 1938 and v. Mises 1938 show the shared appreciation of Mach as an enlightenment thinker; for Neurath's own admiration of Mach see his 1915b, of Popper-Lynkeus 1918; on the relation of Otto to Wilhelm Neurath see Uebel 1991b.

11 Thus Neurath's *index,* which on a first reading may appear as a terminological disciplinarianism of the crudest verificationist variety, thus turns out to be an attempt to resolve the disabling dilemma of authoritarian fundamentalism and laissez-faire relativism.

12 R. S. Cohen, long fascinated by the work of Edgar Zilsel, had explored a variety of theories about 'external' factors in science in his contribution to Frank 1956 and was pointed in Otto Neurath's direction and introduced to Marie Neurath by Carnap (M. Neurath 1973).

13 Eds. H. L. Mulder, R. S. Cohen, B. McGuiness, Reidel/Kluwer, Dordrecht, 1974ff..

14 See Scheffler 1967, Oberdan 1979, Hanfling 1981; in Haller (ed.) 1982e see Chisholm 1982, Davidson 1982, Hempel 1982, Lehrer 1982, Tennessen 1982, and particularly interesting Wartofsky 1982; also Hempel 1969, 1979, 1981. Typical for the neglect of Neurath is Suppe's by-now classic 1977 who disregarded him completely but for a

misinterpretation of an anecdote derived from Carnap (where his attempt to simply physicalise psychoanalytic discourse is taken as an attempt at its axiomatisation; p. 64 fn.).

15 For an interpretation of Carnap as an 'analytic' neo-Kantian see e.g. Coffa 1977, 1991, and Friedman 1987, 1988, 1991; for (implicit) dissent from a conventionalist perspective see e.g. Creath 1991 and Ryckman 1991a, from that of philosophical minimalism Dreben 1991.

16 E.g. Leinfellner 1979, 1982, Friedman 1983, McGuinness (ed.) 1985, J. Lewis, Ryckman 1991b.

17 Compare Watkins 1974, Esser 1984 (reviews of the English Neurath collections) and Freudenthal 1989 with Cartwright 1990, Cat, Chung & Cartwright 1991, Galison 1990, Jacob & Otto 1990, Uebel 1989, 1991a, 1991b, 1991c. (From Quebec see Rosier 1986.)

18 From Italy see Barone 1952, 1982, 1984, Statera 1967, Fistetti 1985, Zolo 1986 (English ed. 1989), 1990, D'Acconti 1986, from France Clavelin 1980, P. Jacob 1984, from Sweden Bergström 1982, and from Switzerland Lauener 1982, 1983, Lüscher 1979.

19 See e.g. numerous contributions in Haller (ed.) 1982e, Gadol (ed.) 1982, Sebestik & Soulez (ed.) 1984, Dahms (ed.) 1985a, Nyiri (ed.) 1986, and the monographs and articles by Rungaldier 1984, Sauer 1985, Haller 1985b, Proust 1986 (Engl. ed. 1989) and 1987.

20 For other overviews see Hofmann-Grüneberg 1988, pp. 12-16 and Nemeth 1990b; I follow Nemeth in discerning a difference in orientation between the German and Austrian critics.

21 Eds. H. L. Mulder, R. Hegselmann, A. J. Kox, F. Stadler, Suhrkamp, Frankfurt, 1986ff..

22 Mohn 1978, 1985, Hegselmann 1979a, 1979b, 1979c, Beckermann 1979, W. R. Köhler 1979a, 1979b. Note also the collection Schleichert (ed.) 1975.

23 In Dahms 1985a see Dahms 1985b, 1985c, Hegselmann 1985 and Heidelberger 1985; Hegselmann 1987; Koppelberg 1987, 1990, Hofmann-Grüneberg 1986, 1988.

24 Haller 1979a [ch. 2]. For a bibliography of Marie Neurath's work see Stadler 1982a, p. 237; for the view from Reading, see Kinross 1979, 1982, 1984, Twyman 1982.

25 Haller 1979c [ch. 3], Rutte 1979 [ch. 13], Rungaldier 1979.

26 Haller 1982c [ch. 15], P. Neurath 1982, Fischer 1982, Rutte 1982b [ch. 14], Wallner 1982, Nemeth 1982a [ch. 21], Stadler 1982c [ch. 12], Dvorak 1982b.

27 Stadler 1985, Dvorak 1985, E. Köhler 1985 [ch. 10], Müller 1985.

28 Rutte 1982a [ch. 6], E. Köhler 1982 [ch. 8], Haller 1982a [ch. 9], Fleck 1982 [ch. 16], Dvorak 1982a [ch. 20], Nemeth 1982b [ch. 22], Stadler 1982d, Fischer 1982b, Müller 1982a, 1982b, Glaser 1982, Weber 1982, Pfoser 1982, Stadler-Türk 1982.

29 Haller 1985a [ch. 7], 1986 [ch. 3], Rutte 1986 [ch. 11], Köhler 1983, Müller 1989, 1991a, 1991c [ch. 18], Nemeth 1990a, 1990b, P. Neurath 1984, 1991 [ch. 17], Stadler 1984, 1986, 1989a [ch. 19] (and passing comments in Janik 1989).

30 Of this 'school' only Haller 1982b has been so far available in English—apart, that is, from Haller's writings on Wittgenstein (1988a).

31 It may be added here that, in his *Survey of Symbolic Logic* of 1918, C. I. Lewis ranked Neurath & Hahn 1909 among those "that are considered the most important contributions to symbolic logic" (as noted in Menger 1980, fn. 2).

32 For Haller's own preferred solution to the problem of epistemology see his 1988b.

33 Note that his reported 'liberalisation' in exile (cf. Neider in Haller, Rutte 1977) did not dampen his enthusiasm for social engineering and collective planning (central to his thought from early on according to Nemeth 1982a, 1982b [chs. 21, 22]); see e.g. his 'International Planning for Freedom' (1942b).

34 Neurath's later embrace of Marxism cannot be held responsible for this, since his still later 'liberalism' did not bring with it a more conciliatory line vis-a-vis the new 'received view' orthodoxy.

PART I

THE CASE OF NEURATH AND THE HISTORY AND HISTORIOGRAPHY OF SCIENCE AND PHILOSOPHY

RUDOLF HALLER

ON OTTO NEURATH*

It seems that the time has come for Austria to remember its more recent past, to discover *that* which may with pride be called 'the past'. People cannot afford to let their riches rot: whoever investigates the cultural and intellectual history of the first two decades of this century will discover that these riches are overwhelming, that particularly the Vienna of that period presents a density of creative forces such as we know only from the Florence of the Renaissance and the Weimar of the turn of the 19th century. I believe that it is not much use if only a few experts recognise this and if that recognition remains limited to esoteric circles.

Well, at this juncture at the latest, I should tell myself: he who sits in a glass-house should not be throwing stones! For when, after World War II, we turned to the philosophy of the more recent past, noted the Vienna Circle and engaged ourselves in its continuation in Austria, then it was first of all the *philosophical* problems of neo-positivism which attracted and fascinated us. The call for a scientific world-conception, the prompting to analyse scientific sentences and delimit them from other, non-meaningful sentences, the definition of philosophy as criticism of language—all these were themes we *re*discovered, and we had to spend much time determining whether and in what form these problems had been further treated and developed in the English speaking world since the glory days of the Vienna Circle. It was not until much later that the historical perspective began to open up, or an appreciation for what one may call Austrian philosophy. And only in the last instance it has become clear how much work will have to be done if we want to present a comprehensive picture of Austrian philosophy.

What kind of role would Neurath assume in such a picture? What secures his place within the history of Austrian philosophy? And what stature does he possess outside of it?

The pursuit of these questions can only make us aware of *one* side of a universal mind, but it cannot do justice to the manysidedness of this man. The present exhibition brings to life an altogether different side; and if we delve deeper into the Neurath phenomenon, we will be surprised again and again by the universality of this mind that contradicts the stereotype of the philosopher in his ivory tower with his own life and actions.

T. E. Uebel (ed.), Rediscovering the Forgotten Vienna Circle, 25–31.

Allow me at least to single out a few things—without claiming to name even only what is of greatest importance.[1] Otto Neurath was born December 10, 1882 in Vienna. He studied there and in Berlin, mainly economics, philosophy and history, and gained his Ph.D. under Eduard Meyer and Gustav Schmoller. In the house of his father, Dr. Wilhelm Neurath, an economist, he had begun to school himself at an early age and to develop his perception and imagination. After his studies he fulfilled his military service and became a teacher at the New Viennese Commercial Academy (*Neue Wiener Handelsakademie*), a position held until 1914. In 1912—after his first wife Anna, born Schapire, had died—he married Olga Hahn, the sister of Hans Hahn, the later Viennese professor of mathematics and co-founder of the Vienna Circle.

Already in the years 1911 to 1913, supported by the Carnegie Foundation for International Peace, he undertook research trips to Galicia, Bohemia and the Balkan states in order to conduct sociological studies about changes in social and economic conditions due to the impact of war.

During World War I, he serves first on the Eastern front, later, stationed back in Vienna, finds the time to prepare for an academic career and in 1917 habilitates in political economy at the University of Heidelberg. Still prior to the end of the war he is named director of the war museum in Leipzig. In the confusion of the year 1919 we find him in Munich as the head of the central economic council, first under the Hoffmann government and then under the Bavarian soviet government (*Räterepublik*), always concerned to realise his socialisation ideas in practice. Like many later members of the Vienna Circle, Neurath was a socialist; the reform of both the outer conditions of life and of education was from early on a practical aim of his economic and social theories. Neurath understood his work in the Bavarian soviet republic not merely as the duty of a civil servant, but also as the possibility to effect the immediate realisation of his ideas. As Ernst Niekisch rightly says, there was little reason to identify him with the soviet government itself. Despite this he was arrested, only set free due to the intervention of Otto Bauer, and deported from Bavaria to Vienna—something which cost him his German university position.

From 1919 to 1924 we find him again in Vienna, not as a teacher of economics, but in the settler movement which was influenced by Adolf Loos, where he finally becomes general secretary of the Austrian Housing and Allotments Association (*Verband für Siedlungs- und Kleingartenwesen*).

In 1925 he founds the Social and Economic Museum of the City of Vienna (*Gesellschafts- und Wirtschaftsmuseum der Stadt Wien*) whose director he remained until 1934. In this museum Neurath developed the so-called Viennese Picture Statistics, collaborating with a number of others, including his later, third wife Marie Reidemeister. As the present exhibition makes vivid and clear, the idea of picture statistics involves a new 'language', more precisely, an auxiliary language which, as Neurath put it, was *not* to take *the place of words* or even number words, but instead contained immediate representational and informational content as a *pictorial sign*.

In its most general form the main demands of such a language can be traced back into the 17th century, when in particular Leibniz, but also many others, supported and developed the program of a universal language of human reason. Signs were divided into those which possessed designative character without the *vehiculum* of the word, and those of which this is not true. About the former, Francis Bacon had already remarked that they exhibited a certain similarity with the things they designated. Signs of most different types thus were interpreted as a *principium cognoscendi*, on the basis of which a thinking being deduces the thus designated thing or fact.

It is true that the reasons which were first put forward for the program of a *lingua universalis* reemerged with the philosophers of the Vienna Circle for *their* program of a universal language, in so far as Carnap, Neurath and the others followed Wittgenstein's propaganda for the symbolic logic of Frege and Russell. But picture statistics shares, as far as I can see, only a few central demands of the program for a universal language, amongst them in particular that for common intelligibility and for univocal designation. Dominant, however, was its universal character in another respect: "Pictures unite, words divide." And Neurath wanted to unite. He wanted to give the same information to children and to workers as he gave to his colleagues. And to his colleagues too, he wanted to give new information, namely, to be able to *see* what they knew. Particularly when it comes to the *comparison* of data, to the recognition of *relations*, the relation of similarity is not only an auxiliary tool, but the very ground on which the information is built.

Of course, the human need for vivid representation had always returned to pictorial symbols. But there was no codification, no unity, to say nothing of a syntactic system of pictorial symbols.

The aim of Neurath's picture statistics was to alleviate this lack by searching for firm rules, those which were the most suitable and rational for the main purpose of the communication of information, the understanding of the cognitive content. For that, one needs first of all an alphabet which secures the *unique* assignment of sign and designated object. Such statistical hieroglyphics, which should not be of a perspectival character, were the building blocks whose combination had also to be ordered by firm rules. But the construction of this pictorial language was also to be one that was appropriate to the psychological laws and regularities under which their users operated. Thus it became important to design a system of rules which brought the following into harmony: density of information with understanding, readability with surveyability, content with form. And of course Neurath was aware of the experimental character of such an enterprise: only by ever new experiments could one hope to find the the most suitable hieroglyphics.

The appropriation of the term 'hieroglyphics' makes clear that Neurath was aware of the different types of visual languages. He soon discovered the following methodological principle for the construction of such a language: what is common to a type or species—that is, a property of all members of a class—need not and may not coincide with what is peculiar or characteristic of that type or species. The combination of these two types of properties, however, is essential to hieroglyphics. Merely stressing one property cannot suffice therefore to picture that which at the same time typifies.

As the natural languages do, the artificial languages present different co-existing systems with different traditions. It was Neurath's aim to reduce these different systems to one, to bring nature and intervention into one unity. This is basically an aim analogous to that of unified science, which wanted to reduce all different scientific (language-) systems to one, as expressed in the program of physicalism.

Neurath, this practical thinker *par excellence*, was interested in the pragmatics of this *instrumentarium*, quite apart from the syntax and the semantics of such a pictorial language. Thoroughly in the spirit of the tradition of the Viennese workers education he wished to bring 'statistics and proletariat' into one context, in order to serve one of the main aims of the Vienna Circle: the "affirmation of the scientific attitude to the world as a social solution" (Kolakowski) and therewith the elimination of ideological, irrational and fanatical attitudes.

In the combative phases of the Vienna Circle everything which resisted integration into the scientific attitude, or even contradicted it, was designated by the slogan and invective 'metaphysics'. It had to go. In this the inventor of Viennese picture statistics agreed with the inventor of truth tables: metaphysical sentences must be made to disappear. But Neurath also occupied a realistic point of view by pointing out how difficult it would be to complete this anti-metaphysical work, and by realising that the unified language could not also be an ideal language.

It was also Neurath, who developed in all clarity the program of a hypothetical naturalism and insisted that within unified science there existed no statements and laws which could not be changed. As a consequence, even basic or protocol sentences were viewed as revisable. With this, one had moved another step closer to the actual practice of natural science and had departed from the metaphysical "absolutism of primitive sentences" (Carnap). Of course, the price of this departure was not small: coherence steps into the place of the adequacy of a system whose endless correction finds its expression in Neurath's picture, made famous by Quine, of the sailors "who must rebuild their ship on the open sea, never able to dismantle it in dry-dock and to reconstruct it there out of the best materials." (Neurath 1932d, p. 92; cf. Haller 1974.)

Accordingly, the truth or falsity of an empirical sentence is not determined by its relation to some reality, but in relation to the entire edifice of science: if a sentence does not agree with this, it counts as false, otherwise as true. But Neurath also noted the operative basic principle of this relation: there are always two possibilities to establish such agreement, namely, either by changing the relevant sentence *or* by changing the system, that is, the edifice of science itself. And even though the concept of truth suffers a devaluation here, the concept of a system gains in precision. It is in this that, in my opinion, the merit consists of a holistic conventionalism as it was developed by Neurath and Quine, according to whom a confrontation of knowledge and reality does not make sense. In their conception the adaptation of a system to sentences is made more plausible than in other metatheories.

We know from Carnap that it was Neurath who propagated the basic idea of the unity of science in the Vienna Circle and with this aimed for the broadening of the Descartes-Leibniz program of a *scientia universalis* (Carnap 1963a, pp. 22f.). For according to the principle of unified science all knowledge, and so all science, is of a unitary nature and it is only practical purposes and needs that make a separate pursuit of scientific disciplines advisable. And it is interesting to see how and with what steps this program, which of course also was Carnap's, was developed. The institutional pendant of these theoretical considerations of reduction was the series of 'Unified Science' and the grandly planned encyclopedia of scientific empiricism, which was founded by Neurath as the *International Encyclopedia of Unified*

Science and later was continued by the Institute for the Unity of Science, led by Philipp Frank.

It was also Neurath who in the discussions of the Vienna Circle pointed again and again to the holes in unified science, through which the old metaphysics could and did threaten to invade again. Also for him there existed no 'world-puzzles', but to explain the reasons for this fact in the fashion of Wittgenstein was for him as much unacceptable metaphysics as Schlick's search for the foundations of knowledge. In this we can see clearly the enlightenment attitude, which pervades the entire life work of Neurath, not only his philosophical efforts. Forever he inveighs against mystification—and metaphysics was its paradigm for him—forever he calls for humanisation and socialisation—and here too metaphysics appears as the real hurdle. Neurath did not see his ultimate goal in the fight against metaphysics as the administrator of all obstacles to the liberation of humanity, but he found in it the mightiest tool for attaining enlightenment.

In Neurath's view, the institutionalisation of ideas belonged to pragmatics, to practice. And no one was better able to invent such institutions and develop them than Neurath. From the founding of museums to that of a series of books, from the planning of settlements to that of hieroglyphics, from the founding of associations like the *Verein Ernst Mach* to the unique bringing together of philosophers in the Vienna Circle, from the education of children to that of adults: Neurath was always able to *do* what was necessary, not only to think it. And Neurath was also convinced that, if one wanted to take seriously the ideas of an enlightened social humanism, what matters first and last are actions.

With the typical overestimation of one's own possibilities and the mistaken estimation of the opposition, Neurath indulged in the illusion that one was 'on the way'; the final sentences of the programmatic pamphlet of the Vienna Circle drafted by him read:

> We witness the spirit of the scientific world-conception penetrating in growing measure the forms of personal and public life, in education, upbringing, architecture, and the shaping of economic and social life according to rational principles. The scientific world-conception serves life, and life receives it. (Carnap, Hahn, Neurath 1929, pp. 317/8.)

Neurath's optimism overlooked that the pathway to the scientific world-conception, ever and again threatened by obscurantism, is full of stones, wearisome and infinitely long.

Karl Franzens Universität, Graz

NOTES

* Lecture given at the opening of the exhibition 'Otto Neurath's Bildstatistik' at the University of Graz, 22 June 1977. First published as 'Über Otto Neurath', in R. Haller, *Studien zur österreichischen Philosophie*, © 1979, Rodopi, Amsterdam, pp. 99-106. In the present version the first two paragraphs have been dropped; translated with kind permission of Editions Rodopi and the author by T. E. Uebel.

1 Cf. the excellent collection Neurath 1973 which contains a selection of biographical and autobiographical sketches and a bibliography of Neurath's writings.

RUDOLF HALLER

HISTORY AND THE SYSTEM OF SCIENCE IN OTTO NEURATH*

Neurath must be rediscovered! In what follows I shall show how certain ideas, which stand in the center of the methodological debates in the theory of science today, were first formulated by Neurath, and in what measure his conception of unified science and his critique of atomism and phenomenalism influenced and promoted the later development of scientific metatheories.

Of course, Neurath rarely presented his ideas with the scrupulous precision which characterises analytical essays. But he possessed what matters in the end, the insight and the overview, as well as the energy and the perseverance, which were needed to promote the realisation of a research program like that of the *Encyclopedia* or to give a new direction to a philosophical movement, like the 'Vienna Circle' (christened so by him).

The two problems which are briefly investigated here concern, on the one hand, the idea of a scientific system and the question of its truth, and, on the other hand, the question of the historical location of such a system and the ways in which it changes over time.

As for the effect of Neurath's ideas, we may rest assured that the friends and foes of the Vienna Circle took note of the few volumes of its journal *Erkenntnis*. This alone explains much of their enduring influence. Then there is Quine's repeated Neurath-motto that had to point everybody in the right direction (cf., e.g., Scheffler 1967). Of the effects Neurath had on the Vienna Circle and in our day I shall not speak here, however (cf. Haller 1979a [ch. 2]).

> For the most part it is not a good sign if scholars are too eagerly occupied with the foundations and the history of their discipline instead of producing new statements concerning the subjects treated by them.

So begins Neurath's 'Ways of the Scientific World-Conception' which was published in the first volume of *Erkenntnis* 1930/31. And Neurath thinks that it is the representatives of the exact natural sciences rather than those of other disciplines who can afford a jaunt into the past without incurring the suspicion that such 'historicising reflections' would slow the progress of their research.

T. E. Uebel (ed.), Rediscovering the Forgotten Vienna Circle, 33–40.

Neurath says of the representatives of the scientific world-conception that they behave like physicists:

They are active and close to the present time even if they move in abstract spheres. They care less for the history of their trains of thought than for new insights ...

He continues with feigned understatement:

But occasionally it is advisable to seek the historical conditions of a view and look around and see how it fits in sociologically, what its connections are with other spheres of life and science, in the interest of the aim of a unified science. (Neurath 1930a, p. 32.)

Like Gilbert Ryle later on (with whom he shares the interpretation of disposition terms) Neurath does not see a *primary* task—"not a task for a workday"—in metatheory and its history, but neither is it something we do not have to worry about at all. On Sundays (or Mondays) a representative of the scientific world-conception may well concern himself with such questions.

The common picture of neo-positivism in its concern for the dynamic of science and scientific theories constitutes an exemplification of Comte's theory of stages. Auguste Comte's famous law of three stages says that both individuals and the human species go through three phases in the development of theoretical thinking: a fictitious or theological stage, an abstract or metaphysical stage, and finally a realistic or positive stage. The realistic or positive stage is the era of modern science which is characterised by an *explanatory* and a particularly *predictive* methodology and builds on universal empirical laws. This phase, beginning with Galileo, becomes the *norm* and *aim* of the development of all those disciplines which have not yet reached the positive stage (cf. Haller 1979e).

If the history of science is considered against the background of this teleological developmental schema commonly ascribed to positivism, then it follows that all *pre*-scientific phases merit no systematic interest. It is *this* picture that has commonly been framed of the canonical conception of science and of the leading conception of logical empiricism.

If, however, my opinion is correct that Neurath was one of the motors of the Vienna Circle—insofar as his opposition to the dominant ideas of Wittgenstein and Schlick led him to fight against phenomenalism, atomism and anti-naturalism, and to inaugurate their overcoming—then there is no reason to be surprised that his picture of the history of science also radically diverges from the cliché of the cumulative increase of knowledge. He did not miss the fact that one cannot force historical changes into the stages schema of the beginning and end of fictitious-theological and abstract-metaphysical stages in the development of knowledge.

Neurath shares with Paul Feyerabend the view that basic elements of the scientific world-conception are already contained in magical practices, and that these could be revived under certain external conditions; but Neurath already published these views nearly 50 years ago [1978].

What distinguishes the magical phase from the theological-metaphysical phase is that the explanatory principles of the former refer to testable empirical events whereas no such control is possible for religious explanations. Rather, what divides and distinguishes science from magic is primarily "the missing systematic connections and systematic control by experience." (Neurath 1930a, p. 37.) Despite this, sorcery and magic as revived in the witch-trials are *more closely* related to science than theology is.

Hearing this, one is tempted to draw the false conclusion (or ascribe it to Neurath) that scientific research would develop particularly intensively during anti-theological phases. By no means, answers Neurath. And he goes on to show this nicely—if only, as usual, by allusions—in the case of the Epicureans' philosophy of enlightenment. They entered the fray as enemies of theology and metaphysics, but had an anti-scientific attitude, whereas the *real* progress in mathematics was effected by the theologising Platonists. With Neurath we must change the primitive picture of a "naive empirical conception of history"—in short, the picture of a linear, cumulative development of science and theory for a new conception which points us towards the much more complicated connections between progress and regress in scientific change (cf. Haller 1977).

Such an approach also allows us to avoid a mistake often made in the camp of naturalistic epistemology, namely, to see in physics the only scientific discipline that is to be taken seriously and to disregard all others, especially the sociological disciplines. As much as such a one-sided approach is understandable if it is rooted in an empirical outlook, it does, nevertheless, encourage narrowing and restricting the domain of science. By contrast, Neurath urges "a more historical approach" even within the framework of physicalism, an approach which makes clear that the modern mechanisation of the picture of the world constitutes also a certain historical fact for which we cannot claim the constancy of physicalist relations.

This aspect of Neurath's conception of science too had to rest unnoticed for a few decades before it re-emerged in the writings of the anti-positivists, only this time without mention of the original author, and perhaps also without direct reference to him.

Similarly with the other principle of change in scientific theories, the principle of the totality, and the priority of the system over individual sentences. There is no doubt that it is Neurath's merit to have been the first to represent the kind of 'holism' in philosophical foundational inquiries which today we associate mainly with the name of Quine, just as it is his to have been the first to attack the exceptional position of protocol sentences in the epistemology of science, and also the 'ideal language' built up from 'clean atomic sentences'. Finally, it also seems to be the case that the idea of physicalism—i.e. the possibility of reducing all empirical-scientific statements to those of physics, or of a physicalist language—goes back to him, even though both Wittgenstein and Carnap claim 'priority' for this conception.

Of all this I can speak here only as far as it concerns the relation between system and the change of a system. For again there is no doubt that it was Neurath who first pointed to the problem of theory change in the Vienna Circle—and of course before Popper.

What then are, in Neurath's perspective, the relations between scientific systems (theories) and history?

First of all, it must be recognised that the principle of scientific investigation has always been a principle of unification or, if you like, simplification, namely, to attempt to reduce the infinite number of possible judgements about the world to ever fewer principles and axioms. The principle of selection between possible systems is that of Occam which recommends economy in the postulation and acceptance of objects and entities, and selects that decision which assumes or presupposes the least number of entities while retaining the same explanatory power and presumably confirmation as its rivals. Concerning knowledge transformed into a system, this holds without exception: every sentence which is integrated into the system is either consistent with the sentences of the system or not. It is inconsistent if it, or the class of its consequences, contradicts a sentence of the system.

Of course, in this form the demand is a very strong one, for it presupposes that all sentences are of the same semantic structure, and it also seems to presuppose that the mutual compatibility of the sentences, i.e. the consistency of all sentences in the system, is provable.

Such a demand is not even satisfied by all the sentences of formal systems, as Gödel has proven. This means only that it is senseless to envisage mechanical decision procedures for empirical sentences, especially since the 'proof procedures' are totally different from those of mathematics. For, as Hempel put it already in 1935, there must be characteristics by which we can distinguish the true sentences of science from the "false sentences of a fairy-tale" (Hempel 1935a; cf. Neurath 1932d and Rutte 1979 [ch. 14]). If there is "no formal or logical difference" between the protocols of a story consistent in itself and a scientific system—as also stressed there by Hempel, summarising Neurath and Carnap—if, therefore, experience and experience alone can be employed as a criterion of decision, then the substantial question remains: what determines the experience involved in a system distinguished as true? The answer which Neurath and Carnap gave was that it was the "*historical fact*" that the majority of scientists indeed accepts a certain class of protocol statements.

This sociologising interpretation anticipates the contentious core thesis of Kuhn's explanation of theory change: "There are moments in the history of science", so Neurath once wrote, "when it matters to know that it is always the totality of hypotheses which can be doubted." (Neurath 1936e, p. 776.)[1] This expresses in all purity the position of holism: when we ask for the validity of a world picture or of a disciplinary matrix, it is not the truth of an individual, isolated hypothesis that is questioned. It seems that Carnap's and Neurath's brand of neo-positivism also followed Mach who regarded the individual observation, the protocol, neither as knowledge or error, i.e. interpreted it neither as true or false, and for this reason had to offer the acceptance or rejection of the whole system as the final decision.

If we use Stegmüller's characterisation of the core sentences of (moderate) holism (Stegmüller 1973, pp. 217f.), we can see at once that Neurath not only provided the central metaphor for Quine's version of holism, but that he also proposed, as it were, the same 'core sentences': first, that a scientific system (a theory) gets accepted or rejected as a whole, second, that there can be no rejection of a system from a failed *experimentum crucis*, and third, that a sharp separation of empirical data (protocols) and system is impossible.

Now, this structure of so-called holism provides the metatheoretical framework for the analysis of the system and for the history of the system, because the question of the acceptance or rejection of a system presupposes alternative systems. And these systems are naturally conceived of as systems of statements, because otherwise the question of truth or falsity could not even arise. Alternative systems can be considered in succession or in coexistence. The former perspective is that of the historian, the latter that of the systematic thinker.

Neurath considers the problem under both perspectives and his most important insight is the following. If there obtains an incompatibility between a sentence and the scientific system, that is, ultimately, the entire edifice of science—whether this happens in the course of the change of one and the same theory or in the comparison of different theories about the same subject matter—then there are always two possibilities for restoring harmony, the agreement of sentence and system: either one changes the relevant sentence *or* one changes the system (cf. Haller 1979a [ch. 2]). I propose to call this principle of the adaptation of scientific systems the '*Neurath principle*'. It explains in the most simple fashion which possibilities are open for the change of a system; at the same time, it makes clear that the decision concerning how to uphold the consistency of the system depends on the circumstances of the case at hand and is pragmatically determined. This principle then leaves us in no doubt that its application lies in the hands of the members of the republic of scholars and that its application is therefore determined by circumstances which determine and cause the decisions of individuals and collectives. Later this principle was elaborated in its structural respects by Quine and in its sociological respects by Kuhn, but of Neurath nothing more was remembered than the motto of the sailors who had to rebuild their ship on the open sea.

The sociologist Neurath was fully aware that scientific research is the process of production by a collective, a collective which is determined by the socio-economic situation and by history, in particular its own history. "Social changes put their stamp on intellectual changes"—the echo in Neurath of the materialistic conception of history (Neurath 1930a, p. 47). It is the repeated re-emergence in this historical development of the old ideas and of concepts of metaphysical origin in new clothing, which makes imperative the repeated attempts to combat metaphysics. One such metaphysical rest seems to be the reference to a self which is the subject of the observational '*constatations*' or protocols, whereas for Neurath, as we saw, there are no empirical designators but those strictly fixed by time-space coordinates. The collective has the critical role of a 'cleansing machine' (Neurath's term) which checks the compatibility of scientific statements with the system and which keeps it free of bugs (i.e. contradictions). In other words, Neurath views the collective of scientists as just that institution which overcomes the subjective caprice of methodological solipsism and is intended to secure the objective status of statements by their intersubjectivity.

Why such a collective, i.e. the republic of scholars, decides in favour of one *particular* conception amongst possible ones, why it choses *this* and not another version of the agreement between sentence and system—that, so Neurath claims, cannot be justified by logical means. And this leads us to the very point at which Kuhn's theory seems like a mere continuation of Neurath's thoughts, just as earlier on Feyerabend's interpretation of magical periods came into view as a direct descendant of Neurath's ideas. For both Kuhn and Feyerabend, the choice of a new theory which is taken after a crisis-like period of insecurity consists in an act of decision for which we may provide a rational *reconstruction*, but not a rational determination. To say how a world which did not become a fact would have developed is possible only from certain limited perspectives and can never encompass all consequences. What seems to count and what history is based on are the *actual events*, 'the necessity of the factual', as it were, in contrast to the infinite realm of freedom of that which does not exist. For this reason it may strike us as grotesque to attempt to simulate the *success*, or even only the real possibility, of a method or theory which is considered wrong or obsolete from contemporary standpoints. But this does not mean that the factuality of the empirical brings with it the necessity of its truth. In this respect Neurath represents a considered semi-scepticism and holds that a final decision in the comparison of theories is not possible:

> Our thinking is a tool, it depends on social and historical conditions. One should never forget this. We cannot act as prosecutor and defendant at the same time and in addition sit on the judge's bench. We confront our present thinking with earlier thinking, but we have no possibility of taking a judge's stand on a point outside. Checking statements with the events is itself part of the characteristic method itself. (Ibid., p. 46.)

Everybody who reads this passage with understanding will realise that Neurath here expressed core sentences of the historical theory of science of our day. It would lead much too far to investigate these theories themselves for their deeper meaning and truth content. It seems to me to be sufficient, and only just, to have laid bare one of its intentionally overlooked sources.

Karl Franzens Universität, Graz

NOTES

* First published as 'Geschichte und wissenschaftliches System bei Otto Neurath', in *Wittgenstein, der Wiener Kreis und der kritische Rationalismus. Akten des 3. Internationalen Wittgenstein-Symposiums, Kirchberg am Wechsel, Österreich 1978*, Hrsg. H. Berghel, A. Hübner, E. Köhler, *Schriftenreihe der Wittgenstein-Gesellschaft Bd. 3*, © 1979, Hölder-Pichler-Tempsky, Wien, pp. 302-307. Translated with kind permission of Verlag Hölder-Pichler-Tempsky and the author by T. E. Uebel.

1 This sentence is preceded by: "If physicists discuss a problem amongst themselves, they often agree about basic questions to such an extent that they do not always note that in the end it is the totality of their hypotheses that is at issue and not the small sector which they currently emphasise."

RUDOLF HALLER

ON THE HISTORIOGRAPHY OF AUSTRIAN PHILOSOPHY*

I

When I reopened the question of an Austrian philosophy in the mid-1960's, I could refer to the manifesto of the Vienna Circle from 1929 (Haller 1966; cf. 1968 and 1975). It was presumably Neurath who composed its first version, and while it was revised by Carnap with help from Feigl, Neurath and Waismann, in the end it did receive the name suggested by him: *The Scientific World-Conception. The Vienna Circle.* Without doubt then it is also the merit of Neurath, inventor of the 'Vienna Circle', of physicalism and the *Encyclopedia of Unified Science*, to have sketched the basic line of the development of Austrian philosophy. Indeed, it speaks for the design he worked out between 1929 and 1935 that my own attempts, and those which went in parallel or succeeded them, followed in the steps of Neurath's view of the pre-history of the Vienna Circle.

Before I come to the main theme of my contribution—the historiography of Austrian philosophy—I want to present very briefly the basic structure of this conception. Its professed aim is to render intelligible the emergence of the Vienna Circle—not its organisation, its outward image, but its grounding in the history of Austria-Hungary—and so to throw light on the *history* of philosophy in Austria as a *pre-history* of *logical empiricism.*

As his interpretative key, Neurath uses the difference between the character and the development of philosophy in Germany and Austria: from the fact of the common language of the two countries and their close historical and cultural connectedness, one cannot conclude that their philosophies developed in parallel as well. Rather, the development of philosophy in Austria represents a chapter in the intellectual history of *Europe*, predominantly of Central Europe. For Neurath, events in Austria generally ran in parallel "more to the events in Warsaw, Cambridge and Paris than to the events in Berlin. And in Vienna and the centres in contact with Vienna, we find philosophical trends which remained unsuccessful in Germany and were abandoned." (Neurath 1936a, p. 676.)

The main characteristic of the philosophy which Neurath's propagandistic writings set out to advertise is the combination of logic and

T. E. Uebel (ed.), Rediscovering the Forgotten Vienna Circle, 41–50.

empiricism. Together with an anti-metaphysical sensibility, this supposedly has led to the unification of previously separate approaches to philosophy. Thus Neurath's repeated advocacy of a label he took from *Itelson* and with which he sought to oppose metaphysical rationalism: "empirical rationalism". But fire and water do not mix well, and I don't wish to talk of the long and troublesome search for a name for the Vienna Circle whose members did not wish to be called 'positivists' (Haller 1984a).

Austrian philosophy—so still Neurath—"saved itself the trouble of the Kantian intermezzo". Instead it opened wide fields of inquiry to the philosophy of Leibniz and subsequent traditions. Similarly, the English and French empiricists are deemed a significant influence, whereas the triumvirate of Schelling, Fichte and Hegel is said to have remained without lasting impact in the countries of the Danubian monarchy. Of course, we must understand this *cum grano salis*. For naturally there existed in Poland more than a few professors who had studied in Germany and spread the teachings of Schelling and Hegel; in the Czech literature we also find Hegelians, as we do in Hungary. But with the advance of the Brentano school, the last exponents of this philosophy lost their influence.

As a historian of philosophy, of course, Neurath has received even less recognition than as a theorist of science, and that means, plainly speaking, none at all. And for those interested—it must be admitted—Neurath did not leave behind any traces of his own path, nor indicate the sources which would show us *where* Austrian philosophy was investigated and *which* lines of development had been established before his own time.

So the question arises: since when does the very *topic* of an Austrian philosophy exist, and in what fashion and from what points of view was it explored? Is it the case that a line of development runs from William of Occam all the way to modern, and logical, empiricism? Does the Viennese nominalism play the same role as the Parisian, and may we assume Buridan's influence in Vienna to parallel that of the Cracow nominalists on the development of Polish positivism? Is it correct to view the idealistic systems as indeed "successors of the scholastic systems"? (Neurath 1936a, p. 682.) Questions multiply—questions which make it transparent that even the superficial parallel between my picture of Austrian philosophy and Neurath's only concerns *central* aspects.

But despite the shortcomings caused by the generosity of Neurath's sketchy design, one cannot discount its merit—a merit which rests in having formulated the question of the Austrian philosophy, and having done so from the fruitful vantage point of seeking a historical explanation for the emergence of the Vienna Circle. Admittedly, it is a common fate of historical explanations to substitute meaningful contexts for the causal connections of events, and it is well known how such a search anticipates the fulfillment of one's wishes: acceptance of such an horizon as the background for one's search leads to the glorified transfiguration of one's own past. It is for this reason that a critical analyst had better follow not the *topos* of glorification but that of excavation, and it was in *this* spirit that Neurath formulated his research program, even if he occasionally did not manage to escape the temptation of transfiguration in his anticipation of its result. What is expected—so Neurath wrote in 1936—is a historical "overview which comprehends the evolution of the main factors of logical empiricism and all its particular implications." (Ibid., p. 685.)

Without a full awareness of the Neurath program I chose the same starting point some 30 years later. And quite naturally the difference from the philosophy in the other German speaking countries was pursued in order to sharpen the contours of Austrian philosophy and search for the inner commonalities of this tradition.

In Neurath we find already an explanation for the remarkable gulf between the philosophical development in Germany and in the Danubian monarchy.

> The church and the court decisively rejected the teachings of Kant and of speculative idealism, which were ready to rule in Prussia, as products of the French Revolution; it was the opponents of Kant who found their favour. For Austrian propagandists this provided the opportunity to consolidate the position of the followers of Leibniz's teachings in school and university. (Ibid., p. 677.)

Werner Sauer was able to enlarge on this externalist approach and confirm its result in the extremely detailed analyses of his thorough study of the early history of Kantianism in Austria (Sauer 1982).

So much to the point of departure for our question. In the next section I wish to show how and since *when* the question of an Austrian philosophy was considered and treated as an independent topic.

II

In February or March of 1855 a 'Report about a so far Unknown Manuscript in the Philosophy of Law by an Austrian Author' was read to the Royal

Bohemian Society of Science (*Königlich-böhmische Gesellschaft der Wissenschaften*) by Robert Zimmermann, then *Professor Ordinarius* at the Charles University of Prague for the previous three years. He complained that "less has been done for the history of philosophy than that of virtually any other science, both for the history of philosophy in general and for its particular pursuit in Austria." (Zimmermann 1855, p. 4.) Of course, he went on, this "nearly total lack of effort" towards a history of philosophy in Austria was due more to external conditions than to the lack of interest or talent, even less to the lack of a subject matter. And to glorify this embarrassing fact he pointed to the supposedly innate modesty of Austrians and their fear of literary publicity.

In fact, however, there is little reason for glorification, as will be discovered when the succession of philosophy teachers in the old empire before 1848 and the appointments in Vienna and Prague, Graz and Innsbruck, Cracow and Olmutz are considered. For nearly two centuries, the empire, this magnificent edifice of the unification of the peoples of Central Europe, left the philosphical training of its youth in the hands of the Jesuits. In addition, the court controlled these studies even *after* the dissolution of the Jesuit order and arranged the study of philosophy according to principles concerning "the disadvantages which might arise for the order and peace of bourgeois society".[1] In this way the court continued a tradition which bore little philosophical fruit, the tradition of subordinating philosophical and general studies to the aims of the state and its sovereign without granting to research its absolutely necessary autonomy. It is no wonder then that a proverbial new age dawned with the university reform of Count Thun-Hohenstein in the years after 1848, but even more so, that an unfruitful period of Austrian philosophy ended. In the judgement of the Imperial Ministry of Culture and Education itself, we can accordingly read that, prior to the reform undertaken, lectures tended to restrict themselves to "the clarity of superficiality".[2]

Philosophy before the revolution of 1848 was confined to the tight circle of those textbooks which the censor of the court's study commission permitted to be used in teaching. As an example of these I can mention the Latin textbooks by Franz Samuel Karpe, who taught in Vienna from 1786 to 1806, *Compendiaria philosophiae moralis institutio* and *Institutiones philosophiae dogmaticae perpetua Kantianae disciplinae habita* (Vienna 1804), the textbook by Joseph Calasanz Likawetz, *Elementa philosophiae in usum auditorum philosophiae adumbrata* (Graz 1820), and finally the textbooks by Johann Peithner, Ritter von Lichtenfels (1795-1866), *Auszug des Wissenwürdigsten aus der Geschichte der Philosophie* (1836), *Lehrbuch der Logik* (1842), *Lehrbuch der allgemeinen Metaphysik und der Metaphysik der Religion* (1845) and *Lehrbuch der Moralphilosophie* (1846).[3] Peithner's realism followed in the steps of Friedrich Heinrich Jacobi[4] and was aggressively against Kant's 'pure subjectivism' which turned humanity into the godhead by conceiving of the supra-sensual as merely moral. Both Wolff and Kant, as well as Schelling and Hegel, were thus classified as mere advocates of speculation. Peithner's textbooks were used in the mandatory three-year courses in all the German speaking Schools of Philosophy of the universities of the Empire except for Prague, and only accredited textbooks could be used for the mandatory philosophical courses of instruction in the high schools.

Zimmermann, who himself was the author of one such textbook, of course knew those of his predecessors. His textbook, *Philosophische Propädeutik*, published in its third edition already in 1867 and translated into Dutch, Italian and Hungarian, remained in use for half a century. His judgement that Austrian philosophy had remained without its historiography covers the textbooks as well. In none of those named, and neither in others, do we find a chapter on Austrian philosophy, for there basically was no such thing. But also philosphy *in* Austria was not treated of in these histories. There was no history of philosophy in Austria, much less a history of Austrian philosophy. Thus we read in Carl Siegel's chapter 'Philosophy' in the *Deutsch-Österreichischen Literaturgeschichte*: "Hardly any history of philosophy in Austria reaches back further than the 19th century." (Siegel 1936) And since it is not besides the point to add a warning, let me not suppress the following passage from Siegel's review of Austrian philosophy:

> Even though it remains a remarkable fact for every Austrian that both a mature work of the philosophy of late antiquity (the self-reflections of Emperor Marcus Aurelius) and a work typical of the German intellect of early modern philosophy (Leibniz's *Monadology*, written for Prince Eugen of Savoy) were written on Viennese soil, it also remains the case that these two, as it were, accidental facts have nothing to do with a history of philosophy in Austria. (Ibid., p.17.)

This squarely puts to rest any superficial attempt to employ association as principle of origin in the history of philosophy. For nobody—with the exception of Kampits (1984)—would want to go so far as to include Marcus Aurelius, Nicolas of Cusa or Leibniz in the history of Austrian philosophy, even though the philosophy of Leibniz must be credited with a considerable influence in that history. Of the German philosophers only Johann Friedrich Herbart had a similar effect, and despite the fact that he himself never taught in the empire.

So what were the sources, one wants to know, from which Neurath, Siegel and, for example, Heinrich Gomperz arrived at their historical lines of derivation? Well—there are predecessors, and the reviews of those named could rely on them. Thus Siegel's own exposition begins by lamenting the fact of the state's prohibition of Kantian philosophy. And it continues with a brief description of the theories of Bolzano and his student Anton Günther and next describes the transplantation of Herbartism, after it had already been established in Bohemia, to the entire Danubian monarchy through Franz Seraphim Exner's reform of studies in Austria. It was the Czechs Franz Cupr (1821-1882), J. Boleslav Kvet (1825-1864), Josef Durik (1837-1902), Ottokar Hostinsky (1847-1910), and the, as Siegel calls them, 'German-Austrians' Robert Zimmermann, Josef Wilhelm Nahlowsky (1812-1885) and the said Franz Exner (1802-1853). Exner established Herbartism in Bohemia, and it was the German-Austrians who brought it to Vienna and Graz. Exner had worked in Vienna from 1830 to 1831 as *Supplent* [academic assistant], became Peithner's successor in Prague from 1832 to 1845, was elected to the Court's education commission and finally, together with Hermann Bonitz, worked as a Ministerial Councillor of the Ministry for Culture and Education on the university reform. It was just this very powerful position—after all, he declined the post of Minister—which enabled him to consolidate Herbartism (with which Bolzano's teachings were often contaminated) throughout the realm.

This historical fact has long been established in our century. And so Siegel remarks rightly that also Aloys Riehl (1844-1924), the most important Austrian neo-Kantian ever, originally came from Herbartian circles, as did Ernst Mach. And because Siegel regards himself to be closest to Riehl, he gives a particularly meritorious mention to the Austrian Kantians from Hugo Spitzer (1854-1937) to Robert Reininger (1869-1955). Only after *their* treatment do we find a review of the school of Brentano, in recognition of the fact that it is very influential "still today".

Nevertheless, Siegel gives a characterisation of Austrian philosophy which is worth considering. He says of it that it

> shows in nearly all its forms a strong attraction towards objective metaphysics and particularly to realism which more or less accomodates the so-called common sense. (Siegel 1936, p. 41.)

Similarly, he remarks at the end of his overview that objectivism "may be designated the main characteristic of 'Austrian' philosophy in its development as whole." (Ibid., p. 48. See also Haller 1984b.) It is to be noted that Siegel puts scare-quotes around the adjective "Austrian". But he still uses it as one would use it without them, by referring to the domain of Austrian philosophy.

Again and again we meet this characterisation of Austrian philosophy as 'objectivist' and just as frequently one which calls it 'realistic'. This does not mean only realism in regard of the problem of universals, and not only objectivism in the question of the dependence on truths in themselves, facts and objects. Someone like Friedrich Jodl also sees himself in opposition to the neo-Kantians when he writes in 1896 that his own study of causality does not satisfy him, but that it nevertheless has value "if it throws proper light on the *punctum saliens* of our *realistic* and positivist world conception against the neo-Kantians of all sorts." (Jodl 1920, p. 162.) In the same place—the biographical description of Jodl's time in Vienna, published shortly after the end of World War I—we find that Franz Brentano is called "the founder and the head of the so-called 'Austrian philosophy'." (Ibid., p. 163 fn..)

In Siegel and Neurath—at the beginning of the '30's of our century—we can see clearly the historiographical concept of an Austrian philosophy. And neither of them fails to try to determine the properties of this object.

It may be thought that these attempts have to be viewed, like the delimitation against German philosophy and philosophy in Germany generally, as connected with the first steps toward the self-definition of a new nation, the First Republic of Austria. But that is not so.

III

Already in the old Empire a number of authors confronted the question whether the prejudice was justified that in its intellectual development Austria limped behind the other European nations, particularly Germany. Thus Josef Klemens Kleibig in his 1914 study of Bolzano:

> There was one area in which the old Danubian Empire did count as dependent and little productive, namely, in the area of philosophy. In the brilliant line of thinkers from Leibniz to Herbart there is hardly anyone who hailed from its soil or who gained his laurels there. (Kreibig 1914.)

The point which Kreibig makes is that it is particularly Bolzano who contradicts this appearance. Entirely in the spirit of Brentano's theory of phases, Kreibig sees the cause of why a thinker of the greatness of Bolzano was so "arrogantly ignored" and neglected to lie in the contemporaneous age "of dialectical intoxication, of blinding paradoxes, of mystical phrases". And he concludes with a relevant passage from the work of Bolzano, in order to highlight the contrast between these two phases: the implicit demand for clear and intelligible discourse is contrasted with the "semi-darkness of expressiveness" intended to mask common thoughts as deep wisdom. His final sentence makes plain that Kreibig does not think that the gap between philosophy in Austria and Germany has been bridged, for he quotes Bolzano's exclamation:

> Germans! When will you finally return from the confusion [meaning the philosophy of speculative idealism, R.H.] which makes you unpalatable and ridiculous to your neighbours?

Kreibig adds only one sentence to the words of Bolzano: "And this call of the noble philosopher still remains to be heeded in our day!" (Kreibig 1914, p. 15.)

The only historiographical source mentioned by Kreibig himself, apart from works on Bolzano's philosophy, is Robert Zimmermann. And indeed, already in the last decades of the last century Robert Zimmermann returned again and again to the topic he pointed to in the earlier mentioned report of 1855: the history of Austrian philosophy. The extensive study 'Philosophy and Philosophers in Austria' was published nearly a hundred years ago, and, like the few that followed, begins with a description of the philosophy of Karl Reinhold (1757-1823) who fled from Austria to Weimar and Jena. Yet Zimmermann regards as the most important break the dissolution of the Jesuit order in 1773 with which the liberation of science and philosophy began, a liberation primarily from the censorship of the church. But, as is known, that censorship was replaced by one of the state, similarly strict, which soon saw a danger for the continuation of its authority in the enlightenment and in the philosophies which descended from it or were sympathetic towards it: all progressive philosophising was suspect of revolutionism. The victims of this mentality were first of all the French philosophies, and amongst the German ones those of Kant and his followers—but ultimately also the philosophy of Bolzano which appeared no less revolutionary, though for entirely different reasons.

These interventions effected, as it were, a standstill in philosophy in the Danubian monarchy which only strengthened the stereotypical picture of a culture backward in comparison with Germany and the rest of the continent. What weighed most heavily was the suppression of any public influence of the most important Austrian philosopher of the first half of the 19th century, Bernard Bolzano. But nobody could say which other intellects were frightened off their individual paths. Thus it took nearly another half century until the real possibility of a scientific philosophy independent of church and state could begin to be realised.

Since the *history* of events does not precede their happening and since historical research is *a posteriori* like all empirical investigations, the newly emergent philosophy in Austria had to wait a few decades before it had grown into a proper object of historical inquiry. One could only take note of Austrian philosophy *once it existed.* But once it did exist, the child received its name—in the works of Zimmermann on the history of philosophy. But the *idea* of this name is even older.

Already in the letters of Michael Josef Fesl to Anna Hofmann, the host and friend of the great Bernard Bolzano, the name can be found. And this leads me to the conjecture, as yet unsupported, that the concept of an Austrian philosophy was not only born in the Bolzano circle—from which, after all, Robert Zimmermann came—but that it was used by Bolzano himself. Others may test this hypothesis. But even it were to prove untrue, it nonetheless correctly focusses on the tentative terminus of our genetic history: the *circle around* Bolzano. Concerning its work, Fesl emphasises clarity and perspicuity (Winter 1970)—just those attributes which *Brentano* was to praise as characteristic of the blooming optimistic philosophy in his 'Die vier Phasen der Philosophie und ihr augenblicklicher Stand' ('The Four Phases of Philosophy and its Current State') which was dedicated "to the academic youth of Austria-Hungary", and which he was to apply in his own philosophy as a methodological principle.

According to Brentano's structural schema of the phases of philosophical development, there is neither cumulation nor non-directed revolutions, but only the ascent or descent of research programs. Brentano proposed two characteristics to be definitive of the ascending phase: firstly a "lively and pure theoretical interest" and secondly a "basically naturalistic, though in many ways still to be refined, method." (Brentano 1895, p. 10.) Of course, such a definition is not sufficient in order to identify a phase; but the *approach* can be used to place the *new beginning of philosophy in Austria* which did obtain historically. According to this schema, Bolzano—of whom Brentano claimed to have read with profit only his *Paradoxes of the Infinite*—must be counted within the ascending phase, whereas all more modern attempts to darwinise Kant have nothing to do with Kant if only for this reason, that his *a priori* judgements "do not derive from any experience, and thus neither from that of our ancestors".

Despite the objections of Sauer (1982, p. 11ff.; for the reply: Haller 1979d), it cannot be denied that not only the rejection of the philosophy of speculative idealism but also of Kantian philosophy remains for long stretches the common denominator of Austrian philosophy, with Bolzano as with Brentano, with Mach as with his students and finally—*pace* Sauer—with the Vienna Circle. This fact is not affected by the somewhat different development of the philosophy of Carnap who does not come from the Viennese tradition and who, in addition, had studied under the Kantian Bruno Bauch.

Well, I shall resist the temptation to present another defense of the thesis that there exists an Austrian philosophy. My point rather was to investigate in what way and since when the *topos* of a history of Austrian philosophy has been in use. And one of the results turns out to be that the object of such a history is not only conjured into existence by the need for identity in a new state, but that such a history had already been recognised to exist in the late flowering of Habsburg empire, when the first steps towards its investigation were taken. Of course, this history could have created its own object, like a fairy-tale or an arbitrary invention, which is just what is assumed by the opponents of the *topos* and fact of an Austrian philosophy and its history. I believe there exist no justified doubts as to its object; as to its name, one may object due to the seemingly nationalistic connotations of its predicate. But it has won its right of acceptance after nearly 150 years—needless to say, a frighteningly short period in comparison to English, French or German philosophy.

Karl Franzens Universität, Graz

NOTES

* First published as 'Zur Historiographie der österreichischen Philosophie', in *Von Bolzano zu Wittgenstein. Zur Tradition der österreichischen Philosophie*, Hrsg. J. C. Nyiri, *Schriftenreihe der Wittgenstein-Gesellschaft Bd. 12, Teil 2*, © 1986, Hölder-Pichler-Tempsky, Wien, pp. 41-53. In the present version, the first paragraph and the appended extensive bibliography on Austrian philosophy has been dropped; translated with kind permission of Verlag Hölder-Pichler-Tempsky and the author by T. E. Uebel.

1 Franz Hammer, 'Von dem ordentlichen Unterricht in der Philosophie', quoted after Rüdegger 1967, p. 17.

2 *Die Neugestaltung der österreichischen Universitäten*, edited by Alois Flier (Wien 1853), quoted after Rüdegger 1964, p. 221.

3 Peithner was Professor of Philosophy in Innsbruck 1819-26, in Prague until 1831, and from 1833 in Vienna until becoming *Emeritus*.

4 Ziemgiebel says that through Peithner the philosophy of Jacobi had become a kind of "philosophy of the state". (Ziemgiebel 1867, p. 318, quoted in Bauer 1966.)

FRIEDRICH STADLER

ASPECTS OF THE SOCIAL BACKGROUND AND POSITION OF THE VIENNA CIRCLE AT THE UNIVERSITY OF VIENNA*

1. Preliminary Remarks

Unlike its philosophy, the Vienna Circle's historical context in the cultural and political life between the two world wars has so far been investigated and retold only very sporadically. Even the relevant fragmentary (auto-) biographical works remain rather in the spirit of recalling 'the history of the house' and rarely go beyond descriptions of an immanent philosophical and cultural-historical nature, and so leave unconsidered the 'how' and 'why' of analyses directed to the immediate historical context.[1]

The following contribution seeks to establish certain basic facts which a historical re-presentation or reconstruction of the Vienna Circle as a philosophical movement and world-view must take account of if it is proceed according to the principles and methods of an 'externalistic' history of culture and ideas. I will attempt to show by examples *that* a conservative-clerical, German nationalist and national-socialist front fought against the Vienna Circle and similarly minded currents because of their liberal-to-socialist, and anti- or non-metaphysical, empiricist orientation, and *how* it did so under the banner and with the catalyser of anti-semitism. It is this opposition that was the cause of the Circle's minimal reception and its dissolution and emigration since the begining of the 1930's.

Restricting the investigation to the University of Vienna brings with it the disadvantage of a merely aspectual analysis of detail, but it allows for a more systematic and pointed treatment. The present perspective is therefore but *one* piece in the mosaic of a comprehensive history of the Vienna Circle, which is to be complemented by others that take as their foci of investigation the documentation of the defamatory press campaign on the occasion of the murder of Moritz Schlick as an indicator of the general social position of the Vienna Circle in the *Ständestaat* (étatist state, a state divided into estates); the record of the remarkably busy, idealistic, organisatorical and practical-political activity by members of the Vienna Circle in the field of people's education (in

T. E. Uebel (ed.), Rediscovering the Forgotten Vienna Circle, 51–77.

the *Verein Ernst Mach*, at adult education institutions, in the school reform movement, through Otto Neurath's Social and Economic Museum) and in the social democratic movement (particularly through Otto Neurath, Hans Hahn, Edgar Zilsel, Philipp and Josef Frank). Only a synopsis of these points of view, together with a historical-genetic problem history of neo-positivism and its importance in a contemporary perspective, would allow a complete understanding of the Vienna Circle as an intellectual and cultural phenomenon.

2. *The Position of Scientific Philosophy*

Despite what several valuable historical-philosophical descriptions may appear to imply,[2] the empiricist, anti-metaphysical and language-oriented 'typical Austrian philosophy' did not dominate the philosophical life of the University of Vienna. For at the same time, adherents of such different currents as, e.g., German idealism (in particular neo-Kantianism, Herbartianism), of natural law scholasticism, of Christian world-view philosophy (*Weltanschauungsphilosophie*), of neo-romantic universalism, were active there (cf. Topitsch 1949, Johnston 1971). Nearly all these currents viewed philosophy as the speculatively synthesizing 'queen of the sciences'. Empiricist and language oriented sub-currents may have existed amongst them, but the smallest common denominator was the sharp division between philosophy and the empirical disciplines on the one hand and, on the other, the newly founded logic and mathematics of Frege, Russell and Whitehead—as well as occasionally attempted amalgamations of philosophy with political world-views. It is a fact of the sociology of science that all these currents which moved outside of the field of neo-positivism (itself mainly characterised by empiricism, logic, analysis of language, and a methodological proclivity toward the natural sciences), were academically constituted as the traditional metaphysical *school philosophy*—as Edgar Zilsel (1930) noted at the time. The social function of this scholastic philosophy, deeply anchored in the bourgeoisie, and endlessly creative of new disciplines, was to stabilise the established system and legitimise authority, as will be shown below.

Switching from an immanent philosophical point of view and a comparative history of the reception of the Vienna Circle tradition (according to contemporary selection criteria) to a concrete historical analysis of the philosophical scene at the University of Vienna in its entirety, we can discern a polarisation of two directions already existing during the Habsburg monarchy. In the First Republic this polarisation is strengthened and turns very much to the disadvantage of scientific philosophy, both in terms of personnel and in terms of lecture themes. This admittedly somewhat simplifying 'two-camps' theory is complemented at the level of world-views: the camp of scientific philosophy is dominated by democratic tendencies (enlightenment, liberal, socialist), whereas the other camp embraces nearly all forms of anti-democratic attitudes, from neo-romantic conservatism to fascist totalitarian excesses. This suggests that the philosophical life should be placed in the context of the party-political cultural struggle of the time between the bourgeosie and the workers movement, though we would not wish to insist on a rigid causal-deterministic explanatory schema.

It would be worthwhile to ask what objective functions were fulfilled by the purist and anti-metaphysical scientism in providing the ideological and theoretical foundations of social democracy, given its self-understanding at the time. (A relation of affinity is undeniable after the emergence of empiriocriticism.) This would provide a useful complement to the already existing investigations between political Catholicism, natural law and authoritarian structure of government, as between universalism and fascism. What would have to be shown in particular is how, in this dichotomous cultural landscape, the Vienna Circle's anti-metaphysical enlightenment philosophy (along with many other movements) was placed in an affinitive relation to the workers movement as its intellectual stimulus and instrument of purification and how, in turn, its missionary charisma and social orientation resulted from its objective function as demystifier of dogmatic-speculative ideas. (For this reason, the Circle's emigration also represents an uprooting of its world-view.) This objective function did not emerge directly from the common minimal political platform of the Vienna Circle as a whole, but from the political engagement of individual members and the role set for them by contemporary protagonists. The incompatibility of content was blurred, on the one hand, by the political and world-view related proximity—the value of one's socio-political position rendered the contradictions of propositional systems ineffective—and, on the other hand, by the spontaneous, hardly questioned partial appropriation by social democracy of rational-empiricist science as an ideological tool against fascist irrationalism. (This particular liaison deserves to be investigated more closely in its development and dissolution; a mutual alienation cannot be denied in the post-World War II Second Republic.)

To turn back to the situation of the discipline of philosophy at the University of Vienna (Wieser 1950, pp. 158, 231, 235ff.): Amongst 22 teachers in the School of Philosophy (including psychology) between 1918 and

1938, neo-positivism had three adherents: Moritz Schlick as *Ordinarius* 1922-1936, Rudolf Carnap as *Privatdozent* and *Titularprofessor* 1926-1931, and Viktor Kraft as *Privatdozent* and *Titularprofessor* 1914-1938 (Friedrich Waismann was only Schlick's librarian 1931-1936).[3] This spread was reflected in the frequency and type of courses offered: traditional history of philosophy was taught most frequently (except for the philosophical 'exercises', a category which together with ethics shows the highest attendance). Thus a list of those philosophers to whom specific lectures were dedicated between 1848 and 1938 shows, by far at the top, Kant (50), Schopenhauer (26), Aristotle (20), Plato (17) and Nietzsche (16), whereas, at very bottom, we find Mach (4), Boltzmann, Einstein, and Brentano (2 each). We get a corresponding picture from the dissertation topics of the same period: Kant (39), Schopenhauer (17), Herbart (13), Spinoza (12), Nietzsche and Leibniz (11 each), Plato (10), Descartes (9), Hume (8), Lotze (7), Aquinas, Aristotle, Dilthey (6 each), Mach as predecessor of neo-positivism (5), finally Stöhr (3), Bolzano, Brentano and Meinong (2 each). Nonetheless, the lecture courses of Moritz Schlick counted amongst the best attended, after those of Karl Bühler and Robert Reininger. If we broaden the criterion 'scientific philosophy' and add the teachers of psychology in the School of Philosophy (Karl and Charlotte Bühler, Sigmund Kornfeld, Egon Brunswik), we get a roughly equal division between the teachers of the two camps; leaving out the psychologists, the adherents of neo-positivism and related thinkers (Heinrich Gomperz, Karl v. Roretz, Emil Reich) constituted less than a quarter. (The comparison by number of semesters taught is more favourable.)

While it is true that three of the four professorial chairs were occupied by Schlick, Bühler and Gomperz, the actual distribution and influence of philosophical currents at the University of Vienna is distorted in this perspective. Not only teachers of philosophy *per definitionem* were of philosophical importance, but also a number of other, in the broadest sense philosophical, teachers from different disciplines and faculties. Examples are Othmar Spann as the *Ordinarius* of economics (1918-1938), teaching in the spirit of his pre-fascist universalism, or the *Privatdozent* for government, law and sociology, Johann Sauter, who, as an illegal National Socialist, held introductory philosophy lectures for students of law and government.[4] (It should also be added that the proximity in world-view and political outlook between Hans Kelsen, *Ordinarius* from 1918-1929 at the School of Law, and the Vienna Circle did not bring with it an equal philosophical congruence, since his conception of natural law was rooted more in neo-Kantianism than in positivism (Wielinger 1977). Similar comments apply to Max Adler who, as *Titularprofessor* at the School of Law from 1920-1937, worked very much in isolation on the synthesis of Marx and Kant (despite his interpretation of Marx after the ideal of Mach's natural scientific positivism).)

Like power and influence in historically grown institutions, the discourse of 'philosophy' manifested itself not only in the framework conceived for it, but also in other 'foreign' disciplines; it did so mostly indirectly, unconsciously and independently, on the basis of certain social processes, forms of behaviour and normative systems, which did not themselves appear in the subject matter taught.

One genuinely philosophical institution was the Philosophical Society of the University Vienna (*Philosophische Gesellschaft an der Universität Wien*), which was also the local chapter of the Kant Society. The lecture program shows that amongst the Society's comprehensive offerings from the most varied disciplines, the advocates of scientific philosophy are in the minority, about one seventh, despite the active publication record of Feigl, Hahn, Kraft, Juhos, and Schlick (Reininger n. d., pp. 21-43). A letter of March 2, 1934, in which Schlick protested against the ordered dissolution of the *Verein Ernst Mach* (because of "social democratic activity") in the aftermath of the February events[5], indicates that the members of the Vienna Circle first intended to organise their own subsection within the Philosophical Society to "develop an empirical philosophy oriented towards natural science", but abandoned this "in favour of greater independence of aim".[6] In the course of these efforts to gain an independent profile, the Verein Ernst Mach had been founded in 1928 at the start of the public phase of the Vienna Circle, and the programmatic brochure *The Scientific World Conception. The Vienna Circle* was published on the occasion of the First Congress for the Epistemology of the Exact Sciences in 1929 in Prague.

A similar spread of disciplines as in the Philosophical Society is shown by the Viennese International Academic Courses (*Wiener Internationale Hochschulkurse*) amongst whose lecturers Moritz Schlick figured as the sole member of the Vienna Circle (Anon. n. d., pp. 8, 14). Similarly for the august and esoteric holy grail of the university, the Academy of Science (*Akademie der Wisenschaften*): the mathematician Hans Hahn was the only one of the neo-positivists who qualified as a corresponding member (Krestan 1972). Whether other Circle members made personal efforts to gain admittance remains an open question. Nevertheless, the complete absence from the Academy of any philosophy oriented towards the natural sciences—

philosophy was represented by Robert Reininger and indirectly through Othmar Spann—can be considered indicative of the history of the reception of the Vienna Circle.

3. The General Intellectual and Political Situation at the Viennese Institutions of Higher Learning

A large part of the philosophical and ideological scene just described becomes intelligible only after a short description of the politics of the intellectual situation in the universities. Such a description also renders intelligible the fate of the Vienna Circle by illuminating its role in the historical events of the day.[7] Generally speaking, the political life of the institutions of higher learning at the beginning of the First Republic was marked by the predominance of German nationalist (*Deutschnationale*) and Christian-social (*Christlichsoziale*) forces with the related, increasingly popular student society *Cartell Verband*, and by the first appearance of national-socialists displayers of the swastica. Compared to the time of the Habsburg monarchy, this marked a strengthened anti-liberal and anti-semitic trend, against which the Social Democrats sank to insignificance. The natural opponents of this dominating phalanx (within which position fights were not uncommon) were, amongst others, the advocates and adherents of 'Jewish' neo-positivism, of the liberal theory of marginal utility, of psycho-analysis and of Marxism.

Of decisive importance for this conservative-to-reactionary climate was, first, the acute unemployment amongst academics and the social misery of intellectuals as a result of the latent global economic crisis and, second, the reactionary, ruthless policy of saving on public expenditure by the ruling block of bourgeois parties, which led to incessant competition and protectionism by university pressure groups. An outstanding element in this ideological cultural struggle in the universities was German nationalism (an allegiance to the German 'nation' regardless of the constitutional independence of Austria) which was found in all big parties as a 'remainder of the trauma of defeat'; in addition, a strong revisionism in reaction to the Peace of Versailles and a 'political Catholicism' essentially founded on natural law theory (the Christian-social tendency) was prevalent in the bourgeois camp (Hanisch 1977).

The establishment of the Austro-fascist *Ständestaat* in 1934 brought further hardship to all democratically minded forces and changed this situation only in so far as the new Austria-ideology attempted in vain to supplant the Catholic nationalist (*katholisch-national*) and 'decisively nationalist' (*betont-national*) coalition of academic teachers. (These wings of German nationalism were differentiated by their extremism: the former were the bridge builders between Catholicism and National Socialism, the latter were still more strictly *völkisch*, i.e. racist, in their orientation.) This was a tragic-comical reflection of the hopeless general political to-and-fro of the Dollfuss and Schuschnigg regime in the powerplay between Mussolini and Hitler on the one hand, and the illegal national-socialist opposition on the other—a to-and-fro not eased by holding the (also banned) workers movement incapable of entering a coalition. All the while, anti-semitism continued to be extensively practiced (Maderegger 1973).

The dispersion of the Vienna Circle can therefore no longer be seen as caused by a sudden invasion of irrational antagonistic forces. The Vienna Circle's anti-speculative logical empiricism and the political profile of its members represented a provocative questioning of every *a priori* legitimation of authority in science and politics. On the more concrete level of social interaction, this constituted one of the main causes of the antagonism in the social and intellectual domain.

In order to examine these interactions my analysis focusses on the practice of *Habilitation* and academic appointments. It is here that controversial standpoints and coalitions emerge undisguised and operative criteria for what is or should be science or philosophy become evident; but the social function of certain concepts of science and philosophy also emerges as an issue of debate. My results exemplify the ideological world-view- and discipline-related preferences of the decision making institutions of authority in the university and make evident the strategies and procedures of exclusion as they bear on interest-determined paradigms of science and philosophy. The predominant political currents manifested themselves at the centres of bureaucratic decision processes.

4. The Personnel Politics, particularly the Practice of Habilitation and Appointment by the Ministry of Education and the Professoriat 1922-1938

The Austrian university system was characterised by a rigid form of organisation and a relatively stable personal continuity in the academic committees. This brought with it a 'guiding thread' in personnel decisions.

The leadership of the Ministry of Education guaranteed at the minimum a conservative line, for the post of Minister was occupied exclusively by clerical-conservative, German-nationalist, even pro-nazi politicians from 1920-1938.[8] The problem of the much discussed autonomy of the university thus turned into a two-sided sword: it was not so much the neo-humanistic ideal of education that the slogan of academic self-determination referred to, but the conservation of privileges and the pursuit of the interests of the estate or elitist caste of professors of science bound by a world-view.

4.1 The Appointment of Moritz Schlick in 1922

The events surrounding the appointment of Moritz Schlick to the Chair of the Philosophy of the Inductive Sciences, once held by Ernst Mach and Ludwig Boltzmann, already show some characteristic elements of the subsequent intellectual atmosphere. In the same year, student riots (secretly condoned by the administration) led to the closure of the universities, and the question whether knowledge or racial descent should be decisive of *Habilitation* was discussed seriously in the right-wing press.[9] It was during this time that Hans Hahn, *Ordinarius* in mathematics since 1921 and one of the first ever academic interpreters of the *Tractatus*, started a signature campaign for Schlick amongst the mainly conservative professoriat, against much resistance from the representatives of traditional school philosophy.

After lengthy negotiations for the three vacant chairs of philosophy, the following order of hopefuls emerged from many domestic and foreign applicants: Schlick, Brunswik, Pichler for the Chair of Philosophy of Nature; Karl Bühler for the Chair of Psychology; and Robert Reininger for the Chair of History of Philosophy (Wieser 1950, pp. 48-50). Subsequently the committee of professors asked Emil Reich (who had taught practical philosophy and aesthetics since 1890 as *Privatdozent* and then as *ausserordentlicher Professor* and was very involved in people's education) and the botanist Richard von Wettstein—both of whom also had spoken for Schlick—whether Schlick was of Jewish descent (Zehetner 1972). After the application of this criterion of qualification—clearly a disqualification of the appointment procedure itself—had failed to show the hoped-for effect, the teacher of Old German, Rodulf Much, and the historian Alphons Dopsch fielded a memorandum against the appointment proposals for discussion. This memorandum claimed that attending to two philosophical specialities (i.e. philosophy of nature and psychology) ran counter to the needs of the university, for "the youth was to be led in the spirit of an idealistic world-conception".[10] In addition, "representatives of total areas ... much wider in range than Mach's thought" were needed. Both vehemently demanded that an Austrian applicant be chosen and proposed the already nominated, 'all-overtowering' Reininger. Their proposal was supported, surprisingly, also by Alois Höfler, who held Schlick to be more suitable as an *Ordinarius* in physics.[11]

This initiative was intended to prevent the diversity of opinion that would have been guaranteed by three chairs by combining them in the one and the sole appointment of Reininger, and so to secure the monopoly of idealistic philosophy. Yet Reininger, Bühler and Schlick—the latter judged in the committee's report as an "original and independent thinker"—were appointed despite this opposition.[12] Afterwards, the dissenters expressed the hope that one would not have the same experience with Schlick as with Mach who always wanted to remain a physicist, or with Boltzmann who once called school philosophy a sort of bad sporting activity.[13]

These episodes confirm that massive reservations existed against an anti-speculative philosophy which took account of the methods of modern natural science, logic and mathematics—Schlick was one of the first philosophers to comment on relativity theory—and that an institutionally anchored anti-semitism functioned as a social instance of control.

4.2 The Attempted 'Habilitation' by Edgar Zilsel in 1923/4

The analysis of Edgar Zilsel's unsuccessfully attempted *Habilitation* is also revealing,[14] for it introduces the issue of the genuine object domain of philosophy. Zilsel, who applied for the *venia legendi* for philosophy as a whole, submitted his book on the religion of genius, his papers in physics, natural philosophy and the didactics of philosophy and, as his *Habiltation*-dissertation, the two-part *Beiträge zur Geschichte des Geniebegriffs* ('Contributions to the History of the Concept of Genius').[15] The *Habilitation* commission (Reininger, Bühler, Schlick, Meister, Gomperz, Schlosser, Wegschneider, Ehrenhaft) informed Zilsel, via Schlick and Gomperz, that he should withdraw his application since it probably would not be supported by the majority of the School, particularly because of the lack of work in the philosophy of nature and epistemology.[16] Zilsel turned down this suggestion on the grounds that he did not want his research interests to be determined by extraneous considerations and that in his opinion the result of eight years work could withstand scientific criticism.[17] After Reininger had declared the dissertation to be philosophically inadequate, and Richard Meister, the *Ordinarius* in pedagogy, had claimed to have discovered fundamental mistakes (amongst them, that "everything focussed on economics"[18])—but with Schlick and Gomperz having already voted for Zilsel—it was decided to ask for an independent professional opinion whether Zilsel was qualified to be a philosophy instructor, based on the philosophical merit of his work in particular.[19]

The commission did not object to the candidate on personal grounds.[20] Although the requested evaluations by Adolf Dyroff and Heinrich Scholz had not arrived yet, the application was discussed in another meeting, as confirmed by the School of Philosophy.[21] Then, in a letter to the Dean of June 3, 1924, Zilsel announced the withdrawal of his application (despite the fact that the majority of the members had accepted the 'scientific value' of his work), since he considered himself unable to submit further "philosophical writings in the strict sense". Thus the divergence between philosophy and science, which lay at the bottom of the criteria of evaluation, was cemented in writing. When the evaluation by Ernst Cassirer, who had "the best of impressions",[22] arrived not long after, it was no longer relevant. In the end, Zilsel wrote once more to the Dean in order to justify his motivation and scientific intentions: he did not approach philosophy

> by coincidence from the history of literature, but ... tried to develop ideas in natural philosophy and the history of philosophy in close connection with the facts of physics and of the humane sciences [*Geisteswissenschaften*], believing that this serves philosophy better than narrowly delimitating it from the fertile ground of the individual sciences.[23]

In addition, he thought it improbable that he would "radically change the methods of investigation".

Now, one may disagree about the philosophical relevance of Zilsel's *Habilitation* dissertation, for it certainly cannot be clearly placed within the schema of classical works in philosophy. But these events show a distinct trend against a conception of philosophy oriented towards natural science and sociology, and for the absolutisation and normativisation of the criteria of the current speculative-idealistic conception of philosophy. One can justifiably claim the proceedings to have been suspiciously tendentious. The commission disregarded the fact that at the time of his application Zilsel was already lecturing at institutes of higher adult education on ethics, natural philosophy, epistemology and history of philosophy (amongst others on Kant), and planned to do so at the university.[24] This only confirms the present diagnosis: exclusion based on motives grounded in political differences and divergences in world-view and in the understanding of science.

4.3 The Appointment of Hans Eibl and Viktor Kraft in 1924[25]

The Ministry of Education also intervened massively in personnel decisions and was supported in this by most of the professors. At this time lists about the racial origin of university teachers began to circulate. After the death of Wilhelm Jerusalem, *ausserordentlicher Professor* for Pedagogy and History of Philosophy—he was also the founder of the sociology of knowledge and of the Viennese Sociological Society, and active in people's education—the two *Privatdozenten* Hans Eibl and Viktor Kraft were considered as successors. Eibl, whose philosophical interest lay in patristic and scholastic philosophy, later advanced to become a 'bridge builder' between Catholicism and National Socialism in line with the 'Catholic nationalists' (together with Josef Eberle, Edmund von Glaise-Horstenau, Oswald Menghin, Karl Gottfried Hugelmann, Othmar Spann).

It is well-known that, beside agreeing with the basic positions of the Vienna Circle, Kraft's philosophical position was characterised by a relatively original constructive epistemological realism and methodological deductivism (cf. Rutte 1973). Politically he was a liberal, although he rejected every political tendency within the philosophical efforts of the Vienna Circle (Topitsch 1960). He had worked in the university library since 1914, gained the *venia legendi* for epistemology with his *Weltbegriff und Erkenntnisbegriff* ('The Concept of the World and the Concept of Knowledge'), and taught mostly epistemology, theory of science and theoretical philosophy.[26]

Eibl complained about the "historical responsibility of the Jews for Bolshevism", advocated revising the Treaty of Versailles, and expressed his sympathy for National Socialism by his engagement for a "common Christian-humanistic" program and the agreement between Hitler and the Vatican in 1933 (Weinzierl-Fischer 1963, pp. 425, 435ff., 502, 505ff.). He had studied classical philology and philosophy in Vienna and worked as a teacher in a *Gymnasium* (secondary school leading to university qualifications) before World War I. At the start of the war he habilitated with his work *Metaphysik und Geschichte* ('Metaphysics and History'), but only lectured after 1921, on the philosophy of antiquity and the middle ages and on questions of *Weltanschauung*. In addition, he taught philosophical propaedeutics for graduates of *Realschulen* (the middle tier of secondary schools) until 1938.[27] By way of a parenthetical contrast, one may mention Kraft's criticism of vitalism, as opposed to Eibl's demand for greater concern with the domain of biological-vital values in Catholicism ; Kraft also welcomed the Paris Peace Congress "despite threats by the Federal Chancellor". (Ebneth 1976, pp. 71, 100-104; Grümm 1950.)

To return to the appointment procedure: agreeing with the Ministry of Education, the committee of professors favoured the appointment of Eibl to *ausserordentlicher Professor* (simultaneously granting the same title to Kraft), because Eibl represented a philosophy "which is needed in the University". In response some professors, including Schlick and Reich, raised a minority vote protesting the partial infraction of the university's autonomy by the ministry. The motion included the following commentary: Eibl was not a specialist in history of philosophy, but mainly interested in problems of religion and metaphysics; anyway, there were already five professors of whom two lectured mostly on history of philosophy; apart from this, theologians were already teaching patristic philosophy, "some of them with a more liberal attitude than Eibl".

This vote was ignored by the School and the accusation of ministerial intervention was rejected. Then a commission (Arnim, Bühler, Ehrenhaft, Hauler, Meister, Reich, Reininger, Schlick) decided, with a majority vote, to petition the School for another discussion of the Eibl/Kraft case, preceded by the appointment of Heinrich Gomperz, and then the adjournment of the whole matter. The School rejected the appeal for adjournment; a vote brought a relative majority for Eibl and a proposal for the appointment of Gomperz as *Ordinarius* for history of philosophy. This apparently unstoppable appointment of Eibl (to keep him in Vienna at all costs, for he had been proposed as *ausserordentlicher Professor* for History of Philosophy in Prague), brought to four the number of professors at the School of Philosophy (Reininger, Gomperz, Karl von Roretz, and Eibl) who lectured on the history of philosophy, despite the School's financial and economic difficulties (Wieser 1950, p. 54).

The Eibl/Kraft case documents *in nuce* the preference of the Ministry of Education and most professors for a metaphysical, Christian-scholastic *Weltanschauungsphilosophie* (world-view philosophy).

4.4 The 'Habilitation' of Rudolf Carnap in 1926[28]

The *Habilitation* of Rudolf Carnap in 1926 proceeded with fewer difficulties. Before 1914 Carnap[29] had studied philosophy, mathematics and physics in Jena and Freiburg (with Frege amongst others); during the war he was a soldiers' representative, a member of the USPD (Independent Social Democrats) and a physicist; after the war he remained in Freiburg as a teacher until his guest lectures to the Vienna Circle in 1925. He is considered one of its strictest anti-metaphysicians, and like Otto Neurath had sympathies for the freethinking *Monistenbund* (Monist League) and for socialism. The First World War effected Carnap's pacifist and anti-monarchist attitude, and his criticism of the existing social order from the standpoint of scientific planning. It was he who provided the strict theoretical elaboration and representation of the many philosophical ideas of the Vienna Circle; he was, so to speak, the successful 'transformer' of the various fragments of ideas into a logical and systematic conceptual edifice, as in his *Der logische Aufbau der Welt* (*The Logical Structure of the World*).

During the proceedings for his *Habilitation* it was of course discussed whether 'logic and epistemology' as a discipline belonged in the university in the first place, and there was a dispute over whether the *venia legendi* was to be awarded for the field 'theoretical philosophy' or 'logic and epistemology' (Wieser 1950, p. 55). Before the *Habilitation* was finalised, however, Carnap's knowledge of religion had to undergo a test of fire when a member of the committee asked him whether he knew the doctrine of the omnipresence of Christ; Carnap answered that he did not, but fortunately was able to guess the correct answer.[30]

4.5 The Appointment to the Chair of Heinrich Gomperz after 1934

The final establishment of Austro-fascism after February 1934 was reflected also in the universities, namely, by a total, legally secured reorganisation in the spirit of the Austria-ideology, combined with a one-sided policy of economy which was turned against all "liberalistic and individualist tendencies"[31] and the overpowering 'national opposition' of 'decisive nationalists' (advocates of a Greater Germany, national-socialist sympathizers) and 'Catholic nationalists' (like Hugelmann, Eibl, Spann, Nadler and Srbik) and the illegal National Socialists (Wandruzska 1954, p. 414).

One of the first victims was Heinrich Gomperz—a complicated case which requires a separate treatment (c.f. Gomperz 1953, Topitsch 1967). Of interest in the present connection is the subsequent process of appointment: it is exemplary for the *Ständestaat*.[32]

Gomperz was retired early and his chair was officially transformed into an extra-ordinary professorship to suggest economies. The government intended to appoint in his stead Dietrich von Hildebrand, the anti-nazi Christian world-view philosopher who emigrated from Germany in 1933. Hildebrand had studied philosophy in Germany before the war, received *Habilitation* in Munich in 1918 and taught there as a *Privatdozent* until 1933. Before his call to Vienna he was *Honorarprofessor* at the School of Theology in Salzburg.[33] He advocated a personalist, Christian-metaphysical ethics and philosophy. Politically he was against every kind of collectivism, left or right; National Socialism and Bolshevism were his enemies. He accepted an authoritarian system in 'extreme situations' and wrote for an 'Austrian Man'. Just as his journal *Der christliche Ständestaat* ('The Christian Estatist State'), he fought against the anti-semitic prophets of 'Catholic national socialism' in the Christian camp—e.g. against Spann, Eibl, Bishop Alois Hudal—and against Josef Eberle's weekly *Schönere Zukunft* ('More Beautiful Future'), even against the *Reichspost* (a nationalist paper). This brought him strong criticism and an ever increasing isolation (Ebneth 1976, pp. 76f., 100ff., 104ff.). After the *Anschluss* of Austria to Hitler's Germany he managed to emigrate to the USA, following an adventurous escape.

The appointments committee—dominated by German and Catholic nationalists and led by Heinrich von Srbik (whose lectures in Berlin were followed with suspicion by the Austrian government)—considered von Hildebrand's proposal unacceptable "on scientific grounds" and recommended, in the following order, Alois Dempf, Viktor Kraft and Karl von Roretz. Dempf had received his *Habilitation* in Bonn and taught there as a *Privatdozent* and, from 1933 until his later call to Vienna as *Ordinarius* in 1937, as *Titularprofessor*. His philosophical interests were medieval scholastic philosophy, ethics and philosophy of culture.[34] Politically and in his world-view he was a conservative thinker, but as a convinced Catholic he clearly differed from National Socialism (which provided the reason for his suspension in 1938).[35] Karl von Roretz had studied law, medicine and philosophy and worked since 1922 as *Privatdozent* for the history of philosophy and at the national library. He represented an anti-metaphysical sensualist positivism similar to Mach's and, in addition to critical epistemology, was concerned, with problems in the psychology and philosophy of culture.[36]

As a compromise the appointments commission accepted admitting von Hildebrand as a lecturer to the university. It is revealing that the proposal contained the additional comment that Hans Eibl was a "universally recognised representative of the Christian world-view" which would justify his promotion to *Ordinarius*! Despite this opposition von Hildebrand was appointed *ausserordentlicher Professor* by Schuschnigg at the end of 1934 (but when he started to lecture he had to request police protection against attempted disturbances from German nationalist students). Dempf was rejected as a foreigner, Kraft and von Roretz failed as well. Both were designated as competent scholars, but, in comparison to Hildebrand, who possessed "in scientific respects a very good reputation", were held less suited for the specialist field of "philosophy with special regard to *Weltanschauungslehre* (theory of world-views)".

Now, if indeed the concept of *Weltanschauungslehre* is elevated to the status of a criterion, then the justification of the rejection of Kraft is plausible and based on a realistic assessment of the applicants. At the same time, however, this makes plain that it was impossible to enage in exact 'non-Christian' philosophy and, once again, the political and world-view-determined struggle over the paradigm of 'scientific philosophy' is played out before us.

4.6 The Appointment to Schlick's Chair 1937 - Remarks on Mathematics

The violent death of Moritz Schlick gave occasion for a massive anti-semitic and anti-positivist defamation and hate campaign for the press of the authoritarian *Ständestaat* (the laudable exception in the Catholic camp, as noted, was Dietrich von Hildebrand in his journal *Der Christliche Ständestaat*).[37] The appointments commission for his former chair in 1937 (Bühler, Knoll, Koppers, Meister, Mewaldt, Praschniker, Reininger, Srbik, Thirring, Versluys and Winkler) quickly concluded that there was no one suitable to succeed Schlick.[38] In addition it proclaimed programmatically that "the history of philosophy is to constitute the real substance and essential task of the study of philosophy". This meant liquidating the Chair for the Philosophy of Nature which had existed since Mach. The by now hardly surprising prejudiced pronouncement on the way and task of philosophy as an academic discipline was designed for the benefit of Hans Eibl, though Heinrich Gomperz was also discussed.

The final proposed ranking of three candidates was Eibl, Dempf and Friedrich Kainz. Kainz had been *Privatdozent* since 1925 and taught mostly philosophy and psychology of language.[39] After the *Anschluss* in 1938 he apparently had no difficulties to become Head of the Institute of Psychology.[40] For Eibl, who was counted as a supporter by the German ambassador von Papen, there was an intervention and pressure on Schuschnigg for his appointment by Bishop Alois Hudal who fought for his vision of a national-socialist state in the spirit of Catholicism (Weinzierl 1963, p. 498). These efforts, however, went unrewarded for in the end Dempf was appointed as successor to Schlick,[41] apparently because as a devout Catholic he was more acceptable to the clerical-fascist Austria-ideology than Eibl, who was politically active in coalition with National Socialism. This was more or less confirmed a year later by Eibl himself in a petition (seconded by Menghin, the Minister of Education) to the *Reichsstatthalter* Seyss-Inquart (the national-socialist govenor of Austria), in which he sought compensation through appointment to the professorship for which he had been recommended twice.[42] His reasoning speaks for itself and for the spirit of the Catholic nationalists and thus merits a brief mention. Eibl complained that he was overlooked by the Schuschnigg regime because of his nationalist standpoint and that, for this reason, his appointment as *Ordinarius* would only be just from the standpoint of the Greater Germany—particularly since he had been following all demands for cooperation from the National Socialists with conviction. Finally, Eibl boasted that he

> revitalised courage and belief in the future ... when through the unfortunate July of 1934 the atmosphere in our circles had been depressed ... through lectures for the National Socialist student group, in particular through a lecture to the extended leadership of *SS-Standarte 89* about the Third Reich and the statesmanship of the Führer.[43]

As is well known, the last traces of neo-positivist thinking were erased after the *Anschluss*, which was welcomed by the majority of the academic teachers who over the years had contributed to the intellectual *rapprochement* with fascism through their own ideological efforts. However bitter the forced emigration, retirement or retreat was for the members of the Vienna Circle, it did not come as a great surprise. Just such a surprise apparently was experienced by those who out of perhaps naive and deluded motives had worked for years toward the fraternisation of Catholicism and national socialism, and who saw their ideas and hopes disappointed by their wished-for redeemer.

Amongst the mathematicians, the resistance to certain appointments was not as evident as with their philosophical collegues. This may be explained by the fact that in mathematics it is possible to decide much more clearly and with less dispute about suitability or unsuitability based on relatively constant criteria of selection. Because of this, the danger of ideologising was much smaller than in philosophy, where every member of the appointments comission felt able and competent to pass a definitive judgement. In spite of this, it should be noted that, except for Hans Hahn, it was impossible for either Karl Menger or Kurt Gödel or Kurt Reidemeister to become *Ordinarius* in Vienna.[44] A policy of economy was also followed with respect to the chairs in mathematics: the third chair was left open in 1935 after the death of Hahn, an application to reappoint was rejected in 1937 and in its stead—as in the case of von Hildebrand—an *ausserordentliche Professur* was created.[45] Another characteristic piece of the mosaic is the fact that Menger's and Gödel's acquaintance with Hahn gave the occasion for anti-semitic displays when both were tested for their "patriotic integrity".[46]

5. *The Vienna Circle in the Wake of Political Forces*

5.1 The 'Deutsche Studentenschaft' and the 'Deutsche Gemeinschaft'

Not only Zilsel, but apparently also other social democratic and liberal applicants were the subject of one-sided application of the norms of *Habilitation* and appointment procedures and strong anti-semitism.[47] Responsibility for this accrues to the overpowering *Deutsche Studentenschaft* (German Student Federation)—an umbrella organisation of all clerical-conservative and German *völkisch* students up to 1933—and the *Deutsche Gemeinschaft* (German Community) (Rosar 1971, pp. 29ff., 37f., 43). The latter was an ideological and active coalition for the advancement of 'Germanhood', which was formed by Catholic and German nationalist officers and academics and distinguished itself in the fight against the so-called 'anarchic tendencies' ('*anarchische Ungeradentum*') of liberalism, Marxism and Judaism. Organised as a registered society from 1919-1930, this protection posse—counting among its members e.g. Engelbert Dollfuss, Emmerich Czermak, Othmar Spann, Oswald Menghin, Alphons Dopsch,

Rudolf Much—was disbanded when the nationalists split in reaction to the ascent of the *Cartell-Verband.* The circle around Spann, the *Ordinarius* in economics, nevertheless continued to be influential.[48] Spann's neo-romantic-universalist teachings of state and economy served as the ideological superstructure for the fascist *Heimwehr* (Home Defence) and who intended so to lay the foundation for the national-socialist building process. His supporters were also active in the *Fachgruppe Hochschule* (Professional Section of Academic Teachers) of the *Deutsche Gemeinschaft.* The so-called 'yellow' list of 'tendentious' professors of December 1925, for example, which, next to Karl Grünberg, Hans Kelsen and Sigmund Freud mentioned Moritz Schlick as well as Otto Neurath in a discriminating context, originated in this group (Siegert 1974).

Karl Grünberg[49] was Professor of Law and Political Science from 1909-1924 in Vienna and represented Marxist economics as a '*Kathedersozialist*' (academic socialist). He also edited the *Archiv für die Geschichte des Sozialismus und der Arbeiterbewegung* and can be regarded as a predecessor of Austro-Marxism. When Grünberg received a call to Frankfurt, Dean Wlassak (together with Menzel, Hupka, Löffler, Goldmann, Pisko, Brasloff) put forward a motion intended to encourage Grünberg to stay. Schwind (together with Jörs, Glaispach, Spann, Hold, Köstler, Hugelmann) responded with a special motion to the effect that "it would be good not only for Grünberg, but particularly for the School, if he were to follow the call from afar."[50] Under this pressure Grünberg abandoned the idea of staying on and went to Frankfurt. Hans Kelsen[51] was Professor for Government and Law of the State, Administration, and of Philosophy of Law from 1919 to 1929, and also the creator of the Austrian constitution. He was the founder of the *Wiener rechtsheoretischen Schule* (Viennese School of the Theory of Law) on the basis of a strict anti-natural law position. His engagement for the *Habilitation* of Max Adler had already brought him the bitter enmity of the *Deutsche Gemeinschaft.* He was fired in the course of the constitutional reform pushed through by the bourgeois block in 1929. Sigmund Freud, too, was able teach at the University of Vienna as *Privatdozent* and *Titularprofessor* only in the face of great resistance. In other meetings of the *Deutsche Gemeinschaft.* in 1926, Spann, Much, Glaispach, Hugelmann, Czermak and others prevented Max Adler's appointment as *Ordinarius* (Siegert 1971, p. 30f.).

Given this overwhelming dominance of the right-wing bourgeois camp, the question arises whether and how these—literally attacked—persons reacted, other than by an understandable withdrawal or by passive resistance.

As mentioned, all leftist groups, and all freethinking-liberal forces, remained a minority and their politics had almost no effect. This is true of the students as much as of the isolated professors and other teachers in higher education. This state of affairs was also, however, a result of a wrongly conceptualised political practice which, in conformity with the general political climate of the First Republic, built on agitation and on the verbal radicalism of cultural and ideological struggle.

5.2 Hans Hahn and the 'Vereinigung sozialistischer Hochschullehrer' in the Fight for Student Rights

From the ranks of the Vienna Circle there emerged an organised political activity which is little known or considered nowadays. Hans Hahn was for several years both a leader of the *Vereinigung sozialistischer Hochschullehrer* (Union of Socialist Teachers in Higher Education) and a member of the School Committee of the City of Vienna, and established a remarkable publishing record, even though this did not have the expected results. In particular, he advocated the equal rights of students and teachers and the application of democratic principles in conflicts with the academic administration, which in turn obviously favoured the vandals of the *Deutsche Studentenschaft* and tolerated their riots.

For example, as early as 1922, Hahn, together with his collegues Ludo M. Hartmann, Tandler und Grünberg, lodged a protest with the Rector of the University against one-sided refusals to provide meeting rooms and against the riots against Jewish and socialist students and professors.[52] This unrepentant attitude of the democratic teachers in higher education was answered from the right only by renewed defamation campaigns. Thus in 1924 the *Deutsch-Österreichische Tageszeitung* ('German-Austrian Daily') once again circulated a list of 200 Jewish teachers in higher education, featuring amongst others Hahn and Felix Kaufmann.[53] Unerringly Hahn stuck to his principles and continued to demand unrestricted freedom to teach and study, free instruction and state-sponsored vocational education, and entirely meritocratic *Habilitation* procedures, with a mandatory explanation for all rejections (since a number of social democratic university teachers had become victims of a tendentious personnel policy) (Hahn 1924a, 1924b). This did not at all diminish the German *völkisch* racism, for Hahn's name reappeared with those of 45 other colleagues (including Emil Reich) in yet another campaign of denounciation.[54]

The contemporaneously founded *Verein für sozialistische Hochschulpolitik* (Federation for Socialist Higher Education Policy) remained without success.[55] The economist Rudolf Goldscheid's savage critique under the title 'Die Pflanzstätten der Wissenschaft als die Brutstätten der Reaktion' ('The Field of Science as the Breeding Ground of Reactionaries') was a notable exception (Goldscheid 1926, p. 132). Due to the reaction of the outraged professors concerned, it single-handedly stopped the Vienna International University Courses that had been taught since 1922. The only voice against discontinuing them was the calming voice of Schlick, who for philosophical and ethical reasons took the stance that "something that is good in itself should not be endangered ... through personal motives." (Anon. n. d., pp. 11ff.)

Other criticisms by Hahn, Kelsen, Tandler, and Max Adler against anti-semitic riots in 1927,[56] which ended with a renewed closure of the University, were defamed as "demagogic intrigues of Judeo-Marxist university teachers"[57] by the *Deutsche Studentenschaft*, who were given permission by the Rector to build a shooting range in the University. Following the events of July 1927—leaving 90 dead, 84 of whom were civilians—the same Rector felt called upon to appeal in an official circular for donations "for the victims of the 15th of July in the police corps"[58]; Hahn immediately responded with a symbolic donation by his Union to the aid committee of the Social Democratic Party and Trade Union, signalling the censure of the actions of the police.[59] Two years later, handbills warning against attendance of the lectures by Bühler, Kelsen, Tandler and Freud were again being circulated.[60]

The same year, the *Deutsche Studentenschaft* returned to its favourite topic 'Race and Science' and polemicised "in solidarity with the German professors against the incessant Jewification (*Verjudung*) of the universities ... by means of cursed liberalism".[61] Amongst the 200 professors named were Felix Kaufmann, Moritz Schlick and Karl Menger. (We need not ask whether there can be anti-semitism without Judaism.) Once again, anti-semitism functioned to find scape-goats.

5.3 Academic Events around the Transition to 'Ständestaat'

In 1930, the pro-Nazi Rector Wenzel Glaispach decreed new regulations for students distinguishing between 'nations' in order to disassimilate the Jews and promote the *Deutsche Studentenschaft* (Fenz 1978). This constituted the apex of racist university politics to date. The court of constitutional law annulled these regulations in June 1931. The protests of the right-wing radicals of the *Studentenschaft* and ensuing riots resulted in the closure of all Viennese institutes of higher education.[62] One year later Czermak, Minister of Education, tried again to pass through parliament these regulations for students according to their nationality, but failed due to domestic and foreign policy complications. The bloody brawls of the National Socialists escalated until the dissolution of the *Deutsche Studentenschaft* and the introduction of entirely new regulations in the course of installing the Austro-fascist (Dollfuss) system in 1933 and 1934 (Weinzierl 1969, pp. 18ff.).

In the *Ständestaat*, the legal foundations for the elimination of 'elements hostile to the state' were laid immediately.[63] Examples include new rules of *Habilitation* allowing personnel decisions to be made without giving grounds, and, in the course of reorganising and economising, the empowerment of the Ministry of Education to force unwanted professors and assistants into early retirement without due process. In June 1935 the Austro-fascist Federal Culture Council decided on new laws for the universities, which were designed for total coordination (*Gleichschaltung*) in the spirit of the 'New Austria'. They prescribed mandatory attendance at lectures for the *weltanschaulich* education of the new state's citizens and at para-military exercises and camps.

The crisis of the sciences at the "universities in the new state" (Walter 1936, pp. 10ff., 18, 26) was said to have been brought about by liberal attitudes, particularly their emphasis on the contrast between science and world-views. In order to legitimate anti-democratic politics, it had become necessary to elevate an overall insignificant movement for purposes of ideological shadow boxing. The positivist, presuppositionless world-view, the positive method of natural science, became the enemy in the university and the counterpart to the 'idea of unification' and belief in a *Führer*. The conformity of the universities was to be guaranteed by an all-encompassing ideological world-view, a unified idealistic conception of science after the image of the middle ages, and a common 'basic science' like philosophy. But the authorities remained helpless against the continuing academic unemployment.

5.4 Moritz Schlick and the Dismissal of Friedrich Waismann

Only a few months before his murder, Schlick opposed the anti-semitism of the *Ständestaat* with his vehement protests against the dismissal of his librarian Friedrich Waismann.[64] According to his own testimony, Schlick himself had joined the *Vaterländische Front* (Patriotic Front).[65] As a pronounced opponent of National Socialism—he even intended to publish a book abroad[66]—he chose what was for him the lesser evil with his characteristic political naivité (or perhaps ignorance). For this reason he also supported Dietrich von Hildebrand—with whom he entertained very good personal relations despite their academic differences—against the German nationalist professors (Ebneth 1976, p. 107 fn. 399).

How fatal and illusory this engagement of the basically liberal Schlick for the *Ständestaat* was, was shown not only by the press reaction to his murder, but already in connection with his protest against the consistently anti-semitic and anti-positivist policies of the Ministry of Education. In a letter to this ministry he gave one last characterisation of his situation at the University.[67] He pointed to the fact that in 1929, when he rejected a lucrative offer to go to Bonn "because of his attachment to Austria", he was not promised an increased salary, but a librarianship for Waismann, who already had for a long time worked at his institute without salary.

Already then, in fact, Schlick was disappointed about the indifferent attitude of the Ministry of Education and the Austrian government which—other than in the case of Eibl—made no attempt to keep him in Vienna. Only after long deliberations and with a heavy heart he decided, pressed by his friends and followers, to stay in Austria.[68]

Schlick's references to his own willingness to compromise could not forstall the dismissal of his closest co-worker. For the *Ständestaat*, Waismann was just as unacceptable as any other non-Arian, for example Robert Reininger and Karl Bühler's assistants Amalie Rosenblüth and Else Fränkel. (Incidentally, their appointment had already been publically criticised in 1931.)[69] A last proposal from Schlick, namely to employ Waismann at least as a scientific auxiliary worker, was simply disregarded. In the end, Schlick complained that "the teaching of philosophy ... is currently very difficult". Could he have meant *only* the policy of economy?

Schlick's favourite student Herbert Feigl already made a realistic estimate in 1931. He saw correctly that, as a Jew and a representative of the Vienna Circle, he possessed few chances for an academic career in Austria, although Schlick was convinced he could make him a *Privatdozent*. Thus Feigl applied to several American universities and finally emigrated in September 1931.[70]

6. Conclusion

Schlick's murder in July 1936 effectively spelled the end of the public phase of the Vienna Circle in Austria, though a small circle of students under the leadership of Waismann still met until 1938 when National Socialism put a final stop to the movement. It is unfortunate that the newer historical literature on emigration and exile takes only incomplete and haphazard account of the representatives of Austrian philosophy.[71] Although interested circles know the chronology of the dissolution and the dispersion of the Vienna Circle from 1931 to 1938 (one positive exception: Broda n. d.), it still has not been stressed enough that the essentially politically motivated emigration and exile was a response to the climate which had become intolerable for open minds from the early 1930's at the latest (cf. fn. 1). The background and causes of this cannot be understood without considering the so-called 'conservative revolution' in Germany and Austria, in particular the situation at the universities described above.

Universität Wien

NOTES

* First published as 'Aspekte des gesellschaftlichen Hintergrunds und Standorts des Wiener Kreises am Beispiel der Universität Wien', in *Wittgenstein, der Wiener Kreis und der kritische Rationalismus. Akten des 3. Internationalen Wittgenstein-Symposiums 1978*, Hrsg. H. Beghel, A. Hübner, E. Köhler, *Schriftenreihe der Wittgenstein-Gesellschaft Bd. 3*, © 1979, Hölder-Pichler-Tempsky, Wien, pp. 41-59. Slightly shortened; translated with kind permission of Verlag Hölder-Pichler-Tempsky and the author by T. E. Uebel.

1 Ayer 1956; Carnap, 1963a; Feigl, H., 1969a; Frank 1949b; Haller & Rutte 1977; Juhos 1965, 1967; Kraft 1973; Kraft 1950; Menger 1965; Passmore 1967; Popper 1974; Fuchs 1949; Johnston 1972; Janik & Toulmin 1973.

2 This is suggested already in Neurath & Carnap & Hahn 1929; see also Neurath 1936a; Heer 1957; Haller 1966, 1968, 1977; Rutte 1977; Oberhuber & Weibel 1976;

3 [Translator's note: There is no precise correspondence between Austrian academic ranks and Anglo-American ones. Roughly, *Ordinarius* or *ordentlicher Professor* (*o. Prof.*) is a full Professor with his own chair; *ausserordentlicher Professor* (*a. o. Prof.*) is a Professor, Associate Professor or Reader without a chair; *Titularprofessor* is an Assistant or Associate Professor or Senior Lecturer for the specific field indicated in his/her *venia legendi*; *Privatdozent* is a Lecturer or Assistant Professor; *Honorarprofessor* approximates to an Adjunct Professor. (An *Ordinarius* is a civil servant, an *ausserordentlicher Professor* may be; the others are not.) *Habilitation* is a postdoctoral qualification required to teach in universities in any of the above positions.]

4 Allgemeines Verwaltungsarchiv Wien (AVA), Unterrichtsministerium (UMin) 1848-1940. Fasz. 761, 4Cl, 1937/38; 44.432/1938.

5 [Translator's note: On February 12, Dollfuss's declared martial law and the suppression of all oppositional movements; the Social Democrats unsuccessfully attempted to stage a general strike and offer some armed resistance to the abolition of the Viennese city government and the arrest of leading members. See e.g. Jedlicka 1970.]

6 Wiener Stadtarchiv, Verein Ernst Mach (Abschrift), Letter of Moritz Schlick to Hofrat Ganz, Polizeidirektion Wien, 2 March 1934.

7 On the topic of the intellectual and political situation at the university, see Fischer 1965, Roszenich 1974, Fischer-Kowalski 1977, Siegert 1974, Broda 1967, 1978, Zehetner 1972, Wienzierl 1969, Topitsch 1967, Ardelt 1972, Dachs 1974, Ramharter 1973, Zoitl 1976.

8 From 1920-1933: Walter Breisky, Emil Schneider, Anton Rintelen, Richard Schmitz, Emmerich Czermak, Hans Schober, Heinrich von Srbik; from 1933-1938: Anton Rintelen, Kurt Schuschnigg, Hans Pernter; after the *Anschluss* of 1938/39: Oswald Menghin, Arthur Seyss-Inquart.

9 E.g. 'Wissen und Abstammung' in *Der Morgen* 11 December 1922. Unless otherwise noted, the newspaper articles cited in the following come from the Allgemeine Verwaltungsarchiv in Vienna (AVA): Grossdeutsche Volkspartei (Archiv) RI-36, 6 002/258, 6 002/259, 6 700/304. Sozialdemokratische Parteistellen (SD) Nr. 66, 155, 232. SD-Klub 89, 136. Christlich Soziale Partei 135/250.

10 AVA, Bundesministerium für Inneres und Unterricht (BfIU), Unterrichtsamt 1922, 4 Cl, Nr. 391, Abt. 2; Universitätsarchiv Wien (UA) Personalakt (PA) Schlick Z. 531.

11 Ibid.. Höfler had edited, with Hans Hahn, the writings of Bolzano, was an adherent of Meinong, and for a while the Head of the Philosophical Society. Hahn remarked that towards the end of his life Höfler became a resigned and conservative teacher. Cf. fn. 17.

12 AVA, BfIU, Unterrichtsamt Nr. 16839-I Abt. 2, 3 August 1922; UA, PA, Sclick.

13 Wieser 1950, p. 51. For a differentiated characterisation of Boltzmann's views on philosophy, see Broda 1955, pp. 87-93.

14 UA, PA, Zilsel. Zilsel (1891-1944) studied philosophy, mathematics and physics in Vienna, worked as an insurance mathematician during the war and then entered the public school service as teacher at a *Gymnasium* (secondary schools leading to university qualification). From 1922 he was granted leave to teach at the Viennese *Volkshochschulen* (Higher Adult Education). He took a relatively independent position vis-a-vis the Vienna Circle and considered one of its most valuable critics. The contacts tended to run along the 'subsidiary circle' with Gomperz and Kraft.

15 Zilsel to the professors of the Philosophische Fakultät (phil. Fak.)(School of Philosophy), 10 June 1923. His other works were: 'Bemerkungen zur Abfassungszeit und zur Methode der Amphibolie der Reflexionsbegriffe'; *Das Anwendungproblem* (1916); *Die Geniereligion* (1918); 'Versuch einer Grundlegung der statistischen Mechanik'; 'Der einführende Philosophieunterricht an den neuen Oberschulen' (1921).

16 Schlick to Dean, 22 February 1924.

17 Zilsel to Schlick, 23 February 1924.

18 Minutes Phil. Fak., 6 March 1924. Meister succeeded Alois Höfler in the Chair of Pedagogy; he was also a vigorous public opponent of the school reform movement (c.f. Hahn 1926, p. 437). By contrast, Zilsel was very much engaged in this movement (c.f. his 1921 and 1924).

19 Dean of Phil. Fak., 10 March 1924.

20 Report of commission, 19 March 1924.

21 Minutes Phil. Fak., 19 March 1924.

22 Ernst Cassirer to Dean, 1 July 1924.
23 Zilsel to Dean, 3. November 1924, after the commission under the leadership of H. Übersberger decided on the temporary adjournment.
24 Report of the *Verein Volksheim* for the academic year 1922/23 and 1923/24.
25 AVA, BfIU, 1924, Reg. 4, Fasz. 629, Philosophische Lehrkanzeln.
26 Wieser 1950, pp. 188-191. Publications up to 1924: 'Das Problem der Aussenwelt' (1904), *Weltbegriff und Erkenntnisbegriff* (1912), 'Philosophie und Geschichte der Philosophie' (1915), 'Gegenstand, Aufgaben und Methoden der Geographie' (1914), 'Intuitives Verstehen in der Geschichtswissenschaft' (1929), and *Die Grundformen der wissenschaftlichen Methoden* (1924/25). His *Grundlagen der wissenschaftlichen Wertlehre* was not published until 1937.
27 Wieser 1950, pp. 194ff. Publications until 1924: 'Platons Psychologie' (1910) and *Metaphysik und Geschichte* (1914), *Augustin und die Patristik* (1923). Later he published *Der Ausklang der antiken Philosophie und das Erwachen einer neuen Zeit* (1928, with G. Kafka), *Christus und die Künstler* (1931), *Die Griechen und die griechische-christliche Philosophie des Abendlandes* (1933), *Vom Sinn der Gegenwart. Ein Buch von deutscher Sendung* (1933).
28 UA, PA, Rudolf Carnap.
29 Carnap 1963a, Feigl 1969a, Haller & Rutte 1977. After his *Habillitation* Carnap remained *Privatdozent* until 1930 when he received the title of *ausserordentlicher Professor*. In 1931, through the initiative of T. G. Masaryk, he was called to the Chair for Philosophy of Nature at the German University of Prague, where he taught until his politically motivated emigration to the USA in 1935. Works until 1926: *Der Raum. Ein Beitrag zur Wissenschaftlehre* (1922), 'Über die Aufgabe der Physik und die Anwendung des Grundsatzes der Einfachheit' (1923), 'Dreidimensionalität des Raumes und der Kausalität' (1924), 'Über die Abhängigkeit der Eigenschaften des Raumes von denen der Zeit' (1925), *Physikalische Begriffsbildung* (1926).
30 Communication from Prof. Wolfgang Stegmüller.
31 Walter 1936, p. 26; 'Änderung der Habilitationsvorschriften', *Die Presse*, 8 August 1934; 'Regierungsmaßnahmen an den Hochschulen', *Wiener Neueste Nachrichten*, 8 May 1934.
32 Ebneth 1976, p. 36f.; AVA, BfIU, Gz. 39735, 4C, 20 February 1935.
33 Wieser 1950, p. 225f. Works up to 1934: 'Die Idee der sittlichen Handlung' (1916), 'Sittlichkeit und ethische Werterkenntnis' (1921), 'Zum Wesen der Strafe' (1920), 'Max Scheler als Ethiker' (1923), 'Zum Wesen der Autorität' (1927), *Reinheit und Jungfräulichkeit* (1927), *Die Ehe* (1929), *Metaphysik der Gemeinschaft* (1930), *Zeitliches in Lichte des Ewigen. Gesammelte Abhandlungen* (1931), *Sittliche Grundhaltungen* (1933), *Liturgie und Persönlichkeit* (1933).
34 Wieser 1950, pp. 228ff. Works up to 1937: *Weltgeschichte als Tat und Gemeinschaft* (1924), *Die Hauptformen der mittelalterlichen Weltanschauung* (1925), *Das Unendliche in der mittelalterlichen Physik und in der Kantischen Dialektik* (1926), *Die Ethik des Mittelalters* (1927), *Sacrum Imperium* (1929), *Die Metaphysik des Mittelalters* (1930), *Kulturphilosophie* (1932), *Görres spricht zu unserer Zeit* (1933), *Meister Eckhart* (1934), *Kierkegaards Folgen* (1935), *Religionsphilosophie* (1937), *Christliche Philosophie* (1938).
35 AVA, UMin, Fasz. 761, 4Cl 1937/38. IV-2-37.333, 17 August 1938. Dempf was considered "unacceptable for the School of Philosophy due to his clerical attitude".
36 Wieser 1950, p. 199. Works up to 1934: *Diderots Weltanschauung* (1914), *Bedingt der Weltkrieg eine Umgestaltung unserer Weltanschauung?* (1916), *Zur Analyse von Kants Philosophie des Organischen* (1922) *Religiöse Epedemien* (1925), *Die Metaphysik—eine*

Fiktion (1927), new edition of Eisler's *Wörterbuch der philosophischen Begriffe* (1927-29), *An den Quellen unseres Denkens* (1937).
37 'Nachruf Moritz Schlick', *Der Christliche Ständestaat*, 8 May 1936; D. v. Hildebrand, 'Gegen gemeine Verleumdung eines Toten', *ibid.*, 19, July 1936. C.f. in the same issue the article signed "F.", 'Der Elefant und die Jüdische Frage', which also sharply criticises the *Schönere Zukunft* (12 July 1936: 'Der Fall des Wiener Professors Schlick - Eine Mahnung zur Gewissensforschung. Von Prof. Dr. Austriacus').
38 AVA, BfIU, GZ.4035, 1937 4CL.
39 Wieser 1950, p. 218f. Works: *Steigerungsphänomen als künstlerisches Gestaltungsprinzip* (1924), *Geschichte der deutschen Literatur* (1929), *Differentielle Psychologie und Ästhetik* (1932), *Personalistische Ästhetik* (1932), *Ein Beitrag zur Wert und Leistungspsychologie des höheren Gefühlslebens* (1934).
40 AVA, UMin, Fasz. 761, 4 Cl, 1937/38.
41 AVA, UMin, Fasz. 761, 4Cl, 1937/38. 14384-I,1 24 May 1937, 4Cl.
42 Ibid., Zl. 12309-I/1, 3. May 1938, 4Cl.
43 [Translator's note: The *SS-Standarte 89*—an illegal para-military group of National Socialists, formed in 1933 and integrated into the SS after the *Anschluss*—was responsible for the murder of Chancellor Dollfuss on July 25, 1934 (cf. Jedlicka 1970).]
44 Hahn was *Ordinarius* from 1921-1935. Menger habilitated himself in 1926 in Amsterdam and already in 1929 became *Titularprofessor/ausserordentlicher Professor* until 1936. Kurt Gödel was *Privatdozent* from 19333-38 although he already published his path-breaking investigations at the beginning of the 1930's.
45 UA, Dean of Phil. Fak., Lehrkanzel-Besetzung Mathematik, Z. 410, 1936/37.
46 UA, Pa, Menger und Gödel.
47 Cf. 'Wissenschaftliche Qualität oder Parteibekenntnis?', *Der Tag*, 16 June 1924; 'Der Hochschulskandal', *Arbeiterzeitung* (*AZ*), 22 November 1924; also *AZ* 18 June 1924 and 8 December 1925.
48 Siegfried 1974, p. 69f., Diamant 1965, pp. 209-218, Ebneth 1976, p. 76f.
49 Nenning 1968, p. 19; Leser 1986, p. 177.
50 AVA, UMin, Fasz. 755, sign. 4C/1. Ph: Prof. 1922/23.
51 Wielinger 1977; Metall 1969, pp. 44, 61
52 'Proteste deutscher Hochschullehrer gegen den Rektor', *AZ*, 10 June 1922.
53 'Rasse und Wissenschaft auf unserer Universität', *Deutsch-Österreichische Tageszeitung* (*DÖTZ*), 23 April 1924. Felix Kaufmann, *Dozent* in philosophy of law, was interested in the philosophy of mathematics and of the social sciences (*Gesellschaft, Staat und Recht*, 1931). He was intellectually close to the Vienna Circle and participated in their discussions. In 1938 he emigrated to the USA.
54 Handbill of the *Deutsche Studentenschaft*, June 1925.
55 'Verein für sozialistische Hochschulpolitik', *AZ*, 28 June 1925 and 4 July 1925.
56 'Die Helfershelfer der Hakenkreuzler', *AZ*, 15 June 1927.
57 AVA, Grossdeutsche Volkspartei, 6 700/307; 'Juden und Marx an den Hochschulen', *DÖTZ*, 16 June 1927.
58 'Der Rektor sammelt für die Polizei', *AZ*, 25 September 1927.
59 'Gegen den Bettelbrief des Rektors. Eine zutreffende Antwort.', *AZ*, 1 October 1927.
60 'Die Heimwehrbewegung ist nicht antisemitisch', *AZ*, 10 October 1929 (AVA, SD-Parteistellen, Karton 66).
61 'Rasse und Wissenschaft', *DÖTZ*, 13 November 1929.

62 'Der Verfassungsgerichtshof bringt deutsches Recht', *DÖTZ*, 24 June 1931; 'Sperrung der Wiener Hochschulen aus Protest gegen das Schandurteil des Verfassungsgerichtshofes. (Otmar Spann spricht auf der Kundgebung).', *ibid.*.

63 'Änderung der Habilitationsvorschriften', *Wiener Neueste Nachrichten*, 28 May 1934; 'Die Regierungsmassnahmen an den Hochschulen', *Die Presse*, 8 August 1934.

64 "In accordance with the order of the Ministry of Education of 29 January 1936 (Zl. 2818/I-1) the employment of one Friedrich Waismann, employed since 1930 in the capacity of librarian at the Philosophical Institute of the University of Vienna, is to be discontinued." Friedrich Waismann (1896-1959) studied mathematics in Vienna and also lectured at the *Volkshochschule*. Works: 'Die Natur des Reduzibilitätsaxioms' (1928), 'Logische Analyses des Wahrsxcheinlichkeitsbegriffs' (19330/31), *Einführung in das mathematische Denken* (1936).

65 Letter of Schlick to Sicherheitskommissar Wien, 23 February 1934, Wiener Stadtarchiv, Verein Ernst Mach (appeal against the dissolution of the Verein Ernst Mach); letter of Albert Schlick to Richard Meister, 22 October 1936, UA, PA Schlick (statement against the public attacks on Schlick); 'Nachruf auf M. Schlick', *Der Christliche Ständestaat*, 28, June 1936. A totally different interpretation is given in Hollitscher 1977, p. 199.

66 Communication from Prof. Trotzmüller, Vienna.

67 Letter to Ministry of Education, 29 February 1936 (AVA, BfU, 4G Philos.: Philosophisches Institut 1936, 7894-I, 1).

68 Ministers of education then: Richard Schmitz and Emmerich Czermak; cf. Mulder 1968, p. 388.

69 *Der Wiener Student. Ein Kampfblatt für die deutschen Hörer der Wiener Hochschulen.* ('The Viennese Student. An adversarial organ for the German students at the institutes of higher eduction in Vienna.') Similarly, *Kampfruf.*

70 Feigl 1969. Feigl studied mathematics, physics and philosophy in Munich and Vienna and taught at the *Volkshochschule* 1927-30. In 1922 he wrote a prize-winning essay on the philosophical importance of Einstein's relativity theory and represented the epistemological position of critical realism. His *Theorie und Erfahrung in der Physik* (1929) was very well received by professionals, e.g. it was praised by Einstein and Pauli (see Feyerabend 1966, p.6).

71 Cf. Klusacek 1966, Arbeiterkammer 1977, Oberhuber & Weibel 1976, Fleming & Bailyn 1969, Fermi 1968, Spaulding 1968, Broda n. d.

PART II

NEURATH AS A METATHEORETICIAN: EPISTEMOLOGY AND METHODOLOGY

HEINER RUTTE

THE PHILOSOPHER OTTO NEURATH*

I

It is highly problematical to speak of the 'philosopher' Otto Neurath, because he was a declared enemy not only of idealistic metaphysics, but of all philosophy. This renouncement of philosophy came easy to Neurath—he did not have to defend an academic position. What in his writings is called *the scientific world-conception, physicalism, unified science, logical empiricism, encyclopedism*, was not conceived as philosophy, as super-science with a specific method, or as the clarification of sentences by explicatory pseudo-sentences (as by Wittgenstein and Schlick), but as an undertaking within the sciences, as their *unification and encyclopedic integration.*

Still, as it tends to go with enemies of philosophy, their enmity is articulated as a philosophy itself, and what emerged in Neurath's case was a very consequential, very radical and in many ways very modern philosophy. In all this Neurath was an original, but not particularly argumentative thinker. He rather formulated theses and programs and felt himself to be the agitator, the propagandist of the *Vienna Circle* and the *Unity of Science* movement: he ascribed great social importance to the efforts for a scientific world-conception and for unified science and understood them as essential contributions to a new enlightenment, the transformation of society, the liberation of the proletariat and, finally, the fight against fascism and for democracy. This agitational and advertising style of some of his writings, which was intended to *hammer* certain conceptions *into* the reader and which, e. g. in his *Empirical Sociology,* takes a somewhat irritating form, left many disturbed and unsatisfied. Amongst them the bourgeois, unpolitical philosopher Moritz Schlick, *Ordinarius* at the University of Vienna and Neurath's opponent in the Vienna Circle, who believed that the philosopher only had to serve the truth and that truth would be victorious—without propaganda.[1] But there also exists another Neurath: this Neurath is a thinker with a formidably secure instinct, who clearly saw through the hidden weaknesses, the hidden metaphysics and absolutist claims of the new empiricism, and who, in so doing, often

T. E. Uebel (ed.), Rediscovering the Forgotten Vienna Circle, 81–94.

anticipated the later criticisms of empiricism by Quine, Kuhn and Feyerabend, and who consequently arrived, with imperturbable consistency, at a position of stunning radicalism and relevance.

II

What then are the basic principles of this philosophy?[2]

(1) Every comparison of sentence and fact, language and the world, knowledge and experience, is to be rejected. The language of science can speak about itself, one part about the other, but there is no going back behind language. Thus there is no instance of appeal beyond sentences from which we could pass judgement on the comparison of sentence and fact, and so on truth. Sentences can only be compared with sentences.

(2) Truth does not consist in the correspondence of sentence and fact, but in freedom from contradiction and agreement between sentences themselves. (The *coherence theory*—a term admittedly not valued by Neurath.) Truth does not exist in itself, but depends on what sentences we have accepted: that sentence is true which can be integrated into the accepted collection of sentences. Similarly, the falsity of a sentence lies in its contradiction with the accepted collection of sentences.

(3) Every sentence, be it empirical or logical, can be revised, and every methodological rule can be revised (fallibilism.)

(4) A testing of hypotheses cannot be effected without referring to certain observation sentences (*protocol sentences*). (Empiricism.) These, like all others, require confirmation and can be *rejected.*

(5) Observation sentences, like all statements about events, are neither simple nor non-corrigible, but highly complex hypotheses, full of theoretical *workings up*. They are embedded in a mass of sentences of varying origin and thus are themselves the *rich* but semantically indeterminate sentences which, as it were, represent the whole of science. By contrast, the scientific hypotheses about laws are the simpler sentences. As it happens, however, observation sentences remain relatively stable in the republic of scholars (remain relatively accepted), though there is no epistemological justification for this stability.

(6) Thus there obtains no opposition of theory and observation; a pure protocol beside and before theory is impossible. One already must have a rough theory in order to raise questions about observation.

(7) A reconstruction of knowledge on the basis of *simple* and *original* sentences is impossible. The *immediately given* is to be rejected, just like the methodological phenomenalism which builds on these, or like an epistemologically distinguished I-language speaking of one's *own* experiences.

(8) Never is a single hypothesis at issue, but at all times a whole complex of asserted or used hypotheses. Basically, the whole of science is at issue (holism).

(9) If there obtains a contradiction between a hypothesis and a protocol sentence, then we are free to alter or delete this hypothesis, or to alter or delete the auxiliary hypotheses employed, or to alter or delete the protocol sentence (conventionalism).

(10) The agreement between theory and observation can be effected not only in one but in indefinitely many ways: the very same observational basis can agree with infinitely many collections of sentences which are in themselves without contradiction, but which contradict each other. (*Multiplicity*, *ambiguity*, *pluralism* of theories and of prediction.) But this multiplicity holds also for reports of facts and for observation sentences: the observable and the observed are also interpretable in multiple ways.

(11) This pluralism of sentences is only reduced by the singularity of decision, which cannot be logically justified, but follows from the need for action.

(12) Thus there can be no verification of hypotheses (unique decision of truth) nor falsification (unique decision of falsity), but only their confirmation or 'shaking up' (*'Erschütterung'*).

(13) The use of rules derived from the logic of science cannot lead to the required selection from the multiplicity of collections of sentences, that is, cannot justify the inquirer's preference of one collection of sentences. For this reason, these rules must be replaced or at least supplemented by sociological and historical analyses of the actual practice of science. Science need not have methodological ideals imposed upon itself; instead, its historically evidenced procedures need analysing.

(14) In his further development Neurath rejects particularly the thought of the *system* of science as one such unrealistic ideal, and replaces the *system* by the *encyclopedia*, by which he understands roughly the provisional collection of the accepted knowledge of a period (encyclopedism). There are no logically-semantically structured complete systems and even less so does there exist the one distinguished (true) system. It does not even make sense

to view this as an ideal towards which science moves. Verification and falsification appeal particularly to the *system*.

(15) The concrete encyclopedias which change in the course of history are, by contrast, never complete, they contain inevitably vague terms, are semantically indeterminate and often not that clearly logically structured, that is, they are full of *imprecise unanalysed terms* (*'Ballungen'*) The concrete encyclopedias are formulated in a mixture of everyday and highly scientific languages, and have physicalistic and metaphysical parts. If this language is successively cleansed as far as possible of its metaphysical parts which obstruct understanding, then one ends up with an intersubjectively and interculturally very useful *universal slang* (*universal jargon*), which is to be preferred for the construction of a future encyclopedia of unified science. Those sentences are true which belong to that encyclopedia which is commonly accepted.

(16) The meaning of the traditional terms of the languages of everyday and of science is changed in often imperceptible but continuous ways by the addition and modification of theories and facts.

(17) Encyclopedism does reign in the rule of the coherence theory, for the following holds of concrete encyclopedias: (a) often they do not allow the firm determination of non-contradiction because of the use of imprecise unanalysed terms; (b) they contain contradictions, namely, contradictions between mutually incompatible theories which, when limited to certain fields of application, give good predictions, and contradictions between theories and control sentences; (c) these contradictions are often removed, but also often left in place as '*shakings*' of the theory in cases where they are opposed by significant *confirmations* of the same theory.

(18) It follows that those hypotheses are to be considered as *confirmed* which agree with many observation sentences and contradict only a few of those observation sentences which are not deeply grounded in the *mass of sentences* (are relatively *isolated*), so that challenging them would have no great consequences. Those hypotheses are to be considered as *shaken* which agree with only a few observation sentences and contradict many sentences which are well grounded in the entire mass of sentences (not isolated).

(19) Completely *isolated* sentences are those which neither agree with the other sentences of the accepted encyclopedia, nor contradict them, and which are thus without any truth value. Prominent amongst the isolated sentences are the metaphysical sentences. But under certain circumstances, these can be integrated as scientific sentences in the course of the

development of science, or once were integrated in a historical encyclopedia; often it is impossible to decide whether they can be integrated (used) although people do work with them. In this way, anti-metaphysics (see below) is relativised.

(20) Opposed to encyclopedism are the manifold forms of *pseudorationalism* which aim for a *system*, which, at least in approximation, is to represent reality in true sentences without any ambiguity. Pseudorationalism aims for a unique judgement of truth or falsity, for the unique justification of sentences and decisions, claiming that this uniqueness follows from adherence to logical-methodological rules.

(21) The, as it were, ontological background for Neurath's conceptions is given by the *aggregational attitude*: in general, science does not deal with isolated quantities, but every quantity is an element of a *cosmic aggregation*, an unordered collection of spatio-temporal objects, and every datum implies the connection of a particular quantity with a cosmic aggregation. Since our knowledge is itself embedded into such aggregations, there obtain situations of unpredictability which seem paradoxical: predictions are themselves social factors which may prevent or bring about their coming true; future advance in knowledge and other creative feats cannot individually be meaningfully predicted because such a prediction would already realise them before their time.

I wish to call these basic principles the *epistemological* principles of Neurath, despite the fact that Neurath wanted to have nothing to do with epistemology. It is plain that this epistemological program is far removed from that which is typically associated with the label *positivism*.

But Neurath has some other principles which fit better with the common conceptions of *positivism* and of enlightenment:

(1) As long as they are free of metaphysics, all scientific statements refer to the spatio-temporal order, following the model of physics, and their terms are reducible to spatio-temporal ones. (*Physicalism*, materialism.)

(2) All the sciences are conducted in principle in the same way; methodical differentiations like that between the natural sciences and the human sciences ('*Geisteswissenschaften*') derive from metaphysics. Thus the need to build up a *unified science* which connects the laws and predictions of all disciplines. It uses a physicalist *unified language* which is intersensual and intersubjective and expresses only relations of order.

(3) Statements about experiences are contained, from the start, in the total mass of physicalist statements. They are to be interpreted

physiologically and behavioristically, that is, there is no contrast between the *psychical* and the *physical.*

(4) The social sciences deal with social behaviour, with the connection of stimuli between individuals and with the collective behaviour of stimuli-connected groups, without having to analyse the connections between individuals in detail (*social behaviourism*). Marxism in particular represents an empirical sociology which seeks to investigate correlations between exclusively physicalistically formed entities.

(5) Meaning too must be reconstructed in behaviouristic terms. As physicalist entities, synonymous sentences are stimuli which under identical conditions bring about identical reactions. Similarly, the *formal validity* of logical sentences must be defined by the *change of certain behaviours.*

(6) Metaphysics as secularised theology is meaningless, and unified science is always to be kept free of metaphysical remnants (compare the epistemological principle (19)). Admittedly, the stimulating effect of metaphysical views in the formulation of theories in the history of science cannot be discounted, but it is always the case that the relevant theory is representable in its scientific worth independently of these views.

(7) In contrast to metaphysics, the traditionally based world-view of magic, which historically preceded the theological phase of thinking, is physicalistic and addresses itself to perceptible alterations of terrestrial matters. In so far, magic is closer to the scientific world-conception than theology. Its systematic refutation would be very difficult, due to the possibility of conventionalist procedures, but the falsity of the magical world-conception can be assumed.

Of course, both these programs, which I differentiated, form one unity for Neurath. Indeed they connect at various points, but I believe that the connection is not too close—one could affirm the epistemological principles even independently of the positivist ones. It is historically noteworthy that Neurath affirmed holism and conventionalism and a historically informed perspective on science already at the time of World War I and its aftermath. In the heyday of the Vienna Circle Neurath affirmed, in addition, physicalism, Marxism, empiricism, the coherence theory and anti-metaphysics. Encyclopedism develops after 1934: in this phase physicalism recedes into the background, anti-metaphysics mellows and Marxism does not seem to play a role any longer. (In this reconstruction I can rely here only on the available philosophical writings of Neurath.)

III

Who are the thinkers by whom Neurath's thinking was influenced to a particular degree? (Since Neurath obviously read very widely, I can only list the most important names.)

(1) There are first of all Ernst Mach and Pierre Duhem, who determined his interest in a philosophy of science informed by the history of science. In the case of Mach, what attracted Neurath were his way of practicing the theory of science as an anti-metaphysical enterprise internal to science, the biological-psychological conception of the process of cognition, and the orientation towards the unity of science—less so his sensualism. Like his academic colleagues of the time before World War I, Philipp Frank and Hans Hahn, Neurath grew up in an intellectual atmosphere dominated by Mach. In the case of Duhem, it was of course the holism and conventionalism which Neurath adopted early on. (Mach, incidentally, viewed Duhem's main work as a continuation of his own efforts.) If today one speaks of the Duhem-Quine thesis, one really should speak of the Duhem-Neurath-Quine thesis in order to indicate the historical development of these ideas. Further influences named by Neurath include Poincaré, Enriques, Avenarius, Jevons, Abel Rey, William James, Karl Pearson, and Russell; William Whewell would have to be added to this list.

(2) In logic, particularly Ernst Schröder with his *Algebra of Logic* and Gregorius Itelson, who remained unknown, left their influence on Neurath. His interest in a terminology cleansed by logic and the analysis of language appears to derive from Itelson. (Neurath started an *index verborum prohibitorum* which is likely to have contained just about all those terms which are dear to philosophers.) In this connection, we may also remember J. A. Stöhr's language-based criticism of metaphysics.

(3) Another source of his thinking are materialism and eudaemonism: Epicurus, the French enlightenment thinkers and the English utilitarians (Bentham, John Stuart Mill), and finally Marx. Neurath looked at materialism not as metaphysical and preferred to declare his *physicalism* to be a new form of materialism. His connections to Marxism are not too close; it would be totally wrong to call him an 'Austro-Marxist.' Neurath is more likely of the type of the utopian social reformer: he always declared his allegiance to the work of Josef Popper-Lynkeus. Neurath's comprehensive socialisation plans, which he developed at the end of World War I and thereafter in the spirit of a *social engineer*, need be seen as deriving from this

and other sources. His subsequent manifold activities in Austrian social democracy served the enlightenment and guidance of the proletariat and the youth: what motivated him was the increase in pleasure and happiness in a concrete sense, not Marxist theory, to which he declared verbal allegiance but which he never discussed more deeply. Neurath viewed Marxism simply as the most advanced, that is, the most predictive social science of his day; for instance, it was essential for him that Marxism's attempts to explain were divorced from the concepts of a personal consciousness and personal motivation, as were those of the behaviourism which he valued. Later, in emigration, one even finds ironical remarks against Marxism (e.g. Neurath 1941a, p. 225 and 1944, pp. 48/9 fns. 28, 44); what gains in importance is his concern for the defense of liberal democratic ideals.

(4) Finally, not to be forgotten for the strong imprint they left on Neurath, there are the encyclopedic, even already partially unity of science oriented intentions of philosophers, from Leibniz via the French encyclopedists, Comte and J. S. Mill, to Spencer and Wilhelm Wundt. For it was to the only rudimentarily realised project of an *Encyclopedia of Unified Science* and its foundation in a new encyclopedism that Neurath's philosophical efforts were directed in the last decade of his life.

(5) I should mention also that it was characteristic of Neurath's reception of philosophy that he was never able to establish a positive relation to the typical German philosophy (Kant, German idealism, neo-Kantianism, phenomenology, etc.). To him, German philosophy always appeared like an island of idealistic metaphysics which had emerged out of the spirit of Protestantism and closed itself off from the new empiricist-scientific currents from abroad (a few exceptions aside). Neurath was unable even to see something of value in the neo-Kantianism that was so strongly oriented towards the theory of science, and he remained mostly sceptical even of a thinker like Max Weber, who contributed so much to the scientific world-conception, due to his relation to neo-Kantianism. By contrast, Neurath lauds the Austrian tradition and stresses its uniqueness (e.g. its independence of Kant). Due to its origin in Catholicism and the scholastics, and Leibniz and Herbart, the Austrian tradition remained open to logical analysis, but also integrated foreign traditions, and so shows, all in all, strongly empiricist, anti-metaphysical, anti-idealistical, scientific features. This holds not only for Austrian positivism, for Mach and his epigones, but also to some degree for the school of Brentano and Meinong (Neurath 1936a).

IV

The influence which Neurath himself has exerted upon other philosophers has been little appreciated up to now. After all, until recently Neurath was to be counted amongst the *forgotten thinkers* and, as regards the Vienna Circle, influential authors like Victor Kraft tended to downplay, rather than emphasise his influence.

(1) Certainly it can be said (and Rudolf Carnap confirmed this repeatedly) not only that Neurath was the propagandist and organiser in the Vienna Circle and the *Unity of Science* movement emerging from it, but that he also provided enormous stimulation in terms of substance or content—in the Circle and later. So much so that Popper once called him *the strongest personality* of the Circle.[3] First, it was Neurath who always constructed the historical, social and political connections to the ideas of the Circle; second, it is likely to be primarily Neurath's merit to have effected the turn from sensualism and phenomenalism (the language of experience) to physicalism (the *thing-language*) and behaviourism. (The true working out of this conception was, however—as always—undertaken by Carnap, and Neurath doubtlessly profited from this, just as he early on accepted suggestions from Carnap's *Logical Structure of the World*.) And the increasing conventionalism in the conception of the empirical basis and the related turn away from the classical conception of truth similarly derived from Neurath.

With all this, Neurath strongly influenced particularly Carnap, but also younger followers of the Circle, like C. G. Hempel and Arne Naess. The further development of Carnap and Hempel shows, however, that this influence was a passing one. They remained empiricists who reconstructed the language of science by logical means as a *system* and never truly accepted Neurath's conventionalism and holism which, as noted, tended towards a historical *behaviouristic of scholars* and an *encyclopedia* as an anti-system. In return, Neurath recorded his scepticism of later developments like that of a logic of induction: for him, inductive procedures were an *inventive, active behaviour* that was admittedly of historical importance, but which was a matter of decision and remained unjustifiable by a logical calculus.

Neurath had his problems with what in most portrayals of the Vienna Circle stands most prominently in the foreground, the meaning criterion, probably because he always rejected the related belief in the immediate given. Generally, it seems to me that, while he influenced a number of thinkers in the Circle, Neurath was not himself influenced in any essential way—despite the impression which Carnap made on him (disregarding the differences of later times) and despite the expressions of respect for Philipp Frank, Hempel and others. Of course, Neurath knew how to value the efforts of the Vienna Circle and related groups for a strictly rational analysis of logic, language and science, and it is certain that he learned many a thing from them; however much he stressed the unity of the movement though, Neurath always pursued in this his own path which pointed in a totally different direction from the formalism and theory of meaning of the Vienna Circle and its tendency to begin the reconstruction of knowledge with a few simple building blocks. Thus later on he was interested more in encyclopedist efforts (e. g. Joergensen's attempts), whereas he derived little pleasure from Tarski's influential rehabilitation of the classical conception of truth: Neurath clearly saw in advance what later did become the case in analytical philosophy, namely, that out of the logical-empiricist movement there devoloped 'a new *metaphysica modo logico demonstrata*, a new scholasticism'.

The development of the Vienna Circle which was co-determined by Neurath in significant ways also meant an increasing distance from Wittgenstein's *Tractatus*, from its logical atomism, its rejection of a meta-language, its concepts of the *unsayable* and of the *showing-itself*, and Wittgenstein's conception of philosophy as *elucidation*. All these doctrines were suspect for Neurath, and one can say that Neurath was the true opponent of Wittgenstein in the Circle. Not unjustly so, he always viewed Wittgenstein as a metaphysician who philosophised about the relation between language and the world and who in so doing entangled himself in pseudo-formulations about what can and what cannot be said. Neurath rejected these juxtapositions in principle, because by their means it was intended to speak with language—but in an, as it were, *pre-linguistic fashion*—about a beyond of language, a pre-linguistic state. Once Neurath summarised the metaphysical sayings of Wittgenstein: most of them were central theses of the *Tractatus* (cf. Neurath 1933a, pp. 1/2, esp. fn. 2). Wittgenstein, in turn seems to have ignored Neurath.

(2) Of course, Neurath was involved in still different controversies, e.g. with Schlick, Karl Popper, Ake Petzäll and Russell. Neurath's criticisms of these and other authors derived throughout from his constant effort to reject the *remnants of metaphysical claims to absoluteness* in empiricism. Schlick who stood closely aligned to Wittgenstein, for instance, attempted to defend the classical conception of truth and the empirical foundation of knowlege—

Neurath answered with his conception of coherence, his fallibilism and conventionalism (see Neurath 1934).

Russell had similar problems with Neurath. He attacked the coherence theory of truth as idealism and relativism and defended the idea that statements had a relation to reality and that truth be decided uniquely by experience, whereas Neurath rejected these *ontological misinterpretations*, because he believed that there existed no such presupposed instance of a *judge* with a constant personality in possession of truth. Often he stressed that in the process of testing scientific claims we cannot be defendant, prosecutor and judge at the same time (Neurath 1941a). (Incidentally, Neurath had already previously criticised Russell's *bourgeois* socialism (Neurath 1929a).) Ake Petzäll too, prompted by the Circle's doctrines about protocol sentences, tried to establish an epistemology which discussed questions of the authority and the distinguished dignity of statements—Neurath rejected epistemology and favoured a *behaviouristic of scholars* according to which *holding-true* is reduced to being accepted and there are no *distinguished* sentences indicating the end of testing (Neurath 1936c). Neurath's criticism of Popper's *The Logic of Scientific Discovery* appears particularly anticipative: he accused Popper of an absolutism of falsification and of wrong system thinking. Neurath pointed to the illusion of uniquely determined decisions of falsification and to the *defensive manoeuvers* of scientists—who knew how to defend themselves against falsifying *attackers*, can retain a falsified theory with good reasons, and do not even necessarily remove contradictions in their theories. Similarly, Neurath saw that the concept of an approximation to the *ideal true system* of science is inherent in Popper's system thinking (Neurath 1935b).

(3) Which were the most important after-effects of Neurath's philosophical work?

Amongst the post-World War II philosophers it is W. V. O. Quine who exhibits a particular affinity with Neurath; Neurath is in many ways the predecessor of Quine. (How much actual influence was in play is difficult to say; that it existed was noted by Quine himself.[4]) Like Quine, Neurath is of the opinion that all sentences, including those of logic, can be revised; that what is always at issue is a comprehensive context of hypotheses and that in case of an empirical contradiction, one may revise, under conventionalist considerations, either the hypothesis at issue or an auxiliary hypothesis or the contradicting observation sentence, without there being firm rules of revision; that several mutually incompatible theories may agree with the same empirical reality; that a revision of theory means a revision in the language employed; that an empiricism of the immediate given and of the construction of concepts and statements from the given is to be rejected; that observation sentences nevertheless constitute the preferred instance of testing; that agreement with observation sentences is particularly stable intersubjectively and interculturally (and that relativism is rejected in this respect); that epistemology is only meaningful not as an *a priori* but as a natural scientific, psychological inquiry; that language and meaning are to be construed only behaviouristically in the emprical stimulus-reaction schema.

The influence of Neurath on T. S. Kuhn is unclear. Still, Kuhn comes from the *Unity of Science* movement so inspired by Neurath: at the end of the '40's he belonged to a *Unity of Science* discussion group in Cambridge (Mass.), in which James B. Conant, Philipp Frank and Percy W. Bridgman also participated (Ludz 1979, p. 234), and his famous *The Structure of Scientific Revolutions,* was published in the *Encyclopedia of Unified Science,* originally edited by Neurath. What connects Kuhn and Neurath? It conforms with conventionalism that Neurath stresses the role of the defensive practitioners of science against their opponents (analogous to the distinction between *normal* and *revolutionary* science in Kuhn)[5] and that for him there are no methodological rules or critical canons of acceptance, reaffirmation or rejection of theories to be found, so that decisions are made according to social rules of the republic of scholars which cannot be comprehended methodologically, but only scientifically-historically-sociologically (in the *behaviouristic of scholars*). Further, Neurath stresses that scientific theories are subject to change of meaning and that older succesful theories do not always represent approximations of later more successful ones; similarly, he rejects an *experimentum crucis* between competing theories; and he believes generally that, despite good predictions, scientific theories nevertheless perpetually lead to inescapable contradictions. The thesis of the theory-dependence of observation, stressed by Popper, Kuhn, and many others, was affirmed by Neurath in the clearest terms as well. In short, the agreement with Kuhn's conceptions cannot be overlooked, whatever the actual influences may have been.

Finally, it is worthwhile to consider the question of Neurath's influence on Paul Feyerabend. Feyerabend is a student of Viktor Kraft and Arthur Pap, that is, experts of neo-positivism, and one can presuppose in his case—as with Quine and presumably with Kuhn—a sufficient knowledge of Neurath's works. Feyerabend himself says: "In the Vienna Circle only Neurath had a clear conception of the properties of scientific research. Used to abstract speech, he expressed these particularities in a somewhat strange manner."[6] Indeed, given the close affinity between Kuhn and Feyerabend, Neurath's agreement with Kuhn's theses already means agreement with Feyerabend's theses. And since Neurath negates, or at least delimits in their authority, scientific criteria like verification, falsification, truth, approximation to truth, freedom from contradiction, since he doubts the existence of universally valid methods of induction and testing, and since he, ultimately, hands the acceptance of scientific hypotheses over to the decisions of the investigators, it is easy to develop his ideas further to the total denial of a *proper* scientific methodology and a measure of rationality generally, and to the postulation of mere ad-hoc strategies which are bound to the concrete problem situation. It is unlikely that Neurath went that far, despite the agreement mentioned which extends to details like certain parallels in the understanding of the magical world-conception and the specific attitude towards Mach. For, against Feyerabend, there is in Neurath the tendency to isolate metaphysics from science as far as possible and not to grant unlimited authority to cultural relativism. Similarly, the underlying attitudes seem to differ: Neurath remains a science-oriented thinker who nevertheless wants to view science realistically and not according to schematic models.

V

I hope I have shown that a closer investigation of Neurath's philosophy could free us from several clichés about the Vienna Circle and neo-positivism, which were created by some critically-minded authors, for example Popper, in order to underline their own originality and to let the Vienna Circle appear as a very limited and cranky movement. Already a closer and particularly less prejudiced look at a thinker like Otto Neurath would provide us with much of the riches of this movement. The authors of the Vienna Circle are, once they are actually read, much more interesting than one is lead to believe following the monochrome pictures of its critics and enemies. But word of that seems to have gotten around: the rediscovery of the Vienna Circle is, after all, in full progress.

Karl Franzens Universität, Graz

NOTES

* First published as 'Der Philosoph Otto Neurath', in *Arbeiterbildung in der Zwischenkriegszeit*, Hrsg. F. Stadler, © 1982, Österreichisches Gesellschafts- und Wirtschaftsmuseum, Wien, pp. 70-78. Translated with kind permission of Österreichisches Gesellschafts- und Wirtschaftsmuseum and the author by T. E. Uebel.

1 On Schlick, Neurath and the other personalities of the Vienna Circle, see also the conversation Rudolf Haller and I had with Heinrich Neider (Haller & Rutte 1977).

2 Support for the following summary portrait can be found in Neurath 1981, passim; cf. Haller & Rutte 1981.

3 See Popper's memories in Neurath 1973, p. 55; cf. Carnap 1963a.

4 E.g. Quine 1960 and 1969. It is particularly Neurath's famous simile of the sailors which has always inspired Quine. "We are like sailors who have to rebuild their ship on the open sea, without ever being able to dismantle it in dry-dock and reconstruct it from the best components." (Neurath 1932d, p. 92.) The parable can already be found in Neurath's *Anti-Spengler*: "We are like sailors who on the open sea must reconstruct their ship but are never able to start afresh at the bottom. Where a beam is taken away a new one must at once be put there, by using the old beams and driftwood, the ship can be shaped entirely anew, but only by gradual reconstruction." Prior to this Neurath remarks (in the sense of Duhem) "that every statement about any happening is saturated with hypotheses of all sorts and that these in the end are derived from our whole world-view." (Neurath 1921a, p. 199.)

5 The affinity between Neurath and Kuhn was pointed out previously by R. Hegselmann 1979a, pp. 44f.

6 Feyerabend 1978. [Translator's note: This phrasing in fn. 48 in the German edition of *Science in a Free Society—Erkenntnis für freie Menschen*, Suhrkamp, Frankfurt, 1979—differs in the addition of the second sentence in place of the parenthetical "as opposed to philosophical analysis" in fn. 11 in the reprint in *Philosophical Papers vol. 2*.]

RUDOLF HALLER

THE FIRST VIENNA CIRCLE*

I

The thesis I present for examination is this. Even before the founding of the so-called Vienna Circle around Moritz Schlick, there existed a first Vienna Circle with Hans Hahn, Philipp Frank and Otto Neurath. This circle is of such constitutive importance for the formation of the circle around Schlick that the judgement can be justified that it was really Hans Hahn who founded the Vienna Circle. To draw attention to this I call the one the *first*, the other the *second* Vienna Circle.

Like every theorist, the historian finds no limits for the formation of hypotheses but their intelligibility. As long as his statements are meaningful assumptions, he has fulfilled one of the necessary conditions of research and possesses the right to have his judgement examined. Initially I do not claim more for my considerations.

Before I defend and justify my thesis I wish to say that it is not just *historical* curiosity which suggests or demands an interest in the question, although the current revitalisation of the discussion of the Vienna Circle does also justify such an interest in the history of its origins. Rather, what demands an explanation which goes beyond the mere description of differences and brings genetic aspects into play, is the question of the *philosophical* and *systematic* divergences in the known Vienna Circle. After all, the revitalisation of the discussion of logical positivism does not spring from the investigation of questions of the nature of theoretical and practical rationality, which emerged only later (Mohn 1977, Hegselmann 1979a), but begins at the end of the 1960's (Barker & Achinstein 1969) and finds its clearest testimony in the monumental *Vienna Circle Collection*.

This revitalisation constitutes first of all a parallel to the rising historicism in the new theory of science. Insofar as this historicism was directed against the standard view of scientific theories, as it was supposedly held in the Vienna Circle, one tended, as I must at once stress, to identify the so-called 'received view' with the theories of the Vienna Circle. It would have only been natural to search out these ideas anew in order to become acquainted with their true nature. For in the meantime, due to the quick

T. E. Uebel (ed.), Rediscovering the Forgotten Vienna Circle, 95–108.

development of the analytical theory of science, the picture of logical positivism had become a cliché which rested mainly on a few simplifications. In this picture, the members of the Vienna Circle supported a reductive theory of knowledge according to which all statements were to be reduced to those about *immediate givens*, or at least could be constituted out of elementary experiences; second, the logical empiricism of the Vienna Circle was *ahistorical* and interested only in the logical-analytical and systematic problems of the sciences; third, the Vienna Circle promoted the model of the *cumulative progress* of scientific theories, and thus an unsupportable conception of theory change. On various occasions I have already pointed out that this picture is wrong and requires radical correction (Haller & Rutte 1981, p. viii; Haller 1977a, 1982a [ch. 10], 1982b). The supposed dogmas of logical empiricism were criticised by the logical empiricists themselves, and some members of the Vienna Circle never held theses which would be contaminated by those just mentioned. Bringing this out clearly appears to me to furnish already sufficient reason for renewed concern with the *real* history of the Vienna Circle. A further reason is simply that logical empiricism was for a long time the only philosophical theory which sought to understand the present century's scientific picture of the world given by logic, mathematics, physics, biology and sociology in a unified perspective, and which made the most developed cognitive tool of evolution, namely, scientific methods themselves, the central object of philosophical analyses. In comparison to this epistemological perspective, one of the original motives of the neo-empiricist movement—the anti-metaphysical enlightenment—could not but recede into the background, although it still thoroughly determined the *scientific* world-conception. The role model of Ernst Mach influenced not only the known Vienna Circle, but particularly so the 'first' to which I now turn.

II

In the introduction to a collection of his essays, Philipp Frank described very clearly and in detail how the first Vienna Circle came to be, which themes dominated the discussions, and who were the main participants (Frank 1949a). Frank's report starts from the fact that around the years 1907 to 1912 there existed in Vienna a group of young scientists who met regularly—incidentally, on Thursdays, like the later Vienna Circle—and whose interests were described as essentially methodological and theoretical, but also as prompted by "political, historical and religious problems". The report makes clear that Ernst Mach and the French theorists of science, Pierre Duhem, Henri Poincaré and Abel Rey, were the main authors who formed and influenced the attitude of this group of philosophically interested scientists. These names also emerged, still earlier than in Frank's, in Neurath's descriptions of the history and prehistory of the Vienna Circle. And indeed, it is these names which we meet again as guides for our orientation in our search for the main lines of thought of the first Vienna Circle.

This type of the theory of science was admittedly never wholly excluded in descriptions of the history of logical empiricism, but mostly it was not investigated in the depth it deserved. And what was particularly overlooked was the degree to which the so-called French conventionalists, together with Mach and Popper-Lynkeus, influenced the ideas of the first Vienna Circle. We are in the first decade of our century during which the main works of Duhem and Poincaré were published—and during which the most important of these were translated into German. Thus *La Theorie Physique: Son Object, Sa Structure* was published in 1908 in the translation by Friedrich Adler with a foreword by Ernst Mach, only two years after its first publication in book form in French.[1] It was preceded by the translations of Henri Poincaré's *La Science et l'Hypothèse* and *La Valeur de la Science* . Also in 1908 Abel Rey's *La Théorie de Physique chez les Physiciens Contemporains* was published in German translation by Rudolf Eisler.

I believe one can represent the central thesis of these French conventionalists with the picture which Poincaré gave of the various geometries (Poincaré 1902; cf. Haller 1983). Poincaré rehearses hypothetical models in order to show how the different perspectives which we occupy in respect of objects result in non-self-contradictory structures, structures which we construct ourselves inasmuch as it depends upon our decision which assumptions we make and accept. We construct the frame and fit into it what we agree about. This agreement—the truly conventional moment—is, of course, not a factual one and also not the consensus of an ideal discussion round, but is a determination or definition which is viewed as rational yet cannot be justified by logic and experience alone.

What makes possible this path between (or, if you like, beyond) apriorist rationalism and empiricism? By not attributing to science the aims usually attributed to it, namely, to provide a description of phenomena that is as comprehensive and as precise as possible and to discover their basic laws.

The answer with which Mach opposes the traditional conception of scientific descriptions of the world already moves the value criterion of the conventionalists into the foreground. Science is not *mere* description, but an *economical* description, that is to say, a description of phenomena and their relations which is reduced to simple elements (Haller 1981; cf. Rey 1905 (1908, pp. 68-109)). The imagination, which presents us with the infinite space of possibility, finds itself increasingly limited by the frugality of rational models with respect to a given domain of objects. The 'adaptation of thought to facts', demonstrated by the natural history of evolution, is continued and perfected in theory construction. Of course, never without the price of simplification, schematisation and idealisation of objects.

Often we are offered a wrong picture of the teachings of Mach—not only, by the way, by the opponents of empiricism—as if he had been, first, a pure inductivist and, second, a pure phenomenalist. Both interpretations are totally unsupportable and only serve to confuse. Without any ambiguity Mach stressed again and again the intervention of the principle of economy in the process of ordering experience; translated, this means the intervention of rational operations in sensual experience. Therefore Mach did not recognise the priority of observation to theory, nor that of the subjective to the objective standpoint. So Mach shares with the French conventionalists the point of view that our experiences are not only ordered according to theoretical criteria, but that the invoked ideal of unique determination is afforded only by decisions which we make in the delimitation of the products of our imagination and in simplifications in general. Mach also agreed with Duhem that "a very large number of experimental values may be assigned to a theoretical value", (Mach 1908, p. iv; cf. Mach 1905, p. 349) and that only a provisional validity of our laws of nature already follows from this.

Indeed, this was one of the most essential insights of Duhem: that to one experiential datum one could give various theoretical interpretations, and that the correspondence with experience—the only criterion of truth for a philosophical theory—is determined by, and bears the stamp of, our decisions. Of course, there remains a great difference between the point of view of Mach and that of a natural classification as the reflex of a real existent order, as propounded by Duhem, for nowhere is Mach inclined to accept Duhem's metaphysical superstructure. But in the characterisation of theory construction, in the estimation of the constructive moments of abstraction, of simplification, no difference between the two seems to obtain.

We cannot deny that some of these arguments need be accepted if we do not want to falsify the nature of scientific research. Thus we must admit that without any description at all no fact may be declared to obtain or not to obtain. We refer to facts and events by individuating them by linguistic means, by describing them as 'a so-and-so' or as 'the so-and-so'. And as we know, we can arrive at an infinite series of descriptions for every fact. And, analogously, one and the same fact may be explained by different theoretical sentences. In the pursuit of practical purposes this multiplicity does not interest us very much: we select the simplest and most convenient, depending upon the requirements of the case in question. But we *choose* and can choose. And matters are not different in the theoretical disciplines: only there, we are also interested in the multiplicity of possibilities.

The decisive considerations which Duhem adds to Mach's about the principle of economy are, first, that there is no *experimentum crucis* which could decide the correctness or truth of one of two theories between which we may choose for the explanation of one and the same phenomenon; second, that, in both the testing of an experimental hypothesis and of a theoretical sentence, we confront the *whole* system of observations with that of theoretical sentences.[2] This holism is difficult to formulate theoretically, but it was perhaps the greatest influence upon the first Vienna Circle. And it is this that has been virtually discounted, if not overlooked, in all descriptions of logical empiricism.

It would be, as noted, an exaggeration to think that Mach's theory of elements totally agreed with the theories of the French conventionalists. Rather, it was just from the tension between the positions taken in the theory of science by Poincaré and Duhem on the one hand and Mach on the other that the bundle of questions emerged which determined the problems which occupied the first Vienna Circle.

I believe that the two French theorists of science primarily arrived at a different interpretation of the laws of nature from Mach because they were interested in *other* problems and investigated other constituent elements of scientific theory construction. The starting point of Poincaré's arguments for a conventionalistic interpretation of theories was the comparison of Euclidean and non-Euclidean geometries. Since—presupposing the axiomatic nature of the theories at issue—different axiom systems can be used for the description of the same structures, the essential question had to be concern for the criterion of the selection of theoretical sentences. And the answer was: if the description of a phenomenon can be separated into two components, a theoretical and an empirical one, and if the former depends upon *conventions*, then the selection of a preferred hypothesis cannot be grounded by criteria of coordination, even less so of correspondence. But if logical or ontological rules of preference cannot guide us, then there remain only pragmatic ones, as e.g. simplicity and convenience. Whether a geometrical system is to be preferred thus does not depend only on whether it sufficiently 'approximates' the facts, but whether it possesses properties which make the coordination with observable phenomena more easy than another, alternative system, and whether it can be integrated into the theoretical system so far used.

What we see here is basically another consequence of Mach's idea of economy. For Mach, the simplification or idealisation of a theoretical sentence is guided by the pragmatic criterion of 'adaptation'; similarly, Duhem repeatedly says: "An infinity of different theoretical facts may be taken for the translation of the same practical fact." (Duhem 1906, p. 134.) It is for this reason that a contradiction by experiment does not determine the truth or falsity of a physical hypothesis. The only test of a theory which Duhem declares to be satisfactory is the comparison of "the entire system of the physical theory with the whole group of experimental laws." (Ibid., p. 200.) It is impossible to know which of the premisses of a many-levelled theory has to be revised if an experimental finding contradicts a theory (Poincaré 1902, pp. 151/2). The explanation which Poincaré provides is illuminating. The core of his argument consists in the thesis that it could be possible that the actual relations within a particular domain were conceived of correctly by the theory, but that the contradiction "only exists in the images which we have formed to ourselves of reality." (Ibid., p. 163.)

It seems to me that here an essential aspect of theory construction in general is touched upon: the framework, within which theoretical hypotheses are developed and made precise, forms the frame of an assumption of intelligibility which concerns the total context of the phenomena of the investigated domain. In most cases we are here led by the natural world-conception, the view of common sense, and the pre-formed ways of predication in natural language. But in many areas scientific research 'violates' such a pre-understanding and does not achieve the re-constitution of the natural picture of the world from which it derived its models. That is why the separation of meaning and designation or, in Frege's terminology, of sense and reference, is more fruitful for the reconstruction of these relations than the use of a rough set-theoretic terminology. And that is why, according to Duhem's proposal, one should use and consider in the selection of a hypothesis not only the logical rules and mathematical and physical principles and postulates, but especially also common sense (*'bon sens'*). Since it is in principle impossible to build up a theory from experimental or observational data alone, that is, purely inductively—for the derivation of theoretical sentences always presupposes or implies further theoretical sentences—every observational fact requires also that it be judged with respect to a theoretical framework.

It is in this that the true justification of the holistic conception lies. It is certainly an exaggeration then to hold that for the conventionalists natural science does not represent "a picture of nature but merely a logical construction." (Popper 1935, p. 79.) As I just mentioned, the integration of the conceptual construction is guided by pictures. Rather, what is demanded—and this demand we find in Poincaré as well as in Rey or Duhem—is that a law must always also have an operative sense, if is to be understood as the definition of an object (according to principles of lawlikeness). In other words, the application of a law, e.g. in operations of measurement, decides about its usefulness and appropriateness. And the conventionalists realised, correctly and early on, that not only verification but also falsification cannot constitute a criterion of selection, and therefore of demarcation.

If then there is no logical (deductive) or empirical (inductive) necessity of preference in this most important of steps in theory construction, the selection of hypotheses, if—as in the pragmatic conception of truth—subjective attitudes and circumstances form the criteria, then we must admit the place of randomness and caprice in the formation of theories. Many see in this the danger of a conventionalist interpretation of scientific research. But just this thought was accepted as liberating and used in the first Vienna Circle, and particularly in the philosophy of Neurath. Why?

Before I return to this question in the next section, I wish to consider briefly the third of the earlier mentioned theorists of science, Abel Rey. I do not thereby recollect a pathbreaking thinker, but a synopticist who was important for the members of the first Vienna Circle in a special way, primarily for the reason that he connected a wealth of material from the texts of contemporary physicists with a clear position and a decisive judgement. Many of the themes and—what is more astonishing—many of the theses which were held in the well-known (second) Vienna Circle can be found explicitly in the work of Rey. First, he shares with Duhem the methodological premiss that a theory of science without regard to the history of science is empty, because the change of scientific theories happens only on the basis of their historical forms and can only be reconstructed on this basis. Second, Rey advocates a theory of the *unity* of science which comprehends philosophy within it and determines its tasks.

In Rey's treatise of 1905 we thus meet already the contrast between the cumulative and the non-cumulative interpretation of the progress of knowledge in the exact sciences in the guise of the contrast between mechanists and conventionalists. Rey, who ever anew discusses sceptical objections to the possibility of actual progress in physical knowledge, seeks to show that it is one and the same domain, one and the same reference, to which both these metatheories recur.

The 'empirical rationalism' of which Rey speaks, unites Mach's empiricist adaptation theory of the evolution of cognition with the idea of a new rationalism whose basic thought rests in the fact that objective experience and thought mutually determine each other, or literally, are "mutual functions of each other". (Rey 1905 (1908, p. 363).) Such a convergence of antagonistic directions seems called for already because, in the interpretation of Rey, empiricism leads "necessarily to relativism". (Ibid., p. 308) One reason for this is seen to be the nature of hypotheses themselves: they are revisable by definition. All generalisations, that is to say, all inductive premisses, are supported by experience and can be falsified by experience.

> It follows from this that they do not hold definitively and unchangingly; they are capable of development like the hypotheses, which, however fundamental, are always made more precise, complete, better. - Thus they permanently remain open to revision and to the further delimitation of their domain. (Ibid., p. 319.)

Rey always very clearly seeks to specify the historical location of the perspective at which modern natural science has arrived, namely, one beyond dogmatism and scepticism, the two extreme possibilities to which rationalism and empiricism lead, or sink. Rey therefore sees the essential task of the metatheory of the exact sciences in the development of structuralism, of a theory of relations. He remains a Machian insofar as he reduces the necessity, which seems represented in the fact that certain relations do obtain, to habits which evolution formed in us. The rational and the empirical are thus understood as moments of a historical development.

Also from this quarter then, the members of the first Vienna Circle came to be well-acquainted with the problems of a historical theory of science—a fact which apparently has remained unknown to the anti-positivists of our own day.

III

In order to recreate the spirit of the group around Hans Hahn, it is advisable to make reference primarily to writings which derive from the dependence I here postulate, and which originate from the time before Schlick's arrival in Vienna and before the publication of Wittgenstein's logical-philosophical essay.

Philipp Frank—who in 1912 had been called to Einstein's old chair in Prague—stated in his fundamental essay on 'The Importance for our Times of Ernst Mach's Physicalistic Theory of Knowledge' (1917) that, first of all, the main value of Mach's teachings consisted in the defense of the edifice of physics against attack from the outside, not in its productivity for physical research itself. This is a correct remark about the character of foundational investigations: they do not represent inquiries in parallel with object-theoretical investigations, but *meta*scientific inquiries into scientific research, inquiries also intended to analyse and answer epistemological questions about the type of research in question. This main value of Mach's teachings showed itself most clearly in the attempt to "bring about a connection between physics on the one hand and physiology and psychology on the other that is as free of contradictions as possible", a connection that would dispose of the need to switch the entire vocabulary of concepts when crossing the boundaries of these disciplines. That is, Frank correctly sees in Mach's attempt in the theory of elements—which he interprets phenomenalistically—one step towards the unification of theories, towards unified science. To express this in the lively image of Mach's: "As the blood in nourishing the body separates into countless capillaries, only to be collected again to meet in the heart, so in the science of the future all the rills of knowledge will gather more and more into a common and undivided stream." (Mach 1896, p. 261.) This was one of the reasons for Mach's importance for the intellectual life of the day, which Frank stressed. The other consisted in that Mach literally carried the spirit of enlightenment into the 20th century: it was ultimately his critique of superfluous entities which made Mach doubt the existence of atoms.

Frank, the follower of Einstein, naturally regards Mach's critique of the foundation of Newtonian mechanics, the critique of the conception of absolute space, as the prime example of a critique in the spirit of the principle of economy, and a critique which also was directed against a purely mechanistic physics. We must agree with Frank when he explains Mach's scepticism of atomic theory with reference to his critique of the 'auxiliary concepts' of materialism. Since the original enlightenment was bound up with a theory whose 'auxiliary concepts' like 'matter' and 'atom' had remained impure, it was necessary for the *truly* enlightened mind to criticise these concepts. And that had been the way of Mach's great works in the history of science, of which we need mention only his *Mechanics* (1883) and the early essay 'The History and Root of the Principle of the Conservation of Energy' (1872).

Again and again Mach criticised the *basic concepts* of natural scientific research—space, time, substance, causality, etc.—for he aimed to rob them of their metaphysical-mythological mantle and to determine their place in the framework of the theories of natural science, e.g. psycho-physics. And like logical empiricism later, Mach already used an argument from the critique of language in order to eliminate what he deemed metaphysical concepts, namely, the argument of their meaninglessness. According to this argument, the application of a pseudo-descriptive expression does not result in a false, but a *meaningless* sentence. And for the same reason, even the question at issue can be discounted as a pseudo-problem. It is primarily this latter aspect which guides Frank's characterisation of Mach as an enlightenment philosopher *par excellence*: "The task of our age is not to fight against the enlightenment of the 18th century, but rather to continue its work." (Frank 1917, p. 75.)

These then are the two basic tendencies which Frank ascribes to Mach. But these are also the postulates which are upheld as norms for scientific research; Frank always remained a convinced Machian in this sense. Transcending his own position in the theory of science, Mach became the herald of the natural scientific conception of the world, as it was advocated by Frank, Hahn and Neurath. Enlightenment and, for that reason, liberation from metaphysics, was and remained the first and ever essential aim of the scientific enterprise.

This impression is immediately confirmed if we consult Neurath's writings, in particular the longest of those which he dedicated to the history of the Vienna Circle. There we read:

> Mach's and Einstein's criticism of Newtonian physics and the new conceptual edifice which resulted therefrom had a particularly specific influence in Vienna. Already as a young physicist, Philipp Frank stood in direct contact with Mach and Einstein. (Neurath 1936a, p. 695.)

Like Frank, Neurath viewed Mach's anti-metaphysics as the sharpest tool in the continuation of enlightenment. And in his justification of the basic empiricist attitude, Neurath, throughout all phases of his development, always referred back to Mach.[3] But however early we find invocations of the authority of Mach, there we also find references to the conventionalism of Poincaré and Duhem, in Frank as in Neurath.[4]

Already in his early work of 1907 on the causal law, Frank formulated a thesis which he himself literally designated as 'conventionalist.'

> The law of causality, the foundation of every theoretical science can be neither confirmed nor disproved by experience; not, however, because it is a truth known *a priori*, but because it is a purely conventional definition. (Frank 1907, p. 54.)

Frank proposes the following interpretation of the real content of the law: "If in the course of time the state A of the universe was followed by state B, then whenever A occurs B will follow it." (Ibid.; cf. Haller 1982f.) In other words, Frank accepts Hume's interpretation of causality as regularity and postulates its conventional definition to be a rule of research for natural science.

The ground is here prepared for the model of two languages, which in the course of the development of logical empiricism was varied in different ways. The two languages are identified with the two branches of natural science.

> Whereas experimental natural science describes the properties of bodies as given by our senses, and the changes in these properties, the task of theoretical science is to provide bodies with fictitious properties which first of all insure the validity of the law of causality. Theoretical science is not research but a sort of remodeling of nature; it is the work of the imagination. (Frank 1907, p. 57.)

Here the form of the combination of Mach's teaching with those of the French conventionalists becomes clear. The free postulation of assumptions becomes comparable to the conception of Kant's critical idealism insofar as both claim a presupposed framework of experience; the Kantian movement ascribes this to the nature of reason whereas Frank declares it to be a product of human free choice.

Mach's basic idea that assumptions, schematisations and idealisations are ultimately products of the imagination, and that science therefore derives from the same source as the other cultural achievements of humans, is identified (with an emphasis equal to that of the conventionalists) with the conventionalists' idea that these assumptions etc. are the product of logically unconstrained, free choice. This means that the nature which is the object of theoretical science "is not empirical nature but a fictitious nature". This nature is not therefore only the object of human curiosity, but also the product empirical action, and only because of this it becomes transparent, comprehensible. The constructive traits of a theory such as Frank envisaged, whose definiteness is created and creatable, are of course not mirrored in experiment, in experience. For this reason, experience has more play than theory has. These created pictures of the world are therefore never true or false but, depending on their '*Gestalt*', their structure, they are either simpler or more complicated. As is common in conventionalism, the principle of simplicity steps into the place of truth.

In a similar fashion, Wittgenstein says in his *Tractatus* that we have always the possibility to chose an arbitrary form of description, and that one may be simpler than another. He calls these forms 'nets' or 'systems' of descriptions of the world. Accordingly, the causal law is not a law which refers to objects, but something which *shows* itself in the fact that there *are* natural laws, that is, lawlike connections (Wittgenstein 1921, #6.32 to #6.361). Likewise, Wittgenstein at this juncture seems compelled to stress the accidental aprioricity of such a net as the distinguishing characteristic vis-a-vis the otherwise similar Kantian conception.

In his philosophical main work which appeared 25 years after his essay on the causal law, Frank corrects his reference to similarities with the Kantian position. He now says:

> If I had already recognized then that the doctrines of so-called 'idealistic philosophy', even of 'transcendental idealism', are not empirical statements—because they are neither tautological nor do they say anything about the real world—but assertions about a world beyond all human cognition, then I would have placed much less emphasis on the analogy between the conventionalist and the Kantian conception of causality. The analogy exists, to be sure, namely, in that according to both views the causal law does not say anything about the empirical world, but only about how humans conceive of the world.

And Frank continues in the vein typical of the Vienna Circle: "But if one wants to take a position that is conducive to scientific progress it is important to stress the *contrast* to the *Kantian* conception of the causal law." (Frank 1932, p. 242; cf. Stegmüller 1960.)

It is impossible here to follow up these reasons in greater detail, though a separate investigation of the main difference between a conventionalist and a Kantian interpretation is required. What is required for present purposes was said already, namely, that what gets corrected in the course of further development is the supposed or actual similarity with Kant, not the acceptance of the kernel of the conventionalist thesis that general laws have the character of tautologies which are simple and afford perspicious representation. For this reason they were defined as adaptable to the given state of research and revisable.

Now Neurath formulated the principle of this revision very early on. And although he follows mostly in the steps of Duhem and Poincaré he went further, in my estimation, than the other conventionalists. To be sure, both Frenchmen stressed again and again that a law could not collide with experience—the reason why according to Duhem there can be no *experimentum crucis* (Poincaré 1902, pp. 105, 179, 135 ff., 151/2; Duhem 1906, pp. 188ff., 208ff.). And both nevertheless asserted the revisability of every theoretical system which failed to harmonise with experience. But neither of the two thinkers formulated the dichotomy of possibilities so clearly as Neurath—the reason why I called this principle of adaptation in theory change the 'Neurath-principle' (Haller 1979c [ch. 3], 1982a [ch. 10]). One can comprehend this principle only if one comprehends the holistic interpretation of scientific systems which Neurath apparently takes over from Duhem with all of his arguments. Since there can be no univocal description of the facts of experience with reference to theory, the dignification of an isolated sentence as an instance of falsification must be rejected. Similarly, it is not possible to apply a criterion for the selection of a hypothesis in an *experimentum crucis*. Since in the Duhemian holistic conception of theories it is furthermore only a scientific system *in toto* that is accepted or rejected, the question was bound to be raised *how* a revision is possible in the case of an imaginary or actual case of conflict of experiential sentence and system. The so astonishingly obvious thesis of Neurath's says that either the system can be changed (the opinion of the French conventionalists) or the sentence itself, because both are of the same status. If one interprets this principle through theoretical naturalism and the perspective of Quine, then it follows that revisions at the periphery of the system are often the ones first thought of, but that in the end it is the experiential sentences which, as 'cornerstones of semantics', help carry the system and which may claim a certain privilege.

As has become plain, these problems did not only re-emerge in the famous protocol sentence debate of the early 1930's, but they accompany the interpretation of theory structures up to the present day.

Without doubt it was due to the development of conventionalist empiricism in the years of the first Vienna Circle that the problem of the interpretation of theories was joined again at the very focus of the position of the members of the first Vienna Circle. When Hans Hahn returned from

Czernovicz and Bonn in 1921 to Vienna, the old group, whose members were also very close in their political convictions, met anew. And it has since been often stressed that the Circle around Schlick was constituted in the main by members of Hahn's mathematical seminar.[5]

I am not of the opinion, which it seems fashionable to suggest these days, that the philosophical divisions in the Vienna Circle, as they became apparent in the protocol sentence debate, derived from the political fronts in the Vienna Circle (which are likely to have existed from the beginning). Rather, these divisions derived from the return to the problems of conventionalism, which had receded into the background due to the discussions of the *Tractatus* and its related epistemology. But it has not been the purpose of this essay to speak of the polarisation of the conception of science in the well-known later Vienna Circle (Haller 1982c [ch. 16]). It will be sufficient if it has become clear what constitutes the basic edifice of the conceptions of the theory of science of the first Vienna Circle. Recognised once, it can be recognised elsewhere again.

Karl-Franzens-Universität, Graz

NOTES

* First published as 'Der erste Wiener Kreis', in *Epistemology, Methodology and Philosophy of Science. Essays in Honor of C. G. Hempel on the Occasion of His 80th Birthday, January 8, 1985* (ed. by W. Essler), *Erkenntnis* 22 (1985) 341-358, © 1985, Reidel, Dordrecht. Translated with kind permission of Kluwer Academic Publishers and the author by T. E. Uebel.

1 Remember that the first translation of Duhem into English (by P. G. Wiener) did not follow until 1954! And the earliest contributions by J. Agassi, A. Grünbaum, N. R. Hanson which discussed Duhem did not appear until the end of the 1950's. Amongst the descriptions of neo-positivism the contribution of conventionalism was merited a separate section at least in Kolakowski 1971, however.

2 Compare Duhem 1906, ch. 6; Rey 1905 (1908, pp. 122ff.). Compare also Grünbaum 1960; Giannoni 1967; Harding 1976; Diederich 1974; Schnädelbach 1971: 165ff.

3 See also Neurath 1915b, and the description, rich in historical materials, by Stadler 1982b, pp. 11-125.

4 As one can see from the index of Neurath 1981, Mach, Poincaré and Duhem are the most cited names, aside from himself and his contemporaries.

5 Compare the eulogy for Hans Hahn by Philipp Frank (1934) where Hahn is called the "real founder of the 'Vienna Circle'"; cf. Haller 1982b and Menger 1980.

ECKEHART KÖHLER

ON NEURATH'S WRITINGS ON LOGIC, ETHICS AND PHYSICS*

This paper provides an overview of seven, mostly shorter essays by Neurath on the interpretation of logical algebra, on the measurement problems of utilitarianism, and on the methodology of science with reference to the history of optics. These contributions show like no others the breadth of Neurath's interests and his knowledge of detail, but they also show that he was at his best not in the role of a purely scientific academic writer, for these works clearly exhibit his characteristic overenthusiastic immaturity. In alternating fashion, we can find extremely penetrating judgements and very practical, well-pointed remarks on the methodology and the history of science, but also presentiments, which, however clever, are not yet thought through, and even wrong-headed misdeductions.

The first group of essays on logical algebra falls into his early scientific career before World War I, the same period that saw his founding of war economics. Two reasons presumably moved him: one derived from his early enthusiasm for Ernst Schröder's *Lectures on the Algebra of Logic* (1890-95) which—one of the great experiences of his youth—he read from cover to cover in one night (he started his studies at the University of Vienna in 1902 with mathematics); the other one derived from his wish for collaboration with his wife Olga Hahn—sister of the Viennese mathematician Hans Hahn—who also worked in mathematics.

Neurath's intense interest in Schröder might seem rather strange nowadays, since Frege and Russell had already started to publish their path-breaking works (Frege 1893-1903, Russell 1903) which are partly still relevant today—and which would have been very relevant to Neurath's problems, as will be shown below—and since Schröder has been pretty much forgotten amongst historians of foundational research in mathematics. The apparent strangeness of Neurath's interest therefore stems from the presumption that Schröder was not very original. In fact, however, he was a key figure in the development of foundational theory, in particular of model theory. It was he who collated in an encyclopedic fashion the results of

T. E. Uebel (ed.), Rediscovering the Forgotten Vienna Circle, 109–115.

earlier work, particularly those of the English writers Boole (1854), Jevons (1879), DeMorgan (1847) and Dodgson (Lewis Carroll), and of the American Peirce (1883, on the logic of relations and quantifiers). It was from his school that the most important result of model theory emerged, the Löwenheim-Skolem theorem (Löwenheim 1915); more generally, it was the algebraic Boole-Schröder conception of logic that led, via Skolem (cf. Skolem 1970) and the Hilbert school (Hilbert & Bernays 1934-39), to the metamathematics that is dominant in foundational theory today. Whereas mathematicians ignored the isolated works of Frege and Russell as too philosophical, they judged Schröder to be a well-renowned specialist colleague. He did not even have to suffer the war of the intuitionists Kronecker and Poincaré against Cantor's set theory (cf. e.g. Zermelo 1932). That Neurath worked on Schröder then is not very remarkable, except that he sought to improve the kind of foundational theory most common in German-speaking Europe.

First the following three essays: '*Ernst Schröders Beweis des 12. Theorems: Für die identischen Operationen gilt das "Kommutationsgesetz"*' ('Ernst Schröder's Proof of the 12th Theorem: the Law of Commutation Holds for Identity Operations'), '*Eindeutigkeit und Kommutativität des logischen Produkts ab*' ('Uniqueness and Commutativity of the Logical Product *ab*'), '*Definitionsgleicheit und symbolische Gleichheit*' ('Identity of Definition and Symbolic Identity') (Neurath 1909b, 1909c, 1910b). Following closely the axiomatic structure of Schröder's algebra of logic, Neurath investigates the concept of equality with reference to the set theoretic operation of intersection (which Boole called 'logical product', written 'ab' for the sets *a* and *b*, nowadays mostly written '$a \wedge b$'). Boole noted the algebraic isomorphism between sets and numbers and so used the same notation for both—something Frege criticised heavily. Maybe it was just this lack of a sign for the logical product in Schröder's notation which lead Neurath to the idea to 'symbolically equate' *ab* and *ba*. This idea follows a semantical consideration in the definition of intersection:

$$\text{(i)}\quad a \wedge b =_{\text{def}} \hat{x}\,(x \in a \;\&\; x \in b).$$

Trivially, the clauses around '&' may be commuted. According to Neurath *ab* and *ba* are given by the same definition, and thus are 'the same product.' Thus Neurath claims that the commutativity of intersection (namely, $ab = ba$) need not be proven separately. Neurath fell victim to his own conjuring tricks, for the commutativity of the mentioned clauses around '&' in the definition depends on the commutativity of the logical connective '&' and involves a small metatheoretical step. *After the permitted commutation one still has a different definition*:

(ii) $b \cap a = \text{def}\ \hat{x}\ (x \in b\ \&\ x \in a)$

That these differently defined products are indeed identical requires a proof, whether it be given traditionally, as by Schröder, or whether it be given metatheoretically, as by Neurath who gives a metatheoretical proof that the definitions are logically equivalent. Calling in addition for the identification of logically equivalent definitions, as Neurath does implicitly, in the end amounts to declaring unnecessary the proof of any mathematical identity. So naturally, the identification of two definitions requires (as noted) a metatheoretical proof, one which according to the theorem of deduction must always be available. Neurath even rushed on to urge that one could do without those signs which have a definition equivalent to that of a sign already introduced: possessing *ab* we would no longer need *ba*. But in this case we could no longer express the commutativity of intersection. A very similar train of thought would lead to the exclusion of the commutativity of union, even the exclusion of any mathematical identity. And that would lead to the loss of a central part of mathematics, namely, the ability to prove that different methods of calculation may lead to the same result.

At the end of his third essay, Neurath gives an example which makes his project precise. He introduces the concept of 'symbolic identity' for the case of the commutativity of intersection and uses as his example the equation $b\sqrt{a} = a^{1/b}$. Apparently Neurath thinks that $ab = ba$ is the same as $ab = a{\cdot}b$! One wants to cry and laugh simultaneously. Neurath even admits in footnote 3 of the same paper that an asymmetrical representation of the definitions of $a+b$ and $b+a$ allows for the proof of commutativity, but still does not recognise that the definitions for *ab* and *ba* are asymmetrical after all (see (i) and (ii)). Instead, he closes with the tantalising remark that "[t]he interesting fact that the logical law of commutativity depends on the axiom system suggests an investigation of how far the law of commutativity of other calculi is eliminable by varying the axiom system. A systematic axiomatics is increasingly needed nowadays." This systematic axiomatics was then developed in Hilbert's school (proof theory), by Carnap (1934/7), and in the Polish school (Tarski 1956).

Alas, Neurath would have found a better way if he had read Frege's 'On Sense and Reference' published already in 1892, which delimits the problem much more clearly. Frege also recognises that equations may appear to be superfluous because the expressions to the left and the right have the same reference in case of actual identity. But the cognitive gain consists in that the same reference is given through different 'ways of presentation' (Frege), i.e. different senses. Following Frege, Carnap later introduced conceptions of identity of different strength—like Neurath, though with a better justification—for example, logical equivalence and intensional isomorphism (Carnap 1947). Today, these and similar considerations constitute a central part of modal logic and the theory of intension.

In their '*Zum Dualismus in der Logik*' ('On Dualism in Logic') Neurath and Olga Hahn introduce several changes in the axioms and the proof procedures of Schröder's system in order to "let the nature and the origin of dualism appear more clearly" (Neurath, O. Hahn 1909). The discussion remains throughout in Schröder's original framework. Put into modern terminology, Neurath and Hahn understand the principle of duality as follows. Sentences about the intersection or the union of sets hold exactly there where otherwise identical sentences concerning their intersection or union hold of the complementary sets. Symbolically expressed: Every true sentence yields another true sentence by exchange of '$\cap$' by '$\cup$' iff every set a is exchanged for its complement a'. Example: $a \cap b = \emptyset$ yields $a' \cup b' = 1$. (Note: $\emptyset' = 1$.)

The main contribution of Neurath and Hahn consists in relating duality to complementary sets, whereas Schröder related them to the connected but difficult to handle concepts of the power and extension of sets. They criticise that Schröder did not prove duality, but only affirms it inductively as an empirical principle evidenced by the theories proven so far. Strangely enough, in Neurath and Hahn there is also no proof to be found, but only a careful listing of dual theorems, for both authors do not recognise that duality is not a theorem which could be joined to others, but that it is a derived proof procedure, that is, a metatheorem about sentences, which cannot be expressed in the axiom system itself. Once again Neurath rushes with a wild hunch into the as yet unexplored field of metatheory. It is no wonder that he later was to value Carnap's efforts in syntax! For it was he who was responsible for their basic ideas (Carnap 1934/7, esp. the Foreword).

Neurath's essay '*Das Problem des Lustmaximums*' ('The Problem of the Pleasure Maximum') delimits the following problem. Utilitarians like

Bentham and Mill helped to found political economy, but, remarkably enough, did not use utilitarian principles as auxiliary tools of economics. "What could be more natural than to examine certain institutions, such as monetary economy and economy-in-kind, free competition and mercantilism, to see whether under given circumstances they can realise the maximum of happiness or not?" (Neurath 1912a, p. 113.)

Neurath very cleverly discusses many of the problems of what are nowadays called 'utility scales'. He introduces 'pleasures' as utility functions of humans and of objects, and asks for their comparability. He mentions the related known problems of ordinal and cardinal scales and uses *intervals*, long before S. S. Stevens[1]. He also deals with questions concerning the types of scales—questions nowadays common in measurement theory, like that of permitted transformations, determination of zero point, the Archimedian principle etc.—but without mentioning the relevant literature like Helmholtz or Hölder. Neurath's negative judgement concludes with the lofty remark that the computation of the individual utility functions for a whole population would already be impossibly complicated.

Faced with so many problems, Neurath demands the use of a dictator in the sense of modern decision theory, that is, of an arbiter or referee who independently settles open questions of weighting (commonly according to certain fairness principles).[2] The development of game theory on the basis of von Neumann's axiomatisation of utility theory and of his proof of the unsolvability of the most common games (namely, mixtures of cooperative and competitive games) (v. Neumann and Morgenstern 1944) seems to show that Neurath was right; similarly so, the emergence of Arrow's paradox and the subsequent attempts to get around it with the help of a 'dictator', for instance a legislature; and even more so the development of bargaining theory by Harsanyi according to which the judge's role is taken by solutions decreed from above (Harsanyi 1977).

In two connected essays still topical today—'*Prinzipielles zur Geschichte der Optik*' ('On the Foundations of the History of Optics') and '*Zur Klassifikation von Hypothesensystemen*' ('On the Classification of Systems of Hypotheses') (Neurath 1915a, 1916a)—Neurath undertakes to derive a methodological moral from the history of optics. Optics appears as particularly fitting for Neurath because, on the one hand, optics had been a clearly demarcated field for centuries up to about 1850 and was only rarely influenced by other investigations, particularly acoustics, and because, on the other hand, the development of optics exhibited sufficiently many contrasts which show interesting features of scientific development.

Neurath's main concern is to demonstrate that the development of optics was not so much determined by the 'rough' (see quote below) dichotomy of undulation/emission (wave/particle duality), but instead was governed by a plurality of characterisations by features which do not fit this common dualism. Neurath focusses on the features of periodicity, polarisability, Huyghen's principle[3], and interference, and investigates with references to these the contributions of the most important theorists of optics, like Descartes, Grimaldi, Malebranche, Huyghens, Newton, Euler, Young, Fresnel etc. In doing so Neurath takes account of the relevant works of Whewell (1873), Mach (1883), Duhem (1906) and Poincaré (1902).

What turns these essays into classics of the history and the methodology of science is Neurath's impressive procedure to explain several cases in the history of optics, which otherwise remain puzzling, with reference to these features. For instance, Whewell counts Descartes amongst the emission theorists whereas Whewell's German translator Littrow, in an addendum, calls Descartes a predecessor of the undulation theory. Neurath establishes that Descartes emphasised features of both the particle and the wave theory.

Neurath stresses that a proper classification of features is indispensible for a mature historiography of science. But even more he is interested in the use of classifications as methodological norms. In particular, he criticises that in the construction of several well-known theories not all variations of then already noted features have been taken account of. For example, Huyghens ignored the periodicity of light entirely.

The consideration of the historical development of science makes clear Neurath's pragmatic, Darwinist side. Rejecting simplifying dichotomies, he writes:

> Dichotomies, however, are not only crude intellectually, but also mostly the product of scientific pugnacity. One characterises the opponent as pungently as possible for the purpose of beating him down as forcefully as possible. At such occasions transitions are only troublesome. Thus dichotomies are a result of a war-like spirit. I do not want to investigate in detail here how far dichotomies, precisely through their deficiencies, have a stimulating effect on scientific life, as pointed out by Vaihinger. Even if that were the case, they would be useful for science perhaps, but themselves unscientific. (Neurath 1916a, p. 15.)

Neurath retains these and related views and develops them further during his collaboration in the Vienna Circle, as R. Hegselman stresses in his excellent overview (1979a, p. 40). In the Vienna Circle Neurath emphasises the importance of scientific disputes for the motivation and development of theories. On the one hand, he recognises the importance of 'well-timed' introductions of concepts and theories, but he also notes that their omission would have been possible just as well—even though the omitted theories would have been more 'adaptable', i.e. better confirmed. The direct inclusion of these pragmatic points of view of the history of science into the concept of scientific rationality itself—see his normative conception of feature classifications—places Neurath with today's 'psychological', 'sociological', or 'historical' movements in the philosophy of science represented by Hanson (1958), Toulmin (1953), Kuhn (1962) and Feyerabend (1970). This is all the more poignant since these latter more or less strongly attack the legacy of the Vienna Circle, the so-called 'received view' which equates the property of being scientific with the potential to be axiomatised. What centrally engages Neurath's interest, however, were just those historical changes of concepts and theories, conceived of both normatively and axiomatically.

Institut für Höhere Studien, Wien

NOTES

* First published as 'Zu einigen Schriften von Otto Neurath über Logik, Ethik und Physik', in *Arbeiterbildung in der Zwischenkriegszeit*, Hrsg. F. Stadler, © 1982, Österreichisches Gesellschafts- und Wirtschaftsmuseum, Wien, pp. 107-111. Translated with kind permission of Österreichisches Gesellschafts- und Wirtschaftsmuseum and the author by T. E. Uebel.

1 See the historical representation in the standard work Krantz et al 1971.

2 "The demand that each must keep sight of the happiness of all at once leads to practical difficulties if one is not to allow the decision of one man as mandatory for all, or if one merely gathers together individuals who share the same basic views about the best distribution of pleasure." (Neurath 1912a, p. 122).

3 It says that light is transmitted like a sphere and that every point reached becomes the centre of another spherical transmission.

RUDOLF HALLER

THE NEURATH PRINCIPLE: ITS GROUNDS AND CONSEQUENCES*

After the powerful intellectual eruptions at the end of the old and the beginning of the new century, out of whose stream our thinking and research is still being nourished, the time seems to have come for taking breath, for taking stock and reflecting. There is a search for sources, a return to the fixed stars of the past, only to orient oneself anew in the confusion of the present. This undeniable fact seems to be, in any case, one of the reasons for the renaissance of interest in the Vienna Circle, and even in its direct and indirect ancestors—or, to phrase it according to a thesis of Neurath's which I find appropriate for my own efforts—this fact seems one of the reasons why there is today not only a world-wide interest in the Austrian literature, fine art or architecture of this century, but also in that which can be called *Austrian philosophy*.

From this perspective it is no accident that the last 10 years saw the publication of more of Neurath's works, of translations of his publications, and of works about him, than all previous years. Admittedly, we cannot claim to have started this movement, but we need not be embarassed either. Not a few matters were seen differently here than elsewhere; still, the 'anticipatory resonance', which Neurath also speaks of, was missing. But in the end that does not matter much.

One such point concerns the history of the Vienna Circle itself, whose basic outlines derive from Neurath himself. Neurath always abstracted from the essential differences within the Vienna Circle in his descriptions of it in order to support his thesis of the unified character of its collective work. These differences found reflection only in terminological questions and in the question as to which members of the Thursday circle were to be called 'real' or 'not so real' members of the Vienna Circle. Thus came about the myth of the unity of this group of researchers, a group which influenced philosophy in our century as deeply as no other, and in whose tradition the entire analytical philosophy of the second half of this century is located (cf. Haller 1982b).

With distance, the differences become clearer, and with it also the figure of Neurath. William Johnston describes him in his assiduous, knowledgable,

T. E. Uebel (ed.), Rediscovering the Forgotten Vienna Circle, 117–130.

but not always reliable history of Austrian culture and intellect, *The Austrian Mind*, as one of the most neglected geniuses of the 20th century. And it is true that Neurath was a universal genius: he worked productively in the fields of political economy, economic history, sociology, mathematical logic, philosophy, theory of science, and in the fields of pictorial statistics, political science, town planning, museology etc. He achieved more in any one period of his life than many during their entire lifetime.

We owe the first Vienna Circle to Hans Hahn, Philipp Frank and Otto Neurath. Without this circle there would have never been *the* Vienna Circle. Around the end of the first decade of our century these scientists discussed the problems of Mach's philosophy and the discrepancy between the positivist and the conventionalist interpretation of science. But the discussion was taken up again only after Hahn had returned to Vienna via Cernovicz and Bonn, and Neurath via Leipzig and Munich. Through Hahn's efforts towards the appointment in Vienna of Moritz Schlick, the first Vienna Circle also gained a professional philosopher. Since Schlick was the only one in the Circle who held a chair in philosophy, he almost naturally occupied the role of *primus inter pares*. The role which Neurath played in and for the Circle, however, remains still insufficiently investigated, although it was certainly he who emitted the most dynamic force, who was the main organiser of the Circle and the movement which it spawned, and who even amongst the philosophers exerted a steady and increasing influence—despite the fact that he lacked an academic position (cf. Haller 1979a [ch. 2]).

Already in 1934, in his *The Logical Syntax of Language*, Carnap refers to Neurath:

> In the discussions of the Vienna Circle, Neurath has been conspicuous for his early—often initiatory—and especially radical adoption of new theses. For this reason, although many of his formulations are not unobjectionable, he has had a very stimulating and fruitful influence upon its investigations; for instance in his demand for a unified language which should not only include the domains of science but also the protocol-sentences and the sentences about sentences; in his emphasis on the fact that all rules of the physical language depend on conventional decisions, and that none of its sentences—not even the protocol-sentences—can ever be definitive; and finally, in his rejection of so-called pre-linguistic elucidations and of the metaphysics of Wittgenstein. It was Neurath who suggested the designations 'Physicalism' and 'Unity of Science.' (Carnap 1934/7, p. 321.)

From this passage we can already see what stature Neurath possessed in the Vienna Circle, in which he became the real antipode of Wittgenstein and so also of Schlick. It would be totally wrong to interpret this politically.

Neurath was fully convinced of the task to insure that the *scientific world-conception* should permeate all forms of personal and public life, of teaching and education, of architecture, and that social and economic life should be governed solely by rational principles. For this reason metaphysics was his arch-enemy, and all relapse into it was a reversal for enlightenment though whose victory alone the scientific world-conception could prevail.

In an unpublished letter to Franz Brentano, Ludwig Boltzmann once says of metaphysics that it causes migraine and retching. Thus the task of philosophy was to heal humanity of this migraine. And Ernst Mach called it the biological task of science to provide the fully sensible human individual with an as complete orientation as possible—without any metaphysics, since no point of view possessed absolute validity. These are only two modest testimonies from the background of that period in the history of Austrian science and philosophy from which Neurath emerged, this truly positivist spirit who did not want to be a positivist. Neurath was a social scientist by training, gained his *Habilitation* in 1917 in Heidelberg in political economy, and had published more than 150 essays already before his well-known writings in association with the Vienna Circle. Of these, only a few concern philosophical questions, most treat of socio-economic problems, from the economic history of antiquity to the tasks of war economics.

These writings have been left mostly unconsidered in the literature so far. But one glance though them would have shown not only that a number of methodological questions had already been broached there, but also that the principles and the content of social Epicurean enlightenment were conceived with an initial clarity to regain which was to remain a task for the development of positive science.

Neurath's methodological problems all lead to the *unity* of all sciences. Neurath rightly saw himself as the social science expert, and with Carnap, Hahn and Frank as a representative of the Vienna Circle—in as much as we understand by it an anti-metaphysical movement that sought to lead the principles of the scientific world conception to victory.

Neurath's empiricist and anti-metaphysical attitude is striking already in the earliest works. When a concept is being analysed, Neurath first asks whether the concept itself is meaningful and whether there is an application for it in daily life or in science (Neurath 1912a, p. 114). If a logical analysis does not yield a unique result, then the members of the republic of scholars require a decision in order to reach agreement. Concerning the most basic questions of how life is to be ordered, and concerning the conception of the best life orders, however, we can no longer reach such a rational decision, and cannot expect it either. But even in such a case Neurath still holds that '*tossing coins*'—a wholly accidental choice—is more honest than a choice based on an '*inadequate metaphysical theory*' (ibid., p. 122).

Similarly, the fallibilistic position which Neurath occupies in the dispute about the foundations of knowledge is evident already in these earliest essays. Here too, the picture of the tossing of coins appears in order to symbolise the ungroundedness of a decision. According to Neurath, there are many situations of practical life where there exists no other possibility for a human individual than to make the choice arbitrarily. From the fact that the system of knowledge did not yet exist which would encompass all the things humans could understand, Descartes had drawn the consequence that human actions should in practice be guided by rules which lend security under uncertainty.

Neurath analyses this idea of provisional rules and concludes that under conditions of ignorance it is not only honest, but also rationally justified to decide by lot which of the possible decisions is 'correct'. By contrast, he who thinks that he could trust his insight even in these situations is either fooling himself or is dissimulating—both consequences of a wrongly understood rationalism, called '*pseudo-rationalism*' by Neurath. A pseudo-rationalist is someone who offers reasons, justifications and insight where there are and could be none. At least Descartes did not make this mistake in the practical domain, but he did in the domain of theory, for "thinking, too, needs preliminary rules in more than one respect" (Neurath 1913b, p. 3). In justifying this demand Neurath develops *in nuce* the postulates of a critical fallibilism without falling victim, like others, to pseudo-rationalism:

> Whoever wants to create a world-view or a scientific system must operate with doubtful premises. Each attempt to create a world-picture by starting from a *tabula rasa* and making a series of statements which are recognised as definitively true, is necessarily full of trickeries. The phenomena we encounter are so much interconnected that they cannot be described by a one-dimensional chain of statements. The correctness of each statement is related to that of others. It is absolutely impossible to formulate a single statement about the world without making tacit use at the same time of countless others. Also we cannot express any statement without applying all of our preceding concept formation. On the one hand we must state the connection of each statement dealing with the world with all the other statements that deal with it, and on the other hand we must state the connection of each train of thought with all our earlier trains of thought. We can vary the world of concepts present in us, but we cannot discard it. Each attempts to renew it from bottom up is by its very nature a child of the concepts at hand. (Ibid..)

Three things are notable about this passage. First, Neurath here already expresses very clearly the principle of the coherence of a system which later on will be taken to be decisive for the integration of empirical sentences: the truth of one sentence is connected to the truth of all others. Second, there is no indubitable knowledge, and so no indubitable system of science. There is no *tabula rasa* onto which such a system might be written. For, third, all the concepts which we use are connected. With this, Neurath anticipates the principle of the totality or the wholeness of knowledge which characterises the theory of holism: every sentence of the system and every sentence about the world is connected to all others, and the truth of every individual sentence is also dependent on the truth of all others. Accordingly, it is the truth of the entire system that is at issue, and not that of an isolated sentence. The decisionism which is clearly in evidence already in the essay on 'The Lost Wanderers of Descartes' is retained also in the later discussions on the status of protocol sentences:

> All content statements of science [empirical statements; R. H.], and also those protocol statements that are used for verification, are selected on the basis of decisions and can be altered in principle. (Neurath 1934, p. 102; cf. Hempel 1935a and Rutte 1979 [ch. 13].)

The question of the criteria of such a choice is also answered in 1934 in a form not essentially different from that of 1913: "We call a content-statement 'false' if we cannot establish conformity between it and the whole structure of science." (Neurath 1934, p. 102)

Given the non-agreement between a sentence and the whole system, there exist always two possibilities to effect agreement: either alter the sentence to be integrated or alter the system. I have called this principle here expressed by Neurath the *Neurath principle*, because it forms with convincing clarity the principle of adaptation in the change of scientific research (Haller 1979a, 1979c [chs. 2, 3]). For this principle is also the principle of the change of scientific theories. When considering this change one must start from the basic motive of scientific research to unify and simplify. According to this principle one must try to reduce the infinite number of possible judgements about the world to ever fewer principles and axioms, taking Occam's principle of economy as one's principle of selection. In other words, that decision is distinguished ontologically which assumes or presupposes the least number of entities while providing equal explanatory power (confirmational capacity). Sentences which are incompatible with the system thus lead to the change of the system, and so to their own revision. That is the principle of the creation of two theories.

To the question *whereby* such a change is ultimately determined Neurath and Carnap answer: by the 'historical fact' that the majority of scientists in fact accept certain protocol statements. This anticipates the Kuhnian thesis which later proved so contentious, namely, that it is ultimately the change of mind of the members of the republic of scholars that is responsible for the acceptance or rejection of a theory. It does indeed seem to follow that the acceptance or rejection by members of the republic of scholars must be offered as the *ultima ratio* by someone who cannot accept an individual observation as it is expressed in a protocol sentence as a basis for his decisions—for such statements may also be altered—and who therefore, like Mach, cannot interpret the individual observation or protocol as knowledge or as error, that is, as true or false. Neurath regards the collective of researchers as a 'cleaning machine' which tests the compatibility of scientific statements with the system and keeps it free of contradictions.

Our thinking is a tool, it depends on social and historical conditions. One should never forget this. We cannot act as prosecutor and defendant at the same time and in addition sit on the judge's bench. We confront our present thinking with earlier thinking, but *we have no possibility of taking a judge's stand on a point outside*. (Neurath 1930a, p. 46.)

The logico-epistemological background lies in the insight, derived from Poincaré and Duhem, "that more than one self-consistent system of hypotheses can satisfy a given set of facts." (Neurath 1921a, p. 203; cf. Haller 1979c [ch. 3].) Here Neurath means the indisputable fact that is so difficult to comprehend systematically, namely, that every fact, every event is compatible with infinitely many descriptions and thus admits itself of infinitely many descriptions. The place of the reality with which we might wish to compare a sentence, could be taken by "several totalities of statements that are consistent in themselves but not with each other" (Neurath 1934, p. 102). This is nothing but a consequence of Duhem's principle, transferred to the context of sentence and reality.

One should not conclude from the foregoing that Neurath opened the doors widely for arbitrariness in the choice of theories and in their evaluation. In his dispute with Spengler—his *Anti-Spengler* was published in the same year as the delightfully polemical essay by Leonard Nelson (1921)—Neurath again and again points out that that which at first seems simply incomparable, that which is most alien to us, is still comparable to our system of knowledge in certain respects, and so is intelligible as well. The relativists and anti-positivists of our day limit themselves to the consideration of successive theories and allege that between them incompatibility and incommensurability obtain; by contrast, Spengler asserts in many places that such incomparability obtains for all significant cultural phenomena:

> The Faustian inventor and explorer is something unique. The primal power of his will, the lucidity of his visions, the steely energy of his thought must be uncanny and incomprehensible to anyone of a foreign culture, but it is in the blood of all of us. (Spengler 1918 (1973, S. 1186))

Neurath points to the inconsistency which consists in, on the one hand, wishing to provide a very general theory of the '*Gestalt*' of cultures, while insisting, on the other hand, on incomprehensibility, or in postulating differences which amount to a 'true' or 'false'. Since Spengler asserts that "what is true for us is false for another culture and this holds of every result of scientific thought", Neurath counters with the question: "Are ancient mechanics and geometry false for us? Were they false for the Arabs? Would our physiology be false for Aristotle? Is it false for the Chinese?" (Neurath 1921a, p. 204.) It is by no means the case that Neurath wishes to blur the differences which he knew very well from his own studies. Rather, he poses the sober question, how Spengler could hold himself free of these principled misunderstandings while he strove to prove and know that all others constantly misunderstand each other. Of course, the incommensurability thesis as advocated by Kuhn and Feyerabend does not exhibit the naiveté which makes Spengler's global theory so indefensible. In several respects these latter theorists should, after all, refer to Neurath, not only when the issue is the status of a paradigm accepted and fortified by the members of the republic of scholars—the scientific collective—but also because of Neurath's advocacy of the theory of holism. Nonetheless, Neurath also offers the basis for *criticism* of these views which partly derive from him. The nucleus of this criticism says that a comparison is only possible between objects and constructions if they *share* the same domain of reference. If we replace 'paradigm' by 'cosmos' or 'world-view', then the following holds:

> every cosmos comprises facts that are quite unambiguously connected with facts of other *cosmoi*, and alongside the changing aspects there arises something lasting and common. The facts that the seasons follow each other, that fire burns and wine makes you drunk, are common to all world-views. (Ibid., p. 206.)

Of course, these are primitive examples, but they point in the right direction. That expressions in different theories and world-views are different, and similarly that they possess different meanings, does not necessarily exclude the condition that they have the same domain of reference (cf. Haller 1971). Even if Neurath's arguments do not wholly convince when he argues from analogy that different people have the same sensations, his conclusion may nevertheless be agreed with; statements are based on the same or at least similar facts in those cases where a unique coordination of expressions is possible.

As I remarked above, already in 1913 Neurath knew that historical and comparative theories are not built on a *tabula rasa*. And even more so, he was familiar with viewing science in a dynamical historical perspective, for he applies such a perspective both in his consideration of the historiography of science and in his description of the changes in theories of optics. But he emphasises repeatedly that he does not want to make the mistake of exaggeration or, even more so, of pseudo-rationalism—a mistake of which he accuses Popper, in my opinion correctly so. We may leave it open whether he always succeeded in avoiding this mistake himself, especially since his own formulation lack the precision which they require.

What then is the picture of the development of science which Neurath possessed *before* the Vienna Circle? From what has been said so far it is clear that we are dealing not at all with the canonical conception of science as the accumulation of knowledge. This becomes even clearer if we return to the central passage of the *Anti-Spengler* where Neurath develops his own position against Spengler's supposition that "every man has his own world": it ends in the famous and later re-employed image of the sailors who have to rebuild their ship on the open sea (Neurath 1921a, p. 199).

An essential precondition of all reflection and all theory building is that we begin with a vocabulary and a set of concepts which we find pre-given. The fact that all changes in our set of concepts with which we describe the world are effected within this language depends on this precondition. But once we progress in our thought, or in the explication of concepts, then "the entire structure of concepts is shifted in its relations and in its centre of gravity, and each concept takes a smaller or greater part in this change." (Ibid., p. 198.)

If one wanted to be really precise, one should "repeat what has hitherto been said, but in a different sense". Only in this fashion could one effect a continuous transformation of an old stock of knowledge into a new, for it is impossible to *fully* complete any one part. For this reason one also must use concepts whose 'blurred margins' are quite essential, and use these concepts which themselves are not totally clear in order to increase the clarity in areas of labyrinthine indeterminacy.

The first metaphor which Neurath offers in order to bring out the nature of this partial progress which always reveals only certain aspects, but nevertheless changes the totality of the conceptual edifice, is that of the miner "who at some spot of the mine raises his lamp and spreads light, while all the rest lies in total darkness." (Ibid..) Neurath then compares the plans and sketches which the miner uses with our conceptual instruments. These are not only conventional definitions and constitute not only auxiliary tools in the service of the most economical comprehension of reality, but they can be claimed to be 'cognition', that is, knowledge, in their own right. There is the question, of course, of how the definitional determinations, which fall under the class of conventions concerning conceptual relations, are to be distinguished from genuine descriptions. By referring to his source Duhem, Neurath makes clear wherein his modification of Machian positivism consisted. Here metaphysics—Neurath says 'world-view'—plays a role for all theories which cannot be overlooked. Its definiteness makes it worthwhile to quote in full the passage containing the image taken over by Quine:

> That we always have to do with a whole network of concepts and not with concepts that can be isolated, puts any thinker into the difficult position of having unceasing regard for the whole mass of concepts that he cannot even survey all at once, and to let the new grow out of the old. Duhem has shown with special emphasis that every statement about any happening is saturated with hypotheses of all sorts and that these in the end are derived from our whole world-view. We are like sailors who on the open sea must reconstruct their ship but are never able to start afresh from the bottom. Where a beam is taken away a new one must at once be put there, and for this the rest of the ship is used as support. In this way, by using old beams and driftwood, the ship can be shaped entirely anew, but only by gradual reconstruction. (Ibid., p. 199.)

The context makes clear that the 'old beams' represent our given conceptual system out of which the new one should 'grow'. In this sense there are no revolutions in the progress of knowledge, but only a steady reconstruction according to the Neurath principle—either change the system or change the conflicting individual sentences.

Of course, this fallibilism must be distinguished from Popper's dramatised version. For Popper, the conflict between an individual sentence and a theory constitutes a falsification of the theory if the incompatible individual sentence turns out to be a basic sentence, whose negation could derived from the system. The conception according to which the empirical content of a sentence is equivalent to the class of the possibilities of being falsified, strongly suggests that this content be increased for this heightens its testability.

But if the possibility of falsification increases the content, every *de facto* falsification, however insignificant, destroys the whole system because of Popper's acceptance of the Aristotelian *modus tollens*:

> By means of this mode of inference we falsify the whole system (the theory as well as the initial conditions) which was required for the deduction of the statement *p*, i.e. of the falsified statement. Thus it cannot be asserted of any one statement of the system that it is, or is not, specifically upset by the falsification. (Popper 1935, p. 76)

But such a strict concept of falsification destroys all science, as Neurath and Carnap immediately recognised. Its most notable instance is belief in the *experimentum crucis*, because testing is to decide the proper standing of the theory, namely, which of the competing hypotheses or theories is to be eliminated. Duhem was able to show that, e.g., Newton's gravitational hypothesis was not inductively derivable from Kepler's observation statements, but contradicted Kepler's laws. Yet instead of eliminating Kepler's laws as falsified, he merely proposed to compare each total system of physical theory with the entirety of experimental facts. The reason given in *The Aim and Structure of Physical Theory*:

> To seek to separate each of the hypotheses of theoretical physics from the other asssumptions on which this science rests in order to subject it in isolation to observational test is to pursue a chimera; for the realization and interpretation of no matter what experiment in physics imply adherence to a whole set of theoretical propositions. (Duhem 1906, pp. 199/200.)

Popper does not accept this part of the conventionalist methodology and wishes to restrict the conventionalist solution only to the basic sentences, whereas Neurath above all upholds the holistic perspective and so rejects a narrow theory of experiment and falsification as inappropriate and unrealistic. Since his *Logical Syntax,* Carnap's defense of holism fully followed Neurath's proposals. Accordingly, the following principles hold for the physical language, the language of unified science. (1) "No rule of the physical language is definite; all rules are laid down with the reservation that they may be altered as soon as it seems expedient to do so." (2) Since the logical content of every law exceeds the logical content of any finite set of protocol sentences, there can be no rules of induction. (3) Statements of law thus are always hypotheses. (4) In the strict sense there is no falsification of hypotheses for even if a hypothesis "proves to be L-incompatible with certain protocol-sentences, there always exists the possibility of maintaining the hypothesis and renouncing acknowledgement of the protocol-sentences." For the same reason there cannot be complete confirmation of an hypothesis. (5) And so there is also no testing of an isolated hypothesis. "The test applies, at bottom, not to a single hypothesis but *to the whole system of physics as a system of hypotheses* (Duhem, Poincaré)." (Carnap 1934/7, p. 318.)

It is unquestionable that in all points Carnap here followed the ideas of Neurath who was the first to advocate them in the Vienna Circle (and of course before Popper). And although Carnap was impressed by Popper's theory of falsification (which was described to him mainly in private discussions) and exerted himself to promote this unknown school teacher, when it came to falsification of hypotheses—the decisive part of *The Logic of Scientific Discovery*—Carnap adopted the point of view that there cannot be in the strict sense a refutation (falsification) of a hypothesis—precisely Neurath's view.

Neurath did not, then, wish to grant an exceptional status to any sentence in the system of knowledge. This still leaves open the decisive question, the question which properties a theory must possess for it to be preferred to other possible and 'real' systems of hypotheses. A system's *freedom from contradiction* can only be considered a necessary but not a sufficient condition of the acceptance of a system.

Neurath's answer to this weighty problem places the pragmatic aspect in the foreground and uses the pragmatic argument of the success of a theory. The measure of the degree of success of a theory is its agreement with the protocol sentences compatible with it. In this Neurath sees the justification of empiricism. Even if protocol sentences are not finally fixed or fixable, they are nevertheless "the ultimate beyond which we cannot go" (Neurath 1933a, p. 2). But since protocol sentences are also revisable—"We do not renounce the judge, but he is replacable" (Neurath 1934, p.107)—there is no firm foundation of knowledge, there are "no sacrosanct sentences at all" (ibid., p. 110). Against this, one may object that, given such permissiveness, the system actually used by us could not be distinguished nor could it be brought into any relation to reality, to the facts. Edgar Zilsel formulated it as follows:

> Without an experienced substrate of relation, without any givens to be designated, all symbolic edifices stay hanging in mid-air and no structure may be distinguished from totally arbitrary other structures as the structure of the experienced world. (Zilsel 1932, p. 159.)

If one uses Wittgenstein's distinction between what can be spoken of and what cannot, then the same thought may be expressed also in this way.

> Every definition of science that can be expressed comprehends not only one unique system of sentences, but arbitrarily many systems of sentences that are internally coordinated and mutually incompatible. (Ibid., p. 152.)

Zilsel here uses the Wittgensteinian distinction of what can and cannot be spoken of in order to prepare a decisive objection against the coherence theory. For if what can be spoken of is limited to the connection of sentences with sentences, then one can never get beyond such a system of sentences. From this Zilsel formulates the dilemma of the empiricist philosophy of science as one between what can and cannot be spoken of. Either we restrict ourselves to such systems of sentences, in which case there exist arbitrarily many mutually contradictory ones, none of whom are distinguished, or we distinguish one, unified science, by applying it to what cannot be spoken of.

Faced with this dilemma, which could be called the dilemma of the justification of science, Neurath developed one of its horns, his 'encyclopedism' as he called the later and more mature version of historical conventionalism. In this he was the pioneer of the so-called anti-positivists Kuhn, Toulmin, Feyerabend, etc., who owe more to him than is generally known today. Neurath's proposed solution of the principle of adaptation is also that of W. V. O. Quine, whose thesis of the underdetermination of a theory by (its) data represents a revitalised and modernised form of justifying the Neurath principle. Not only the sociological critique of the standard or received view of scientific systems, but also the overcoming of unhistorical approaches to science are what the contemporary philosophy of science owes to the co-founder of the Vienna Circle. Viewed in this way it seems strange that this merit is credited to the so-called anti-positivists.

The Vienna Circle was much more colourful and dynamic than its descriptions so far would have us believe. Today there can no more be any doubt that it was Neurath who contributed the strongest colours, simply by trying to lead empiricist enlightenment beyond itself—and that means also beyond Wittgenstein.

Karl-Franzens-Universität, Graz

NOTES

* First published as 'Das Neurath-Prinzip - Grundlagen und Folgerungen', in *Arbeiterbildung in der Zwischenkriegszeit*, Hsrg. F. Stadler, © 1982, Österreichisches Gesellschafts- und Wirtschaftsmuseum, Wien, pp. 79-87. Translated with kind permission of Österreichisches Gesellschafts- und Wirtschaftsmuseum and the author by T. E. Uebel.

ECKEHART KÖHLER

METAPHYSICS IN THE VIENNA CIRCLE*

1. Of all the basic conceptions of the Vienna Circle, the rejection of metaphysics was the most penetrating, the most spectacular, but also the most repellent. It was not the case, however, that the Vienna Circle made the elimination of metaphysics its central concern, or that its interest lay exclusively in something like a materialistic criticism of ideology. Again and again—that is, whenever it had become clear that a more tolerant attitude towards the actual practice of science required it—its radical anti-metaphysical slogans were quietly circumvented or openly revised. The Vienna Circle did not share Brouwer's restriction of set theory, logic and classical mathematics to effective constructive procedures; it did not share Mach's rejection of atomism and relativity theory; it did not share Watson's discardment of psychological concepts like introspection and the unconscious. Its attitude towards what Quine rightly called "ontological commitments" was therefore notably tolerant and pragmatic: there was little desire to wield 'Occam's razor'. Moreover, the Circle's interest in metaphysics did not even extend so far as to notice the radical difference between the *logical* concept of metaphysics of Aristotle and the rationalists Descartes and Leibniz on the one hand, and the *mystical, dialectical* concept of metaphysics of the neo-Platonists Plotinus and Proclus and the German idealists Fichte and Hegel on the other.[1] The main concern of the Vienna Circle was rather the elaboration of a 'positive' philosophy of science; the discrimination against metaphysics was a defense against philosophical movements which did not keep up with the scientific development of their time. That is the decisive point of the Circle's concept of metaphysics[2], as I shall show below.

Nevertheless, the attack on metaphysics was the most noticable and probably also the most unifying theme of the Vienna Circle. All important members, from Schlick to Neurath, were passionately engaged in its elimination, and particularly the works of Carnap from the early 30's more or less made it their program to develop a philosophy of science truly free of metaphysics. My concern here will be to show that this program cannot succeed, for it simply overlooks that the two most important topics of scientific metatheory—the observational base and logic—are eminently

T. E. Uebel (ed.), Rediscovering the Forgotten Vienna Circle, 131–142.

metaphysical, for they must hold *a priori* with respect to empirical investigations. Now admittedly it was noted in the Vienna Circle that the observational base and logic must count as metaphysical[3]; any attempt to view this matter neutrally, however, was stopped in its tracks by the firm conviction that, compared to the new logic and science, the metaphysics of the 'school-philosophers' (e.g. Kant's transcendental deduction, Hegel's dialectic, Heidegger's nihilism) simply had to be got rid off. The members of the Vienna Circle showed a boundless anger about obscurantism, irrationalism, uncontrolled speculation: metaphysics was like theology and could not be reined in by generally accepted criteria of research which allowed one to reach definite results by intersubjectively valid methods of testing. The application of their delightfully striking meaning criterion only served to sharpen this insight.

But it was not the case that the Vienna Circle was against everything that was valid *a priori*, as for example the analytical sentences of logic or the observational base. For a sympathiser of the Vienna Circle who is even only superficially acquainted with the history of metaphysics it is outright embarassing to have to note that already Aristotle considered *this a priori* as the essence of metaphysics.[4] Of course, the Vienna Circle did not wish to deal with Aristotle because he could be used so effectively for a rigid scholasticism; and it must be said that of his entire work the metaphysical writings were the most opaque and questionable. Yet a historical judgement of the dispute about metaphysics simply cannot but accept Aristotle's concept of it as basic.[5] Thus it remains a very grave and sensitive question for the interpretation of the logical empiricism of the Vienna Circle to ask how it was possible to eliminate metaphysics and yet defend the (notably anti-psychologistic!) *a priori* validity of analytical sentences. The answer was that *a priori* validity does not refer to any 'reality'—even less a transcendent, Platonic one[6]—but expresses *linguistic conventions* concerning the use of sentences.

This linguistic conventionalism derived from two sources: (a) from the conception, strongly urged by Neurath and Wittgenstein, of logic and observational base as matters of language; (b) from the conception, arrived at by Menger against the background of the dispute between logicists, intuitionists and formalists, that different logics can exists side by side and find justification according to their point of use (Carnap's "principle of tolerance" derives from this (Carnap 1934/7, §17; Menger 1930, 1933)). The latter development in particular was decisive for the turn towards linguistic conventionalism, for, strangely enough, the linguistic turn as effected until then was not at all meant conventionalistically. For the Wittgenstein of the *Tractatus* 'the language' was definite and unrevisable (its function was, after all, to picture possible worlds), and for Neurath logic was not a matter of "decision" (Neurath 1934, p. 104) nor did he ever allow any other than the physicalist observational base. But due to the obvious assumption that language is a human institution, the linguistic viewpoints of Neurath and the early Wittgenstein strongly facilitated Carnap's linguistic conventionalism—despite their apparently anti-conventionalist absolutism.

2. Although I shall focus below mainly on Carnap, it is very important to be clear about the sometimes radically different motivations of Neurath and Wittgenstein with regard to metaphysics. First of all, common to both was the deeply moral character of the elimination of metaphysics from science, in so far as morality for both—and independently also for Schlick—was concerned with happiness. For both, the temptation to assert metaphysical sentences was detrimental for the conduct of a happy life. For Neurath, the detriment lay in mistaken *actions*, particularly political ones, whereas for Wittgenstein, it lay in the violation of his aesthetic *asceticism*. It is clear that these attitudes stimulated mutual ridicule. Neurath thought of metaphysics as the absolutism of a 'pseudo-rationalism' which pretended certain knowledge where there could be none (Neurath 1913b; cf. the discussion in Nemeth 1981, p. 177). This included all variants of absolutism and theology, metaphysics and mysticism, transcendental teachings and all hypotheses about 'reality': all of them were damaging and played a role analogous to Marx's religion, as opium and means of enslavement. By contrast, Wittgenstein held this very mysticism, transcendentalism and religion to be far more important than all science, for they penetrated and gave soul to everything in life. It was just that it could not be described in words, just like the God of 'negative theology' could not be grasped by human reason.[7] Indeed, it was one of the most important insights of the *Tractatus* that logical structures cannot be represented in certain simple empirical languages, that they are 'unsayable' and in this respect transcendent.

By now, the development of metamathematics and semantics has made clear what was missing here: means to talk about entire linguistic systems—means which Wittgenstein himself used all the time in his 'elucidations', although with the proviso of having to discard them later like a scaffolding.[8] The situation was apparently the following (apart from the fact that his incessant emotional turmoil prevented Witgenstein from any systematic development of a theory): given his aesthetic attitude, Wittgenstein despised going beyond the objects immanent to life to *explicit* abstractions. It was as if a master craftsman was suddenly pestered by insistent questions concerning the nature and point of his work, or as if one interrupted one's experience in the middle of an orgy and sought to comprehend it explicitly. Such things are simply not done, they are bad form. Nevertheless, such transcendent abstractions do have their proper place *elsewhere*, and it is even a form of *cultural primitivism* to prohibit their elaboration. It is all the more important to note that throughout the *Tractatus* Wittgenstein himself *did* engage in metaphysics in virtue of his strictly delimited 'meaningless' elucidations. Neurath was absolutely right to protest against them *ad nauseum* when the Vienna Circle worked its way through the *Tractatus* around 1926.

The Vienna Circle's opposition to metaphysics must be judged, in addition, against the background of a cultural uprising against absolutism, a rebellion whose spirit was perhaps most poignantly formulated by Nietzsche. Indeed, the parallels between Nietzsche and the Vienna Circle even pertain to *detail*[9]: (a) the denial of an object for logic; (b) the linguistic conception of logic; (c) the assimilation of mathematics and logic; (d) the connected rejection of Kant's synthetic *a priori*, in mathematics as well as in physics; (e) the decisive rejection of metaphysics; and (f) a proto-existentialist conception of scientific method as a form of life.

3. The Vienna Circle's attitude towards metaphysics was much more radical than the earlier sceptical teachings of the nominalists and idealists, because the latter really only wanted to delimit the field of metaphysics, whereas the Vienna Circle followed Neurath in his rejection of any theses about 'reality'. Characteristically, they did not return to Occam's razor (with the exception of Hahn (1930)) in order to prune metaphysical overgrowth, as was the method of the nominalistically and extensionalistically inclined Quine and Goodman, who explicitly engaged in *ontology*. Carnap and Menger in particular objected in the spirit of the principle of tolerance against the elimination of ontologically demanding theories like set theory, the Leibniz-Wittgenstein theory of possible worlds, and the theory of the continuum.

Instead of using Occam's razor, once intended to allow the better development of what remains of metaphysics, one settled on an obviously much more radical principle: the meaning criterion. Carnap used it to establish the meaninglessness, and so the untenability of theses about reality, motivated by Neurath's radical demand to reject not only transcendent entities, but any thesis about reality (Carnap 1928b, §§9-10, 1932a, 1934/7). His argument runs in analogy to Wittgenstein's observation that the logical structures of a language cannot be topicalised as objects by that language itself, and that therefore one cannot express in this language the fact that sentences of a particular structure are logically true or contradictory. Similarly, Carnap pointed out that the determination of an observational base as real, physicalistic, phenomenalistic, idealistic etc. could not be effected within the same language that uses this base, and that the different metaphysical interpretations of the observational base are irrelevant for the use of a language. This insight is very valuable in so far as it says that such metaphysical theories fall outside of the framework of the expressive means of the language in question.

The matter becomes tricky as soon as one gains access to a *metalanguage*, as Carnap did. In this metalanguage all those transcendent matters of fact can be expressed which cannot be immanently expressed in the object-language—e.g. logical truth, possible worlds etc.—but also whether the base is realist, phenomenalist, physicalist etc. There it becomes possible after all to investigate the different metaphysical interpretations, because they influence the applicability of the language. It turns out, for example, that Carnap's argument in his *Pseudoproblems*, §§ 9-10, only holds given the assumption of the empirical equivalence of the different interpretations of the observational base. (Carnap probably made the even stronger assumption of the *identity* of the different interpretations, thus trivialising his result.) For according to the different interpretations of the observational base *different procedures of observation and measurement* are imputed to the observational predicates of the object-language—all the while supposing that, say, physicalist or phenomenalist procedures of measurement give the same results. (Whether this is technically feasible at all times may be left open.) But although the different interpretations cannot be distinguished under these circumstances in the object-language, it is *nevertheless* possible to distinguish between them on the metalevel with reference to differences in economy, reliability etc.

It is extremely interesting to see how Carnap employed the *metalinguistic* standpoint for the problems of the observational base, once he reached it.[10] It turned out that the distinctions which he rejected as meaningless in *Pseudoproblems* now gained a precise and easily established (metalinguistic!) meaning. Phenomenalist ('K'-) concepts referred to brain processes (the *body*) of an observer, whereas physicalist ('D'-) concepts referred directly to external objects. The (obviously now meaningful) choice of an observational base can now be made with reference to the efficiency of measurement and theory (whose calculation *within* the same observational base would remain circular).

Obviously, different metaphysical standpoints are after all meaningful in this respect. That they are meaningful is something which Carnap admitted more or less clearly from beginning to end. In the *Aufbau*, §§ 54, 58, Carnap asserted the 'epistemological primacy' of the autopsychological base. Later, Carnap was talked into physicalism by Neurath.[11] And finally Carnap claimed that ontology must be judged according to pragmatic criteria (Carnap 1950). Unfortunately, Carnap never became fully aware that these pragmatic criteria have to enter into metatheory itself as epistemological norms, although he once explored ethics as a normative value theory in an extremely interesting investigation.[12]

4. We have seen that it was the consistent application of the meaning criterion which led to the unprecedented radicalisation in the battle against metaphysics. But we have also seen that the step towards metatheory annuls this effect of the meaning criterion, for it is by its means that the interesting, but so far 'unsayable' properties of language (logical structure, observational base) can now be explicitly expressed. It is noteworthy that Neurath kept the meaning criterion at arm's length (maybe because it was derived from Wittgenstein?); perhaps he sensed that in the end it did not help to eliminate metaphysics. In any case, the situation was ripe for a *new attack* on metaphysics. For now the point was to show that a language remained free of metaphysics even when it was described by a metatheory. This would be all the more interesting since the latter makes explicit the 'metaphysical presuppositions' of the object-language.

This great project was attempted by Carnap in his *The Logical Syntax of Language*, the most thorough and intensive effort in the history of philosophy to totally eliminate metaphysics. Its aim was to conceive of language as a physicalist structure of signs and their relations to each other and to *establish* that all otherwise metaphysical properties (logical structures, references to reality) were *nothing but physicalist-syntactic properties*. This would be the definitive realisation of Neurath's physicalist program for the elimination of metaphysics.[13]

The formalist approach of Hilbert which disregarded any semantic interpretation of signs was of great importance here. In this context it was particularly important for Carnap to overcome the deficit in the formalist program which had been shown by Gödel: its incompleteness and the consequent absence of an (effective) proof of non-contradiction. This was also important for the point of view of logicism, which Carnap was still enamoured with, seeking to comprehend all mathematical structures as logical ones.[14] The success of Carnap's syntax program constituted moreover the pinnacle of Wittgenstein's efforts to establish logic and mathematics as meaningless tautologies, something which was indeed achieved by means of the non-effective concept of syntactic analyticity (Carnap 1934/7, §14 and particularly §52).

But what happened to the meaning criterion in the *Logical Syntax*? The meaning of *logical* structures was interpreted by purely syntactic structures. The meaning of the *observational base*, however, was rendered by translations of various theses about reality into the 'formal mode of speech' as syntactical mappings of empirical relations in the 'material mode of speech' (ibid., §79).

Did Carnap then succeed in eliminating metaphysics? No, for two reasons. (a) First of all, already the syntactic metalanguage makes very strong ontological presuppositions. Carnap describes it as the "mathematics and physics of language" (ibid., §73, p. 284), and these are naturally used in a 'material' sense. In particular, it turns out that Carnap uses for his 'criterion of validity' sets of signs of (at least) countably infinite magnitude; moreover, he uses non-effective (i.e. transfinite!) rules of inference (the rules of consequence of Part III) in order to prove the completeness of his 'language II', in which amongst others the continuum and the related integers and differentials are defined for their use in representing physical theories. In addition, he presupposes the common observational base of physics itself in order to describe syntactic structures. (b) Secondly, Carnap repeatedly and in an underhand fashion involves a semantics whose signs are, after all, used as meaningful ones. This becomes apparent on all those occasions when Carnap recurs to *translations* (ibid., §§61-62). It is trivially true that a semantics, i.e. a fully developed theory of truth, can be built up directly on translation rules, for the 'syntactic coordination' which is effected by a translation presupposes the determination of (semantic!) synonymity. At crucial points then, Carnap illicitly imported semantics into his 'pure' syntax. His rather quiet conversion to the Tarskian concept of sematics around 1935[15] finally showed that his sympathies lay with the reality theses all along.[16]

Metaphysics after all! The zeal of old was exhausted. What remained were merely weak echoes from the time of *Pseudoproblems* (Carnap 1963b, p. 933).

5. What then was the point of this entire circus? Did the attempt to crush metaphysics have any point at all? A positive answer to this question can be found in the earlier state of metaphysics[17], for it was indeed regressive, obscurantist, anti-scientific.[18] Neurath in particular always pointed to the connection of metaphysics with superstition, messianism, occultism, irrationalism and mysticism (cf. Nemeth 1981, III, 2). Nowadays, many are inclined to think these intellectual movements unimportant, for they are held to have been overcome in the course of industrialisation, secularisation and democratisation. At the time of the Vienna Circle matters did not look so good, however. Representatives of various forms of irrationalism in politics and art agitated everywhere; yet in metaphysics the irrationalists performed their tricks like in no other field: for this reason, metaphysics could not but appear to the Vienna Circle as the intellectual dumping ground of the Western mind. We need not be surprised then that there was this tremendous outrage against the nonsense that was particularly common in academic philosophy. In *this* respect the uproar caused by the Vienna Circle is to be applauded after all, and remains exemplary for the entire history of philosophy.

This situation in the German-speaking lands was due primarily to German romanticism and its philosophy, German idealism. Already from its beginning this romanticism tended towards irrationalism, was antagonistic to the enlightenment, and presented itself as populist if not outright '*völkisch*'. We can count among its philosophical representatives Hegel whose apology for totalitarianism was justified by his dialectics in which concepts were 'rendered fluid'—a continuation of the occultist and gnostic teaching of Plotinus (Topitsch 1967b, pp. 15ff); also Fichte with his nihilistic, proto-fascist metaphysics of action and his postulation of an absolute ego. This Pandora's box had been opened by none other than Kant with his lamentable incomprehension of logic, his unknowable 'things in themselves', and his nonetheless postulated 'transcendental unity of apperception', according to which all true laws of nature are synthetic *a priori* forms of reason of the ego.[19] Every historian of logic nowadays who dutifully considers Kant (not to speak about Fichte, Hegel or Schelling) cannot be but exasperated when he compares him with the great examples of Aristotle, Leibniz, Frege and Russell! From Kant onwards the level of logical exactness had sunk to such a degree—Fries, Bolzano and Peirce excepted—that one can speak of an age of darkness which was lifted only by Frege and Russell.[20] The wide response to the anti-metaphysical point of view, first to that of Ernst Mach, then to the related one of the Vienna Circle—especially in the blockbuster form of Ayer's *Language Truth and Logic*—can certainly be explained as a feeling of redemption from the obscurities of the metaphysics of German idealism.

Without exception, the Vienna Circle was opposed to Kant's transcendental philosophy and Hegel's dialectic. Against the background of the then current 'schools', this fact can be regarded as a particularly unifying factor. Now it is admittely false to claim that the Vienna Circle only aimed at German idealism with its criticism of ideology: think only of the Neoplatonists' magical abuse of the '*nous*'; I only claim that the Vienna Circle would have never started on this path without the bad example presented by Kant and German idealism

6. Today it is not as if one had to 'resign' oneself to metaphysics, because it is 'inescapable' or 'inevitable'. The fact of the matter is rather that metaphysics simply amounts to applied logic.[21] Founded in Greece, occidental metaphysics reached magnificent achievements: modern science which conquered the whole world. As Aristotle remarked, philosophy begins with wonderment, and we may note: most worthy of wonder is this metaphysics.

Institut für Höhere Studien, Wien

NOTES

* Revision of a lecture given at the workshop 'Die Geschichte des Wiener Kreises,' February 26, 1982, Zentrum für interdisziplinäre Forschung, Universität Bielefeld. The work was undertaken as part of project no. 4517 of the Österreichische Fond zur Förderung der wissenschaftlichen Forschung: 'Wissenschaftsphilosophie in Österreich 1918-1938,' led by Prof. Dr. Edgar Morscher, University of Salzburg, and Prof. Dr. Erika Weinzierl, University of Vienna, with co-workers Mag. Dr. Friedrich Stadler and Karl Müller, Vienna. First published as 'Die Metaphysik beim Wiener Kreis' in *Philosophie, Wissenschaft, Aufklärung: Beiträge zur Geschichte und Wirkung des Wiener Kreises*, Hrsg. Hans-Joachim Dahms, de Gruyter, Berlin-New York, 1985, pp. 190-204. © 1986 Eckehardt Köhler. Translated with kind permission of the author by T. E. Uebel.

1 On Aristotle's concept of logic see fn. 4. The lack of a history of metaphysics from a standpoint close to that of the Vienna Circle was partly filled by Topitsch 1958. Though rich in criticism of ideology, this work unfortunately does not contain a treatment of ontology and of the conceptual apparatus of the mathematical sciences. Particularly the literature on the

history of ontology from the standpoint of modern logic has remained somewhat backward, with the exception of the writings of Heinrich Scholz (see fn. 4) and the areas of Leibniz and Bolzano studies: Weingartner 1983, Morscher 1969, 1973.

2 That the criticism of metaphysics was *not* the most essential aspect is clearly shown by the close relations of the Vienna Circle to the (rather Aristotelian) logicians of Lwow and Warsaw, and to Popper, who never rejected metaphysics.

3 E.g. Schlick 1931 strongly rejected Weinberg's assimilation of logic to metaphysics, pointing to its tautological character.

4 According to Aristotle, metaphysics (always designated as 'first philosophy', sometimes as 'first science') was first of all a science in the sense of his *Posterior Analytic*, i.e. an axiomatic-deductive theory (following Scholz 1930). Aristotle *always* designated it as the theory of the 'axioms' (i.e. of the—logical!—laws of contradiction, excluded middle, identity; at one point he adds the syllogisms to the 'axioms', and sometimes other logical laws, particularly those of the logic of relations) and of *substance* (i.e. the determination of the fundamental types of objects and their constitution, the concept of essence, the abstract concepts of mathematics and the theory of ideas). The axioms roughly correspond to logic, substance to the observational base and higher order concepts, say sets. Repeatedly, Aristotle characterises metaphysics methodologically as applicable in *every* individual science (in particular in every domain of individuals) and ontologically as the science of being as such, i.e. of objects with no particular description. Both of these characterisations can be explicated by the concept of logical universal validity. Theology plays a wholly subordinate role: in one obviously late book of *Metaphysics*, God is defined—without any reference to the other books—as unmoved mover, that is cosmologically and physically, *not* ontologically. Substance is characterised as the (probably technomorphic) composition of matter and form, whereas the (biomorphic) concept of entelechy hardly shows up in the metaphysical books; by contrast, the leading roles are played by the categories, rules of derivation, semantic questions and mathematics. See Köhler 1968, 1969.

5 The school of Lwow and Warsaw around Lukasiewicz and Lesniewski derived via Twardowski from the *Aristotelian* Brentano; see Skolimowski 1967. Leibniz, Bolzano and Peirce also found Aristotelian scholastics fruitful and did not attempt to separate metaphisics and logic; see e.g. Burkhardt 1980, § 5.04, 'Logic and Metaphysics'.

6 Incidentally, it is *impermissable* to complain about Aristotelian metaphysics that it contained transcendental theses; its difference to Platonism lies just in this aspect. Even non-sensible things like numbers and forms are viewed immanently as objects; this led directly to nominalism, in contrast to Platonic-Christian theology.

7 Wittgenstein's letter no. 23 to Ludwig von Ficker is revealing here: "... the point of the book is ethical. ... my work consists of two parts: of the one which is here, and of everything which I have *not* written. And precisely this second part is the important one. For the Ethical is delimited from within, as it were, in my book; and I'm convinced that, *strictly* speaking, it can ONLY be delimited in this way." (Wittgenstein 1969a, pp. 94/5.)

8 Wittgenstein 1921, §6.21. The dispute in the Vienna Circle about Carnap's demand for a metalanguage in the sense of his *Logical Syntax* was characteristic for this attitude. As a representative of Wittgenstein, Waismann objected obstinately, apparently for the sole reason that Wittgenstein thought so. Already in 1914 Russell had proposed a metalogic to Wittgenstein who did not want to know. Neurath shared Wittgestein's agreement with the saying of Loos that "Ornament is crime" (which Neider used to convince Neurath to shave off his beard), but Neurath objected to Wittgenstein's rejection of metalogic; see Neurath 1932a, 1933a.

9 Revealing here the discussion by Beth 1938 and 1965, pp. 176ff.; pertinent also Del Negro 1923.
10 This was undertaken first in Carnap 1932e, Sect. 1, a very important and much too neglected response to Neurath1932d, where the 'protocol sentences' are cleansed of the absolutism of the earlier atomic primitive sentences. What is particularly important is the first differentiation of internal and external ontological questions (here with regard to the non-logical observation concepts), which only much later (and to great applause) was picked up again in Carnap 1950 with reference to logical-mathematical axioms.
11 Carnap 1932b, §§5-6. Bergmann 1954 had long considered logical positivism to be suffused with metaphysics.
12 Carnap 1963b, §32, pp. 1000ff. At issue is the introduction of the modal concept 'utinam' which was to represent so-called 'optatives'. Apparently Carnap quietly took over this concept from Menger 1939, who in turn used the wrongly forgotten 1926 book on ethics by the logician Ernst Mally of Graz.
13 Of course, already the term 'physicalism' indicates what kind of metaphysics was represented by the 'anti-metaphysician' Neurath: a version of materialism. (See Fischer 1982a.) In this context it is interesting that Neurath proposed a (naturally behavioristic) *semantics*: "'Synonomous sentences' are to be defined as stimuli, which, under definite reaction-tests, evoke equal reactions." (Neurath 1932d, p. 95.)
14 Carnap 1935, integrated into the English version of his 1934/7. See the excellent review by Kleene 1939.
15 Brilliantly shown by Coffa 1977. Incidentally, *Semantics* was Carnap's working title for *Logical Syntax*, whose title was prompted by Neurath.
16 See also Neurath's report on the general mood after Tarski's pathbreaking presentation of his semantics at the Paris congress. Most fatal for the syntactical program seems to be the following quote: "Rougier expressed the view of not a few members of the congress when he remarked wittily that he used to think that representatives of logical empiricism would have to be afraid to look on the menu to check whether they received what they ordered when the waiter brought a dish of veal, for they thought it impermissable to compare a sentence with reality. But after the expositions by Tarski and Lutmann-Kokoszynskas every representative of logical empiricism could breath more freely..." (Neurath 1935c, p. 666/7.) Neurath also described his own expressions of his disappointment at the congress about the introduction of ontological issues through semantics; this disappointment led to the later disagreement with Carnap over the latter's works on semantics.
17 A very lively description of the situation is given by Linke 1961. He was no adherent of logical empiricism, although he praised its scientific nature, particular that of Carnap's works. He very knowledgably describes how the 'cleanliness' of the logical method was confounded by Dilthey, Scheler and Husserl (!), finally resulting in Heidegger. J. Kraft 1932 passes an even harsher judgement in the course of his solid discussion of Husserl and Dilthey.
18 Carnap 1928a, §182, rejected the possibility of a scientific but non-conceptual, purely intuitive metaphysics with a telling reference to Bergson's irrationalist philosophy of life (Bergson 1903). Schlick 1913 and 1926 passed a similar judgement.
19 See the magnificent, if too relentless, book on Kant by Aebi 1947. This most thorough book that has ever been written about Kant gained brilliant reviews abroad, but was criminally neglected in Germany and practically suppressed by the philosophical profession, because it amounted to an uncompromising dénouement of the Kantian tradition. What must be particularly lauded is Aebi's description of the transition from Kant to Fichte and Hegel and how the latter picked up the most irrational aspects of the former.

[20]Bolzano was fully aware of the devastating effect of Kant's transcendental aprioricism: "...it would have to become one of his *life's tasks*, by the *dissemination of clear concepts* to give direction to the unredeemed confusion, which Kant, without suspecting it himself, brought about in Germany by his own philosophy." (Quoted after Morscher 1973, p. 33.) Note also that therather careful and neutral historian of philosophy Hans Wagner accused German idealism as follows: "Of that which Kant demanded by way of 'discipline' from pure reason, virtually nothing was left". And he praises a "blossoming and fruitful *empiricism*", "for this intellectual movement brought undeniable progress which a systematic thinker must integrate in some way or other." (Wagner 1977.) And this in the opening address of the German Hegel congress of 1975!
[21]So Scholz 1940, §2.1, as the result of the 'new foundational research'.

HEINER RUTTE

ETHICS AND THE PROBLEM OF VALUE IN THE VIENNA CIRCLE*

I

It is known that problems of ethics and the philosophy of value (understood as theoretical problems) played a small or peripheral role in the Vienna Circle: according to participants like Viktor Kraft and Heinrich Neider, they did not constitute topics in the discussions proper (cf. Kraft 1973, Haller and Rutte 1977). On the other hand, something has been rediscovered which had been repressed for a while in the post-war empiricist, analytical philosophy, or had simply not been talked about, and which, probably due to the 'positivism-dispute' of the mid-'60's, the younger generation did not even notice[1]—namely that, to different degrees and according to their individual temperament, *most* members of the Vienna Circle were politically engaged for social reform, even zealously so, and that *all* of them felt themselves to be free-thinkers and advocates of the enlightenment who intended to educate the broader population (as was recently documented in great detail by Friedrich Stadler (1982b)). So there seems to be a correspondence of sorts (perhaps even of a compensatory nature) between the strict rejection of objective values and the objective justification of values, shared by all thinkers of this group, and their forceful declamatory *moral and political will*, their urge to promote the liberation and increased happiness of humanity. Just this mixture of *destruction* and *engagement* would appear to have been particularly explosive for outsiders and opponents of the group. In the ideologically charged atmosphere of those years it seems to have been received by those who held different political and religious views as aggressive, destabilizing and outright dangerous in a way that is difficult for us to imagine today. (This is proven by the reaction to the comparatively unpolitical, liberal bourgeois Moritz Schlick, whose ethics was defamed, especially after his murder, as the culture destroying-teachings of a profligate sensualist.)

The ethical writings of the Vienna Circle—which I date from the middle of the '20's to the end of the '30's—were nevertheless incidental products when considered from its internal philosophical perspective. It is

T. E. Uebel (ed.), Rediscovering the Forgotten Vienna Circle, 143–157.

characteristic of these writings that they did not constitute a context of discussion, and did not emerge from such a context. The individual authors developed their views independently of each other and from different sources of influence—one may gain the impression that they wanted to speak in these matters more as private individuals than as philosophers, and that the scientific attitude receded in comparison to that of the reformer or moralist, so that these writings occasionally appear rather idiosyncratic. In this sense then, there is no 'ethics', 'moral philosophy' or 'philosophy of value' of the Vienna Circle—no one doctrine or 'orthodox' teaching , which would have provided the object of refutation, overcoming or defense for the philosophical discussion of subsequent generations.

The most important writings (books, essays or manuscripts) concerning ethics and the philosophy of value which were written by the representatives or associates of the Vienna Circle seem to be (in their temporal order): Moritz Schlick *Vom Sinn des Lebens* ('On the Meaning of Life', 1927), Otto Neurath *Lebensgestaltung und Klassenkampf* ('Personal Life and Class Struggle', 1928a), Schlick *Fragen der Ethik* (*Questions of Ethics*, 1930b), Ludwig Wittgenstein's ethics lecture ('A Lecture on Ethics', probably worked out in 1929, published only in 1965), Neurath '*Soziologie im Physikalismus*' ('Sociology in the Framework of Physicalism', 1932a), Rudolf Carnap '*Theoretische Fragen und praktische Entscheidungen*' ('Theoretical Questions and Practical Decisions', 1934d), Karl Menger *Moral, Wille, Weltgestaltung* (*Morality, Decision and Social Organization*,1934), Alfred J. Ayer *Language Truth and Logic* (1936), Felix Kaufmann *Methodenlehre der Sozialwissenschaften* (*Methodology of the Social Sciences*, 1936), Viktor Kraft *Die Grundlagen einer wissenschaftlichen Wertlehre* (*Foundations for a Scientific Analysis of Value*, 1937), Friedrich Waismann's lecture 'Ethik und Wissenschaft' ('Ethics and Science', worked out in manuscript form probably 1938, published 1983). I include Wittgenstein in this connection because towards the end of 1929 and throughout 1930 he and Schlick and Waismann discussed ethical questions with reference to Wittgenstein's ethics lecture and to Schlick's book, which Wittgenstein had read and annotated in the margins (Waismann's relevant notes of these talks are available (Waismann 1979, pp. 68, 115ff.)). Ayer's book similarly belongs in the context of my topic, not only because of the author's personal contacts with the Vienna Circle, but also because it established the connection between the Vienna Circle (especially Carnap) and the philosophy of analysis, as it was practiced by Moore and partly by Russell, and because the views, which were there developed and soon became very influential, serve in a sense as the paradigm for radical empiricist teachings, and so also for the teachings of the Vienna Circle. Besides these writings which topicalised and analysed questions of value explicitly, there also exist in the works of the Vienna Circle, particularly in Neurath, Carnap and Hans Hahn, loosely scattered remarks about the values of and in life which are basic to the empiricist or scientific world-conception.[2] These remarks have a more programmatic character, they express an attitude towards life without really analysing it—and remained not without influence for this very reason.

Most of these writings and manuscripts have recently been edited or newly edited in German or English, most recently Schlick's *Fragen der Ethik*. This provides, if you like, the public occasion of my lecture. Still outstanding, in my view, are only Kraft's *Wertlehre* and Kaufmann's *Methodenlehre*. (In retrospect, Schleichert's collection *Logischer Empirismus - Der Wiener Kreis* (1975) seems to possess particular merit, for it was this volume that moved the forgotten, 'engaged' and 'life-affirming' side of the Vienna Circle back into the light—in contrast with the picture drawn in Kraft's influential book (1950). Kraft apparently seems to have consciously neglected these aspects (which also seem to have made him personally uncomfortable) in order to provide for the re-entry of the Vienna Circle into the republic of scholars in the conservative atmosphere of the post-war years, something he succeeded in doing only after considerable delay.)

Going through these writings in their temporal order makes clear that the early ones—Schlick's *Vom Sinn des Lebens* and Neurath's *Lebensgestaltung und Klassenkampf*—still moved outside of the typical philosophy of logical analysis which characterised the Vienna Circle; they came, as it were, before the linguistic 'Turning Point in Philosophy'. As to Schlick's *Der Sinn des Lebens*—similar thoughts had already been expressed in his *Lebensweisheit. Versuch einer Glückseligkeitslehre* (Wisdom of Life. Towards a Theory of Happiness) of 1908!—this was written in the spirit of Epicurus and represented, in tone and attitude, a 'philosophy of pleasure.' The authors who were used for this renewal of hedonism (Friedrich Schiller, Nietzsche, Guyau, Ruskin) are not commonly cited as defenders of empiricism. It is this Epicuraeism which connects Schlick with Neurath's *Lebensgestaltung und Klassenkampf* (one of the philosophically relevant chapters of this book is entitled 'Marx and Epicurus'), but Neurath, as can already be seen from its title, wanted his book to serve the liberation of the proletariat, differing therein from the more aesthetic views of Schlick who sought the meaning of life in purpose-free play. Neurath intended his to be a thoroughly partisan, political book of struggle; its intellectual home is to be found in the cultural, educational and life-reform efforts of social democracy. But both of these writings develop *ideals* of the new man, humanitarian ideals, and thus have a strongly normative bent, and therefore stand outside of the conception of philosophy of the Vienna Circle (without necessarily contradicting it).

It only in his *Questions of Ethics*, at the beginning of the 1930's, that Schlick tries to present a moral philosophy which gives expression to the principles of the Vienna Circle without neglecting the hedonistic aspect of his theory of happiness. I do not know whether this book summarises the views of the Vienna Circle (or of a part of it) concerning ethical questions, but I suspect that it did not do so. Instead it seems to have provoked critical reactions in the Circle which Neurath even formulated in writing.[3] Schlick claims for his work that it is primarily a *psychological* investigation; he wants to give a psychological answer to the questions "Why do we act morally?", "Why is something recognised as a moral norm?", i. e. give a causal explanation of moral behaviour which proceeds from hedonistic and altruistic assumptions and is able to show that acting morally can become increasingly pleasurable due to the education of the social and altruistic inclinations to be found in humans. At the same time, Schlick provides an extended *metaethical* investigation of the meaning of 'morally good' and finally even postulates a *normative* moral principle. (I will return to both these points.)

What also will be noted when we consider the ethical writings of the Vienna Circle is that the genuinely philosophical *analyses* and *explications* of ethics and the problem of value were attempted not by the 'left' of the Circle, but by the less politically engaged 'bourgeois liberals': Schlick, Kraft, Kaufmann, Waismann. By contrast, Neurath again and again proclaimed—as did Hahn, Carnap and Frank occasionally—the social reformist life-ideals of the empirical scientific world-conception, but he always considered a *genuinely philosophical* ethics simply as residual theology or metaphysics (his tendency in 'Sociology in Physicalism' and his *Empirical Sociology*, also published in 1931). For him, only the investigations of a concrete theory of happiness or 'felicitology' were scientifically meaningful, investigations which proceeded from questions like "Which order of life makes people happier than others?" On the other side, Kraft and Kaufmann wholly limited themselves to a philosophical and psychological analysis of questions of value—the *proclamation* of ideals or values of life was furthest from their mind—and while Schlick often speaks in a normative ethical vein he did not wish to express a socio-political tendency or to make propaganda—this he even found distasteful.

A special case amongst the writings mentioned is presented by Karl Menger's book *Morality, Decision and Social Organization.* His pronouncedly formal investigations only marginally touch upon the philosophical problem of morality, but gained great importance for decision theory, social science and political economy.[4] In respect of his philosophical attitude towards the problem of justifying morality, however, Menger does not differ from the other representatives of the Vienna Circle. Like all others, he *denies* that a scientific or philosophical ethics could provide the justification of a moral system, a justification which would replace moral decisions and reduce the plurality of moral systems

II

I now turn to the consideration of the *problem of value judgements* and its *analysis* in the Vienna Circle. I begin with Schlick's definition of the moral value judgement as given in his *Fragen der Ethik.* The traditional question of ethics, as also expressed by Moore, is the following: ethics is the general question of what is good. For Schlick, a meaningful answer to this question must consist of the definition or explication of the concept, i.e. he understands this question as "What is understood by 'morally good'?" or "What is called 'morally good'?" The definition required by answering the question has a *formal* part which appears trivial to Schlick: those actions are called 'good' of which it is demanded of us that we perform them, and we qualify actions as 'good' if we desire them. The moral judgements thus appear to the acting subject as *demands*, and the morally judging subject expresses a *wish* by this judgement.

The most weighty problem for Schlick, however, lies in the question of *who* demands or desires something, and of *what* is being demanded or desired. To answer this means to give a *material* determination of that which is called 'morally good'. My following reconstruction is intended to bring out the principal considerations which moved Schlick in his *Ethik.* Here goes:[5]

(1) What is called 'morally good' are decisions of the will and subsequently actions and ways of behaviour of individuals which have been *formed by society.* The society amounts to the majority of the human

environment of the individual agent in question, the fellow human beings with whom the individual is in contact through action, speech and writing.

(2) Moral agreement by society consist in the following: (a) society *believes* that such decisions, actions etc. are conducive to its well-being and maintenance; (b) society *desires* its well-being and maintenance and the realisation of everything that contributes to it; (c) the wish of the society is apprehended by its members as a *demand* upon them.

Of course, this way of speaking of *the* society which believes something, wishes and demands something, is ultimately fictitious for an empiricist. Moral values are judged by *individuals*, they are the valuing subjects. It seems that Schlick has in the back of his mind something like Hume's idea that the individual becomes a 'voice of the society' or 'voice of the community' with his moral judgements. Of him who passes a moral judgement, judges a decision of the will, an action etc. as 'morally good', the following holds first of all (I return to the explication of the concept):

(3) *He believes* that the decision (or action) in question is useful for and conducive to the well-being and maintenance of society, and *he desires* the well-being and maintenance of society and the realisation of that which contributes to it.

Two essential elements of moral *approval* or *disapproval*, of moral *praise* or *blame*, are thus circumscribed. But what is essential to moral value judgements (or value judgements in general) does not yet seem to have been captured. True, what is captured are the *content* of a moral value judgement (the decision or action x is conducive to the well-being and maintenance of society), and also the *presupposition* of *every* moral value judgement (namely, that every moral judge desires the well-being of society and the realisation of that which contributes to it). But the semantically specific property of such a value judgement seems to be expressed by the *connection* of the factual judgement with the presupposed desire or demand. For this problem Schlick gives only a pointer when he formulates that something is called 'good' "if it is as one desires it to be", or—for the properly moral domain—that something is called 'morally good' if it is "desired by the human society" (Schlick 1930b Sect. IV.1). The semantic essence of the value judgement seems to me therefore to lie in the *statement of the satisfaction or frustration* of a desire or a demand which the valuing subjects shares. Therefore I add to formulation (3) as follows.

(4) If someone judges x to be morally good, then he *expresses* his desire for the well-being of society and the realisation of everything conducive to it,

and he *states* that *x fulfills* this desire. And this of course implies the statement that *x* is conducive to the well-being of society. Only that is what I would consider to be a moral approval or disapproval, or moral praise or blame.

But the point of a moral value judgement still does not seem to have found a sufficient explication, if we want to do justice to Schlick's intentions. What is still to be shown is that the moral judge proceeds *in harmony with society*, that he is a 'voice of society', that is to say, that he only judges morally if he conforms to the beliefs of society with his own beliefs concerning what is conducive to the well-being of society. Only in this fashion can it be made clear what it means to give a moral judgement as a 'voice of society'. Accordingly I widen my definition as follows.

(5) When someone judges *x* as morally good then he expresses his wish for the well-being of society, he states that *x* satisfies this wish and there obtains a *consensus* between his beliefs and those of society concerning its well-being. With this, the material definition of the concept 'morally good' seems completed according to Schlick.

Without doubt, such a definition is appropriate to the moral valuations of those subjects one tends to call 'useful members of society', i.e. to the moral valuation of subjects who pledge allegiance to the *dominant moral value system*. Yet the reflective moral subject, the 'moralist' in the widest sense, typically does not confess this allegiance to the dominant value system, but opposes it. He does not share the belief of society about what is conducive to its well-being, although he also is concerned with it. His moral valuations do express approval or disapproval and praise or blame in the indicated sense, but he does not represent 'the voice of society'. And talking about 'value systems' leads to another problem of fundamental importance: Schlick's definition of 'morally good', which I reconstructed freely, fits the moral valuation of individual decisions, actions, modes of behaviour, but is not appropriate for the value judgements we make when we postulate or proclaim *moral principles*. What is called a 'value system' is, after all, the logical context of such principles.

Take the sentence 'Thrift is a virtue' as an example of such a principle. Someone who says this does not make a valuation in the above sense. To be sure, he desires the well-being of society, etc., and this desire confronts the subject of will and action very clearly as a *demand of society*[6]—one even gains the impression that value judgements take on the character of a demand on the individual only in such moral principles, that it was somewhat premature to attribute this demanding character already to the valuations of decisions, actions and modes of behaviour, as I did in (2) above. Furthermore, he who proclaims this principle states (in a descriptive sense) that thrift is conducive to the well-being of the society, and this belief of his can be in conformity with society or not. In the former case he would be something like a teacher of morality, in the latter case something like a moral legislator or reformer. But it is not the case that the proclaimer of the principle states the satisfaction or frustration of the desire at issue, i.e. he does not approved or disapprove something, does not praise or blame. He expresses the moral principle in the form of a *demand, not a descriptive statement*—he demands thrifty behaviour under certain circumstances, or he demands that that thrifty behaviour be approved or praised. But his demand only possesses *validity*, if the presupposed factual judgement (that thrift is conducive to the well-being of society) is true (if the factual content is true), and if the further presupposition can be granted that he desires the well-being of society (if the value character is endorsed).

With the latter expressions I have already begun to help myself to the terminology of Viktor Kraft. For it was Kraft who cast in relief these problems of *supra-individual* or *supra-personal* validity of value judgements in his above mentioned *Grundlagen einer wissenschaftlichen Wertlehre*[7], and did so much more clearly than Schlick (though not, as I see it, in contradiction to Schlick if his definitions are explicated as I have above). Kraft distinguished between the *factual content* and the *value character* of value judgements, and he thinks of principled value judgements like 'Deception is despicable' or 'Thrift is a virtue' as supra-personal *instructions for taking a position* vis-a-vis a certain matter, in our cases certain forms of behaviour. Such value judgements are not true or false in a descriptive sense, but they possess validity (comparable to the validity of rules or norms) for Kraft; to claim their validity means to promote their recognition in a general, impersonal fashion. That is to say, one who evaluates makes the claim 'anonymously' in the name of a supra-individual authority and in doing so directs a demand to people in general and requires that they take up an evaluative attitude in response, in which the essential value character lies. This response may range between acceptance and rejection, friendliness or opposition, praise or blame, approval or disapproval, etc.

In general, Kraft distinguishes between value judgements as 'statements of actual attitudes' and value judgements as 'general instructions for the taking up of attitudes', and it is the latter which I have just described. Kraft characterised the former as *ascriptions of a value character* to objects, and they are to be understood just as impersonally as the 'general instructions for taking up an attitude'. For Kraft, such a value judgement expresses "the determination of a [friendly/oppositional, accepting/rejecting etc., H. R.] attitude by an object in general".[8] I do not know exactly what he means, but it seems he is concerned with same problem which the explication above of 'good' bequeathed to us: '*x* is good' is supposed to *state* something about *x* (namely, that it *x* is such as to cause a certain evaluative—'positive' or 'negative'—response) and it is supposed to *express* at the same time such a response on part of the speaker. Above I have tried to characterise this semantic situation in a case of moral approval after Schlick's schema as the 'statement of the satisfaction of a desire which is shared by the speaker'; Kraft, however, speaks of value judgements in general (not only moral ones), and it appears difficult to adequately characterise aesthetic judgements like 'This picture is beautiful' as statements of the satisfaction of desires. In principle, however, it does seem that what is involved in all these value judgements is the statement of the satisfaction or frustration of wishes, tendencies, demands, norms, rules, etc., which are *shared or accepted or simply followed by him who evaluates*, and it is this sharing, acceptance or following which his value judgement expresses at the same time.

In any case, Kraft instructs us to distinguish between two kinds of value judgements. I would describe the lower form as *approval/disapproval, praise/blame* (and I have located the moral approval/disapproval which Schlick had in mind on this level) and the higher form as the *postulation of principles, of virtues or vices*. But I must note the following difference: Kraft characterises principles like 'Thrift is a virtue' as exhortations towards the taking up of an evaluative attitude of approval/disapproval etc., whereas I have characterised them more simply as exhortations to the respective types of action. According to the former determination, what finds expressions is that "thrifty behaviour is to be lauded", according to the latter that "engaging in thrifty behaviour itself is recommended", or something like it. Which analysis is more appropriate I do not know; probably this demonstrates two legitimate uses for statements about virtues and vices.

III

I now return to my basic historical theme. The Vienna Circle is generally held to denounce moral and other value judgements as cognitively meaningless, the *mere expression of feelings* and other psychological states:

emotivism is the favoured label for this. I believe I have shown that this cliché is totally wrong. Schlick as well as Kraft very carefully distinguished what Kraft called the 'factual content' and the 'value character' as the components of value judgements; in other words, they clearly distinguished the cognitive from the truly evaluative component. So it is not even correct to speak of a 'return' of the problem of the validity of values in Kraft—as suggested by Hegselmann.[9] This problem is well recognised by Schlick, and even the other Vienna Circle authors I mentioned do not provide an 'emotivist' analysis that would accord with the cliché of the common attribution. For instance, Felix Kaufmann's analysis is quite confusing in its multidimensionality: he introduces value judgements at one point as analytically true in the sense of expressing a just aim, while at another point he compares them to existential judgements because they are subjected to the same procedures of verification, whereas he deals with imperatives as judgements about the inner behaviour of oneself—but this very plurality of conceptions of evaluations shows the degree to which the author was concerned with the problem of validity, the differentiation of factual and evaluative components (cf. Kaufmann 1936, chs. I, 5 and II, 3).

So-called 'emotivism' entered the debate in a substantive way through A. J. Ayer's 1936 book *Language, Truth and Logic* (and Kraft already refers to it critically, at least in the 2nd edition of his book in 1951), but according to Kraft a similar standpoint had previously been taken already by authors like Braithwaite, Hägerström and others. The particularly emotivist element consists in attempting to analyse the entire value judgement as a kind of *interjection* like 'ah', 'oh', 'bah', etc.. Ayer's analysis of expressions of the form '*x* is good', like his entire book, constitutes the combination of Moore's technique of analysis with the verificationism of the Vienna Circle, in particular of Carnap. Like Moore, Ayer find that every descriptive analysis of the meaning of an evaltive expression like '*x* is good' cannot but fail: if 'good' is equated with any particular property of objects (e.g. 'pleasing'), then it remains meaningful to say, i.e. it is possible to say without contradiction, that the object at issue does posses the relevant property but that it is not 'good' (e.g., something is pleasing but not good). But Ayer does not draw from this the same consequence as Moore, that 'good' is an unanalysable property, but that expressions like 'good' do possess a descriptive character for they have the function to express a special feeling of approval/disapproval on the part of the speaker or to cause such feelings in the person addressed. If then someone describes a theft as injustice, he shows or expresses his moral disapproval of this action; he does not say *anything more* about the theft, and he does not say anything about himself and his feelings, he only *shows* that he has them. Insofar such an expression seems *comparable* to an interjection like 'Theft, bah'.

On the other hand, the very example shows that valuations do have a factual content for Ayer, and he does not wish to exclude a *meaningful dispute* with the moral domain. He only concludes that the seemingly moral dispute in truth is a dispute about this factual content, that is, about *empirical questions of fact*:

> we find that argument is possible on moral questions only if some system of values is presupposed. If our opponent concurs with us in expressing moral disapproval of all actions of a given type *t* [e.g. of the type theft, R. H.], then we may get him to condemn a particular action A, by bringing forward arguments to show that A is of type *t*. For the question whether A does or does not belong to that type is a plain question of fact. (Ayer 1936, p. 111.)

Thus we cannot argue for moral principles themselves, we can only praise them or reject them in the light of our own feelings. It is for this reason that ethics is only possible in a scientific sense as the psychology or sociology of moral behaviour, as casuistry (i.e. as the logical investigation of a given moral or value system), or as the analysis of morally evaluative expressions (metaethics).

I think Schlick and Kraft could agree with all these assertions of Ayer's, and *de facto* did so: what they pursued as ethics thus appears to be a mixture of metaethics, psychology and sociology. The only point on which they could not agree is what might be called 'the Moorean conclusion' which Ayer draws from his Moorean argumentation: that ethical concepts are *pseudo-concepts* and therefore are *unanalysable* (ibid. p. 112). What Ayer himself actually did was to emphasise the specifically semantic function of an evaluation over and against the descriptive function, to make the former available for an anlysis, and to make one proposal for such an analysis. One can develop this analysis and conduct it more comprehensively, as Schlick and Kraft tried to do, but as far as I can see, nothing of the emotivist conception has to be withdrawn in doing so. (What needs to be shown is in what way factual content and value character are connected in evaluations, because only thereby could the logical structure of a value system be made transparent.)

It seems to me to be rather the cliché of emotivism which brings on the confusion. It overlooks that Ayer only *compared* evaluations with interjections, but did not equate them. (Incidentally, Kraft made the same mistake in his criticism of Ayer.) It is sufficient for a true interjection to say simply 'bah' in a particular situation, but the corresponding value judgement would have to go 'Theft, bah'; not to distinguish this constitutes the source of the misunderstanding of emotivism. I think Ayer only wants to say that an evaluation constitutes a *meaningful whole* that can be neither true or false, because an evaluative expression is *constitutive* for it, an expression which cannot be analysed as descriptive, but has an expressive or signaling function that is comparable to an interjection. And it seems very difficult to fault this. Thus I believe that the relevant authors of the Vienna Circle and of logical empiricism agree to a large degree about the essentials of the problem of value judgements and their analysis, and that they tend to complement rather than contradict each other. Often oppositions and controversies are suggested about these matters; this is connected with the fact that the explicit and direct discussion of these problems only begins in a later phase of the movement inaugurated by the Vienna Circle, in the metaethics of the analytical philosophy of the post-war years.

IV

What does require a separate investigation is Wittgenstein's conception of ethics, for here the problem of the *limits* of an empiricist or scientific world-conception finds a clear expression. In his talks with Schlick and Waismann, Wittgenstein gives an indication of his principled and fundamental unease with Schlick's ethical position.[10] For him it is established that ethics cannot have anything to do with facts, that in the 'true' ethics no sentence about the world could appear and that sentences about the world cannot contain a sentence of this ethics. (For instance, the description of a murder would not contain or imply an ethical sentence.) Wittgenstein thinks that the ethical value judgement has an *absolute* character. (This thought is developed in his ethics lecture (1965).) By contrast, Schlick suggests that ethics does deal with facts, that it leads to *knowledge of facts*. And this seems intended not only in the harmless sense that ethics provides a psychological, causal explanation of moral behaviour.

But first note that Schlick does not provide an answer to the question 'What is good?', but wants to answer the question 'What is being called "good"?' Schlick does not say: that *is* good which human society agrees to evaluate positively—as supposed by Kutschera who then objects that in this case presumably nothing is good[11]—he says: that is being *called* 'good' which human society agrees to evaluate positively. In other words, Schlick proposes that 'good' is not a simple predicate which attributes a state to the world. Wittgenstein, by contrast, takes the question of the nature of the good literally: in his ethics lecture he uses the formulation that ethics asks for what is good, what is valuable, what really matters, that asks for the meaning of life and the proper way of life. If such questions are conceived of as ethical and not as empirical questions concerning convenience and means-ends relations, then they possess only absolute answers; and for Wittgenstein these are *meaningless*, without a natural sense: ethics is *supernatural*.

For Schlick the question 'What is good?' becomes 'What is called "good"?' But what about the question of *value*, and of what has value? Schlick does not achieve full clarity here (cf. Schlick 1930b, sects. VI.1, VIII.7). For him, value consists only in relation to a subject that feels pleasure and pain: if there would exist no pleasure or pain in the world, then there would exist no values, everything would count for the same. Thus Schlick identifies the concepts of *happiness* and *the most valuable*, and since the way to the valuable leads via the capacity of an individual for happiness he even pronounces a moral imperative: keep capable for happiness, be prepared for happiness! How then must we understand Schlick's principle which seems to have caused Wittgenstein's scepticism: "What is to count as the highest value must be derived as fact from human nature and life"? (Ibid., sect. I.9.) If 'valuable' *means* something like 'pleasurable', 'pleasure-increasing' or 'bringing happiness' for Schlick, then this principle says merely that it is an empirical *fact* that for humans this or that is most pleasurable, or that they find this or that most pleasurable. But sometimes it seems as if Schlick held pleasure and happiness to be valuable in themselves, simply *worthy of pursuit*, and that it is only on this basis that the imperative 'Keep prepared for happiness' gains a meaning which does not express a trivial means-ends rule. Then it obviously would have to be understood as absolute in Wittgenstein's sense. But then it could be trivially objected that an absolute imperative is not a sentence about the world, and this would end the *logical* discussion rather quickly—which would please Wittgenstein for he did not intend a logical discussion about the derivability of 'ought' from 'is', etc.

For Wittgenstein, his own discussions of absolute value, etc. are logical nonsense: he describes the experience of astonishment that the world exists, the experience of absolute security and the experience of guilt vis-a-vis God—these are *experiences* to which he grants an absolute value due to his own personal experience, in order to then speak in the sense of Kierkegaard: "It is the paradox that an experience, a fact, should seem to have supernatural value." (Wittgenstein 1965, p. 10.) Now he stands at the same juncture as Schlick with his experiences of happiness. Nothing seems to lead beyond this.

But maybe these weighty examples of religious or mystical experience only confound the simplicity of the problem from which Wittgenstein had begun. If someone asks "What shall I do?" then this question can be easily answered by reference to the aims of the questioner, in meaningful sentences about *functional* behaviour. But if these goals themselves are in question for this person, or if these goals become more and more merely intermediate goals, then it becomes meaningful to ask: "What do I really want?", "What should I want?", althought these questions are logical nonsense. (If 'oughts' are the expression of willing, what would it mean that I 'ought to will?') These questions do not express a problem about which of my wishes it would be more convenient to grant priority, and these questions also do not exhort others to prescribe my goals for me, or to order me to do anything, and these questions also do not aim for a discovery on my part of so far hidden movements of the will of my person. In a fashion that remains rather dark logically, such questions rather would seem to appeal to the *autonomy* of the subject.

Schlick confronts fairly analogous problems in his above mentioned *Vom Sinn des Lebens* (and related writings). A life that is ruled by means appears to him like a meaningless runaround, for the *value* of life can only lie in the pleasure of activities that are engaged in for their own sake, in which means and end, action and success are one. Nevertheless Schlick would not want to speak of the 'absolute good' or the 'absolute correct', or of the 'supernatural' character of ethics, just as it was impossible for Wittgenstein to identify the moral good with pleasure or happiness in any form. What we find here are obviously deep divergencies in spiritual *attitude*, but whether these differences could be formulated in the form of a *philosophical dispute* about the question of the meaning of life seems to me to be very dubious indeed. For both Wittgenstein and Schlick are aware that with such questions they transgressed the limits of logically meaningful speech.[12]

If, following the empiricist way of thinking, such delimitations are taken seriously, then not much more could be said at this juncture that what was already said by Waismann (1983, p. 184): "Declaring oneself for a particular morality is a very deep process—I do not know another expression to designate our feeling that the core of our personality emerges or manifests itself in this choice."

Karl Franzens Universität, Graz

NOTES

* First published as 'Ethik und Werturteilsproblematik im Wiener Kreis' in *Von Bolzano zu Wittgenstein. Zur Tradition der österreichischen Philosophie*, Hrsg. J. C. Nyiri, *Schriftenreihe der Wittgenstein-Gesellschaft Bd. 12, Teil 2*, © 1986, Hölder-Pichler-Tempsky, Wien, pp. 162-172. Translated with kind permission of Verlag Hölder-Pichler-Tempsky and the author by T. E. Uebel.

1 Compare the discussion between A. Beckermann, R. Hegselmann and W. R. Köhler in *Analyse und Kritik* 1 (1979) no. 1.

2 See Carnap 1928a, Carnap-Hahn-Neurath 1929, Hahn 1930, Neurath 1931c, and the posthumously published Schlick 1952 and 1962.

3 For example, without mention of Schlick's name in section 6 of Neurath 1932a. Criticism of Schlick's hedomism is suggested also in Menger 1934 and explicitly formulated in Kraft 1937—again without the mention of his name. Compare also the summary in Kraft 1950, 2nd ed., ch. "Values".

4 Oskar Morgenstern already published a remakable appreciation of Menger in this respect in his 1936.

5 The following reconstruction is based on Schlick 1930b, sects. I.5, I.7-11, I.12, IV.1, IV.3-4, IV.6 (1984, pp. 60f., 63f., 66ff., 70f., 74, 112f., 115f., 118, 122f.).

6 That the moral principle confronts the addressee not only as the demand of the speaker but as the demand of society at large does not mean that it really would have to be a demand by society (its majority). The addressee merely believes that the speaker postulates his demand in the interest of society, and if the addressee accepts the principle, then he belives that this demand *is* in the interest of society.

7 The following is based on Kraft 1937 (1951, esp. ch. IV, also chs. II, 1 and III, 4.5). See also Kraft's summary of his standpoint in his 1950, ch. "Values".

8 Kraft 1937 (1951, p. 187: "Bestimmung einer ... Stellungnahme durch einen Gegenstand ganz allgemein").

9 Hegselmann 1984, sect. 4; later on, Hegselmann himself concedes that these distinctions were already drawn by Schlick.

10 Compare his conversations with Schlick and Waismann, published as Waismann 1979.

11 See Kutschera 1982, p. 187. I refer to my reconstruction of Schlick's views in part II of this paper: Schlick bases his explication attempt precisely on the culturally determined antagonistic multiplicity of moral evaluations (Schlick 1930b, sect. 4.4) and thus he stresses that that is called 'good' in a society (or culture) of which this society *believes* that it is conducive for its wellbeing. Thus mutually *incompatible* modes of behaviour are called 'good' dependent on the society, culture and epoch, but the meaning of the expression 'good' nevertheless remains the *same*.

12 Compare my discussion of this problem in Rutte 1985 where I refer to Janik and Toulmin 1973 in which Wittgenstein's ethical concerns are placed in the centre of the interpretation.

FRIEDRICH STADLER

OTTO NEURATH—MORITZ SCHLICK: ON THE PHILOSOPHICAL AND POLITICAL ANTAGONISMS IN THE VIENNA CIRCLE*

In drawing the intellectual and personal profiles of Moritz Schlick and Otto Neurath, the two dominant adversary thinkers of the Vienna Circle, I present the outline of the physiognomies of two individuals as well as the inner physiognomy of a 'scientific community'. The example of these two strong personalities in this heterogeneous circle of scientists furnishes us with the minimum base of doctrinal consensus required to present some aspects of the pluralistic dynamic of theories which characterises this community; it shows the essentials of a 'psycho-sociogram' typical of all scientific communities; and it also provides evidence of the mixture of personal, world-view-related political, and philosophical-scientific elements.

My result relativises the myth (based on the combative programmatic pamphlet *The Scientific World-Conception. The Vienna Circle* (Carnap, Hahn, Neurath 1929)) that the Vienna Circle represents a closed anti-metaphysical school of philosophy, and serves to criticise the ahistorical perspective which describes neo-positivism as a traditional philosophical movement—similar to neo-Kantians and neo-Wagnerians—that was possessed from its inception by a tendency towards self-dissolution and showed disintegrative scientific communication amongst its only contingent forms of organisation (*Verein Ernst Mach, Erkenntnis*).

The following consideration of internal and external factors can sketch the specifically collective aspect of the Vienna Circle, which points beyond its description as a loose scientific community, but only in suggestive contours. The extreme positions in this grouping are here drawn very pointedly, partly because of the remarkable conclusion that there *did* exist what was demonstrated to the outside despite all of the personal and doctrinal differences—namely, the consciousness of representing a common movement which, from the beginning to the end of the Circle, prefaced all divisions by the credo of being an open experimental forum for discussion of the scientific world-conception.

T. E. Uebel (ed.), Rediscovering the Forgotten Vienna Circle, 159–168.

In his *Allgemeine Erkenntnislehre* (*General Theory of Knowledge*, 1918, 2nd. ed. 1925) up to the mid-1920's, Moritz Schlick[1] owed allegiance to non-positivist critical realism—as it has been advocated since by Herbert Feigl, Viktor Kraft, also by Karl R. Popper—but from 1925/6 he presented, with Friedrich Waismann, the strongly Wittgensteinian wing of the Circle with his 'Turning Point in Philosophy' (1930a). It is not the case, however, that one should speak of a congruence of the positions of Schlick and Wittgenstein, as can be seen from the example of their conceptions of philosophy later in the 1930's and their divergent conceptions of ethics and aesthetics. Schlick presented a (methodological) phenomenalism like that of Carnap in his *The Logical Structure of the World*. This occurred only *after* his critique of metaphysical school-philosophy at the height of the second scientific revolution at the beginning of the 20th century; there he summarised the results of the natural sciences within the framework of a programmatic conception of philosophy representing a system of most general principles. In this development, philosophy became a therapeutic activity of linguistic analysis of statements of the individual sciences, an activity designed to overcome metaphysics and to clarify the meaning of statements with the all too well-known and misunderstood dictum that the meaning of a statement lies in its method of verification. So we must conclude somewhat sketchily that Schlick changed, via the 'linguistic turn' à la Wittgenstein, from his early realistic position to a loosened and liberalised standpoint towards the end of his life. The latter found expression in his little noted essay 'The Vienna School and Traditional Philosophy' (1937) in which he distanced himself from the concept of Neurath's radical physicalism as a strictly anti-metaphysical, 'non-philosophical' unified science (later, encyclopedism).

Analogously we can similarly distinguish stages in the development of Neurath[2]—though with greater continuity of content and aim—from the programmatic pamphlet (co-)written by him (Carnap, Hahn, Neurath 1929)—which, as is known, was criticised by Schlick together with Wittgenstein as simplistic and propagandistic—up to the historically referenced self-portrait of logical empiricism in his essay on *The Development of the Vienna Circle and the Future of Logical Empiricism* (1936a). In the latter, the program of 'unified science' becomes the widened encyclopedic program, and the Vienna Circle or the Unity of Science is classified as a movement—provoking amongst other things Schlick's just mentioned repositioning of the Vienna Circle. From the relativising of the meaning criterion by the principle of tolerance (introduced by Karl Menger) on up to his death, Schlick still accorded a positive and constructive role to philosophy—a role which extended to ethics—whereas throughout his life Neurath was convinced of the dissolution of philosophy as an autonomous discipline into the comprehensive empiricist system of encyclopedism. In this, he unintentionally remained indirectly 'philosophical' or philosophically relevant, as can be seen by considering his work (Rutte 1982a [ch. 6], Haller 1982a [ch. 9]).

Neurath recognised the doctrinal divergences in the Circle and described them already at an early stage, still before the controversies around the 'problem of the basis'; nevertheless, he fortified its organisation and stressed in terms of content what was common to the neo-positivist movement against the outside. With certain reservations this harmonising tendency also holds of Schlick, for his internal criticism was objectively blunted to a large degree by his public-institutional role and function, as in his activities in the Verein Ernst Mach, and in particular his struggle against its dissolution in February 1934 (Stadler 1982b, pt. 2). One year earlier Neurath stated somewhat too optimistically and euphorically (and from today's view, wrongly):

> But while Schlick came from 'realism', Frank as well as Hahn and Neurath were already 'free from philosophy' ... Wittgenstein provided the 'Circle,' which has been meeting around Schlick during the last few years, with a powerful stimulus and a fertile ground for discussion by forcing it to take issue with many problems. But it is possible to accept Wittgenstein's theory of truth and truth-functions and his extremely fruitful approach to the radical analysis of language and nevertheless reject decisively and without reservation his attempt to legitimize at least provisionally some form of idealistic, even mystical metaphysics in an indirect way, via preliminary elucidations. ...

Neurath concludes with some Utopianism that he has

> taken the radical position within the framework of the 'Vienna Circle' that one cannot make even preliminary remarks in anything other than a physicalistic language. Any discussion of whether "language or the world is more complex" must be dismissed in advance—or, in short, any confrontation between 'world' and 'language,' such as the one that Wittgenstein suggests. To formulate this as strictly as possible, "*sentences can be compared only with sentences*" ... It seems that this uncompromising attitude to the basic metaphysical sentences of Wittgenstein's *Tractatus* is becoming more and more prevalent, and that it will be possible to come to some kind of agreement about the thesis of 'protocol sentences' as the basis of *logical empiricism*, which is opposed to idealistic half-measures and others like them. (Neurath 1933a, fn. 2 (pp. 274/5).)

It is known that this antagonism was sharpened in 1934 in the debate between Neurath's coherence and Schlick's correspondence theory (cf. Hempel 1981), and that Neurath's hope for a universally accepted 'materialistic' (read: physicalistic) solution of the problem of the empirical foundation of scientific knowledge and truth remained unfulfilled. His vision of harmony had failed and the basis problem of empiricism remained, together with the concepts of physicalism and unified science, one reason for the thorough-going pluralism in the Vienna Circle up until Schlick's death.

Neurath's consequential and unyielding position becomes intelligible only in a greater historical and developmental context, which shows the stunning continuity of his thought and action (Stadler 1982d). Through the influence of his father, his intellectual socialisation was formed by liberalist, social-reformist, economical and visual elements which raised the central question of a just social order; this was to be reached by 'social engineering' through planned economy, socialisation, social reform, picture language, and the scientific world-conception. Working in this sprit, Neurath could not but aim for a conception of enlightenment which comprehended theory and practice. This conception manifested itself variously in his organising, in his scientific and in his popular-educational work in the Vienna Circle, the Verein Ernst Mach, the Social and Economic Museum, and abroad in the Unity of Science movement: we can trace a developmental line from models of scientific philosophy, 'scientific world-conception', to encyclopedism in signs and language.

Neurath criticises the opposition to the scientific enlightenment by means of his functionalist concept of anti-metaphysics (i.e. according to varying analyses of society), a concept which found its positive doctrinal and methodological transformation in the 'unified science' of the early 1930's. It seems possible to demonstrate a systematic connection between his rational-empirical historicism as an epistemological principle and his holistic coherence-theoretical relativism in the theory of science. In one of his last self-portraits Neurath described for us how strongly he was impressed until the end of his life by the idea of collective work and planning in society and science for the purposes of humanisation and democratisation:

> I also stress planning for freedom. I mean by planning for freedom orchestration of variety. Others may prefer other programs. But none can use Logical Empiricism to ground a totalitarian argument. It conceals not a loophole for dogmatism. Pluralism is the backbone of my thought. Metaphysical attitudes lead very often to totalitarianism, but I do not know of any consistent logical empiricist who came thereby to a totalitarian view. (Quoted after Kallen 1946, p. 533.)

The intellectual biography of the grand-bourgeois Moritz Schlick, who, on a superficial view, was also indebted to 'this-worldly' anti-metaphysical enlightenment thinking, was rather different. Coming from an aristocratic background, trained in the natural sciences, this favoured student of Max Planck and friend of Einstein was pre-destined to criticising the speculative-idealistical school philosophy, becoming one of the first who, with Einstein's agreement, approached relativity theory philosophically and who first presented scientific philosophy in his *General Theory of Knowledge* at the end of the World War I. In contrast to Neurath, Schlick always remained within a domain of scientific discourse which he understood to be autonomous. Philosophical and world-view/political differences could not but result from this contrast.

Though I cannot go further into the the Vienna Circle's problem of empirical foundations, I may, however, give a cursory pointer towards the difference between Neurath's Machian, syntactically oriented 'relativism' and Schlick's semantically oriented 'absolutism'. This opposition, characterised in these terms by Neurath himself, also involved the definition and function of anti-metaphysics in the elimination of idealistic speculation from the domain of science which was intended by both. These antagonisms find open and occasionally polemical expression in the correspondence of Schlick and Neurath. Neurath classified Schlick's 'On the Foundation of Knowledge' (Schlick 1934; the opposing position: Neurath 1934) as "philosophical absolutism", "rather worrying" and "sadly mystical" and mocked the "Wittgensteinery" of the "Schlick-sect" in their criticism of ethics[3], whereas Schlick complained, with allusion to Neurath, about the dogmatism of some older members, about the "anti-intellectual character" of some publications, about Neurath's terminology of a "right" and "left wing" of the Vienna Circle, and about the "unscientific and unserious" character of Neurath's book *Empirical Sociology*: "If one declares triumphantly on nearly every page that one can do without God and the angels, then this is very boring for comrades-in-arms, appears dogmatical to opponents, and is ridiculous for both."[4]

Quite apart from the fact that here the well-known personal animosities of the finely attuned, aesthetic Schlick against the combative, vital Neurath found expression, there also was another difference which cannot be discounted, namely, how the envisaged aim of a 'consistent' or 'logicising empiricism' was pursued. For Neurath, the levelling of philosophy and science was to help in achieving a physicalistic (behaviourist, materialistic) unified science and a corresponding encyclopedism, whereas for Schlick a scientific philosophy could coexist in parallel with science. What is decisive for Neurath in this context is the holism and fallibilism of the theory of science and a pragmatically and sociologically understood conventionalism which points to the social conditions of scientific communication and signals the necessity for a synthesis of internalist and externalist approaches—long before the works of Kuhn and Feyerabend. The community of scientists must itself therefore become a scientific problem which cannot but escape the (absolutist) 'pseudo-rationalism' (Neurath 1935b).

Here we touch on their world-view *cum* political differences. Neurath demanded the theoretical and practical unity of thought and action and—like Philipp Frank and Edgar Zilsel—joined the history of science with the sociology of knowledge and of science, whereas Schlick endorsed an autonomist conception of philosophy and science which, complemented by 'questions of life and meaning', amounts to a scientific ethics and a traditional philosophy of culture and society, and correspondingly, to the liberal-bourgeois understanding of politics in the sense of 'lowly' party and day-by-day politics.

A shortened social analysis and political abstinence determined Schlick's aesthetic-moral perspective on socio-political conditions, as can be seen from his posthumously published small book *Natur und Kultur* (Nature and Culture, 1952) which was written against German National Socialism. What becomes transparent in Schlick's self-understanding, in his vocabulary and in his practical proposals for solutions, is a laudable humanism, pacifism and cosmopolitanism, anchored—unlike Neurath's social eudaeumonism—in an individual-hedonist picture of society. The enlightenment postulate of cultural development and the increased sensibility of customs and consciousness thus lead Schlick—with a rigorous criticism against Spengler—to an abstract eudaemonism and libertarianism characterised by an anti-state liberalism which operates with ideal-type constructions like 'state' and 'economy' without references to empirical-causal results. The state merely has the task of the 'protection of life', even if the latter may lack the 'morality' necessary for a happy life. This idealistic understanding of politics by a liberal intellectual in the wake of Europe's descent into fascism could not but lead to the kind of helpless anti-fascism that is characteristic of his proclaimed moral-voluntarist conception of the state. Starting from this world-view with the motto "As little state as possible!" (1962, p. 42) Schlick inevitably arrived at a plea for a political '*Laissez fair*': "Liberalism is the only political form of thought which is adapted to the modern form of life determined by communication and technology." (Ibid., p. 47.)

Against this, Neurath urged the application of the socialist idea of planning on all levels of theory and practice—from war economics, total socialisation, the settlement movement and trade union movement, all the way to the idea of planning for freedom in a humanist world society in which a democratic science was to function according to the form of non-capitalist and collective production process.

This background makes intelligible Schlick's vain struggle against the dissolution of the Verein Ernst Mach, his surprising declaration of loyalty to the Austria of Dollfuss and his critique of the repression of academia in an étatist, authoritarian fashion (Stadler 1979a [ch. 5]), as well as his pointed remarks and confessions in his correspondence. This background also makes intelligible the disappointment of the emigré Neurath who expressed his regret over the role "for Dollfuss against unified science" taken by Schlick (who stylised the Dollfuss strategy as a bulwark against National Socialism).[5]

There also is no more surprise in Neurath's sharpening of the dualism of empiricism and metaphysics (the latter with reference to Plato and Hitler) and in his praise of the time of the Circle as the best period of co-operation despite these disappointments.[6] In the Social and Economic Museum, the Verein Ernst Mach, the Vienna Circle, and as the organiser of unified science in exile, the enlightenment worker, encyclopedist and social reformer Neurath was unable to separate science and politics (given his more concrete political consciousness and the diction of the workers movement), so that the 'publicity work' (viewed sceptically by Schlick) became a necessary correlate to the 'academic playing field'.

And although according to his 'Lost Wanderers of Descartes' (Neurath 1913b) it was the case that, due to the dualism of fact and value, political actions could not be derived as norms from the scientific world-conception, Neurath nevertheless aimed at a concerted action of scientific work and social practice: with monism as the means of empiricist communication and pluralism as an attitude to the development of hypotheses, this aim amounted to a uniform rational-empiricist program.

Schlick's aristocratic context in the area of academic communication and in his private circle—against propagandistic positions with the bourgeois values of value-free science, the Socratic ethos of clarity, cleanliness and love of truth—collided with Neurath's impulsive, spontaneous spirit and organisational vehemence in the context of the Viennese adult education movement. As promoter, innovator and fiery speaker at the 'base' between proletariat and *petit bourgeoisie*, the polyhistor Neurath demonstrated the practical cosmopolitanism of his wide interests in literature, art, politics and architecture through his inter-disciplinariy orientation and his concrete utopias, his popularisations by concrete example, situated between encyclopedism and social reform, like ISOTYPE, 'universal slang', and 'unity of science'.

If in conclusion we must take note of a number of personal and doctrinal differences between Schlick's and Neurath's paradigmatic profiles (which are exemplary for the pluralism in the Vienna Circle), then we must also evaluate the objective dimension of the Vienna Circle—over and above, that is, the commonalities which from their personal perspectives made up for the disagreements. It is in this objective dimension that the lasting influence of the neo-positivist movement lies, after all.

Commonalities show up in their opposition to Plato, Spengler, Kant, in their this-worldly empiricist ethical and scientific attitude informed by criticism of language, pluralism of theories, monism of cognition and method as based on the natural world-conception, common public work in the Verein Ernst Mach despite their internal differences, dualism of facts and values, their nominalism, hypothetico-deductive theory construction, and in the world-view-related and political embedding of rationalism and empiricism within the cultural syndrome of the 'late enlightenment' (Stadler 1981). Both thinkers were in their own way dominant figures in the Vienna Circle and in the Verein Ernst Mach, representing the maximum of differences and a still cohesive minimum of a common consciousness and public self-understanding in their publications and institutions. The latter served to translate their basic values and aims—like science as providing a basic attitude (empiricism, logic, and criticism of language)—into the framework of an implicit program of enlightenment.

Schlick made it possible for the Vienna Circle to gain an academic platform under difficult conditions in the University of Vienna, whereas Neurath's cooperation shows not only in his original scientific work, but also in his work as organiser and populariser in Austria and abroad, in particular for the International Congresses for the Unity of Science. Schlick's personal achievement was the natural scientific and logico-linguistic founding of a scientific philosophy and world-conception based on the natural sciences, against the 'chaos of philosophical systems'. And even though his early tragic death prevented a comprehensive conclusion to his work, his scientific ethos and spirit, his personal authenticity and his professional competence bestowed upon Schlick the central role of *doyen* in the Vienna Circle and allowed him to promote the international reputation of the 'Vienna School' as its ambassador abroad.

Of course, despite the 'Turning Point in Philosophy' which he initiated, Schlick remained a genuine 'philosopher' (see Schlick 1937). In his last retrospective of the neo-positivist movement he drew—against Neurath's 'unified science'—the harmonious picture of an anti-metaphysical, but essentially philosophical movement which employs anti-metaphysics only as a methodological principle in the elimination of philosophical errors. (It is noteworthy here, however, that despite all of his internal criticism Schlick did not renounce the collective concept of a philosophical-scientific movement.) The distinction between 'questions of fact' and 'question of meaning' allowed him to determine the different domains of science and philosophy: the philosopher seeks to clarify the meaning of our statements; the scientist attempts to decide their truth. In the end, Socrates is presented as the true father of philosophy.

Schlick postulated ethics as an equally genuine philosophical task, namely, the clarification of ethical concepts as a more important task than the solution of theoretical problems. With everyday and common sense philosophy understood in that way—oriented contrary to the scientifically narrowed conception of logic—Schlick indirectly meets again with Neurath's encyclopedism, which argued against metaphysics as a relevant political ideology between the green and brown fascisms, but without the help of a genuine philosophy besides or above the individual sciences.

For Schlick, there was no abstract frame of reference like that given by a pluralistic encyclopedia as an open research program, but there was the wholly concrete perspective of democratic science in the socio-political climate of contemporaneous English society. Due to his subjectivist self-understanding as an apolitical thinker, the strictly anti-Nazi 'stay-at-home' Schlick had to sublimate his moral position in a philosophy of culture and society and express it as a timeless philosophical-literary reproach. His dichotomy of 'experience' (*Erleben*) and 'cognition' (*Erkennen*) demanded a high price in comparison to pragmatic-empirical historicism and relativism, namely, that of projecting political forces into a domain that appeared to render them rationally decidable, the domain of philosophical and scientific discourse—with the consequences of neutralising practical action and indifference against everyday politics.

Yet if we consider the revaluation of the Vienna Circle which has begun in the last few years and was stimulated by the Schlick-Neurath centenary—a historical-critical inventory which followed in the wake of the emotional and mythologised 'positivism dispute' and the static criticism of the 'received view'—then Neurath's prognosis, made in 1938 to Herbert Feigl, happily seems to come to true: "Commonalities will remain; differences will pass as temporary phenomena."[7]

Universität Wien

NOTES

* First published as 'Otto Neurath-Moritz Schlick: Zum philosophischen und weltanschaulich-politischen Antagonismus im Wiener Kreis', in *Schlick und Neurath. Ein Symposion*, Hrsg. R. Haller, *Grazer Philosophische Studien* **16/7** (1982) 451-463, © 1982, Rodopi, Amsterdam. Translated with kind permission of Editions Rodopi and the author by T. E. Uebel.

1 On Schlick's life and work see: Schlick 1979a, 1979b; Feigl 1937; Waismann 1938; Zilsel 1937; Rutte 1976, 1977b; Juhos 1963, Parthey/Vogel 1969; Wheeler 1969; Stadler 1982e.

2 On Neurath's life and work see: Neurath 1973, 1979, 1981, Hegselmann 1979a, Fleck 1979, Nehmet 1981, Stadler 1982a.

3 Neurath's letters to Carnap, 7 May 1934, 14 May 1934, 17 June 1934, 13 December 1934, 18 January 1935, 22 September 1935. All in the Vienna Circle Archive, University of Amsterdam. (The publication of the Neurath-Carnap correspondence is in preparation.)

4 Schlick to D. Rynin, 4 November 1933, to W. Köhler 13 March 1934, to R. Carnap 20 January 1935, to P. Frank ca. 1930. Vienna Circle Archive, Amsterdam.

5 Neurath to Carnap , 18 July 1934, Vienna Circle Archive, Amsterdam.

6 Neurath to Carnap, 1 July 1936.

7 Neurath to Feigl, 1938, Vienna Circle Archive, Amsterdam.

HEINER RUTTE

NEURATH CONTRA SCHLICK. ON THE DISCUSSION OF TRUTH IN THE VIENNA CIRCLE.*

My remarks concern the discussion of the problem of truth and the basis of knowledge, which took place in the years 1934/5 between members of the Vienna Circle in the journals *Erkenntnis* and *Analysis*. In his essay 'On the Foundation of Knowledge' (1934) Schlick criticised the theory of protocol sentences held by Neurath and Carnap, which for him amounted to an indefensible coherence theory of truth and an indefensible conventionalism concerning the empirical base. He pleaded for an absolutely certain foundation of empirical knowledge, his 'constatations', by means of which a comparison and agreement of knowledge and reality became possible.

Neurath responded with his article 'Radical Physicalism and the "Real World"' (1934) in which he rejected Schlick's allegations and summarised his standpoint. The controversy was continued in *Analysis* with contributions by Hempel, Schlick and Juhos.

Neurath characterised his position as follows.

> In opposition to Schlick, I advocate the following tenets: (1) All content statements of science, and also those protocol statements that are used for verification, are selected on the basis of decisions and can be altered in principle. (2) We call a content statement 'false' if we cannot establish conformity between it and the whole structure of science; we can also reject a protocol statement unless we prefer to alter the structure of science and thus make it into a 'true' statement. (3) The verification of certain content statements consists in examining whether they conform to certain protocol statements; therefore we reject the expression that a statement is compared with 'reality', and the more so, since for us 'reality' is replaced by several totalities of statements that are consistent in themselves but not with each other. (Ibid., p. 102.)

By contrast, Schlick's constatations are 'sentences' which cannot be written down and which concern the obtaining of immediately given, private data of experience in one moment of experience (e.g. 'red patch here now'). They are ego-dependent, egocentric sentences, non-hypothetical (incorrigible), but they possess this absolute validity only in that moment of experience itself. All empirical knowledge leads to these momentary endpoints concerning which we may speak of the direct, complete and final verification of a sentence

T. E. Uebel (ed.), Rediscovering the Forgotten Vienna Circle, 169–174.

by the fact itself (comparison of sentence and fact). Neurath's protocol sentences are instead hypothetical from the start and take an objective form; in a greatly simplified formulation they say that such and such a person at such and such spatio-temporal location observed such and such an object. I consider a few arguments for Neurath's standpoint.

As a physicalist, Neurath naturally does not deny that there are real facts. But he opposes the philosophical way of speaking of sentence and fact and of their agreement in the case of truth. For this way of speaking implies a comparison of sentence (proposition) and fact. To engage in such a comparison which would constitute absolute verification, we would have to formulate a sentence about the sentence and a sentence about the fact, and these in turn would have be true in the sense of the comparative relation: for this to be established, (1) we need another comparison and (2) the truth of the sentence about the fact must be presupposed—which was supposed to have been established by the first comparison (as verification). For this reason Neurath holds that we can only compare sentences with sentences.

For Schlick, however, the comparison of sentence and fact is given with the constatations: I have a hypothesis, e.g. that the cathedral has two towers and then I look at the real cathedral and make the constatation (Schlick 1935a). On closer analysis this is only a comparison between two sentences with the same meaning, the comparison of a hypothetical first-person sentence about the content of experience, which is derived from the hypothesis, with a non-hypothetical first-person sentence about this content of experience. A decision about this identity of meaning (e.g. of the two 'I'-expressions) can only be a hypothetical one, and so it requires another verifying comparison, etc. So there can be no verifying comparison of sentence and fact by constatations which would finalise a test.

For Neurath, the coherence of a system of statements means, besides freedom from contradiction, the conformity of statements of a theoretical context with statements which stem from other theoretical contexts, in particular with protocol sentences. Schlick's demand, which goes beyond this to the conformity of protocol sentences with reality, could only be served by a pointless repetition of the protocol sentence. The addition of the predicate 'true' to a sentence (defined according to the correspondence theory) does not say more than the sentence itself (Ayer 1936, p. 88).

The objection that due to the distinction between truth definition and truth criterion one can understand that a sentence is true correspondence-theoretically without making the comparison, can be met on Neurath's grounds: if we assert a matter of fact *p* in the sense of the correspondence theory, then this means the same as the assertion that the sentence '*p*' is true; that is, by means of the sentence we assert the relation of correspondence itself, we make the comparison.

By contrast, if truth is not a property of sentences in general, then the problem of truth only becomes an issue if someone asserts a sentence and it is not known by us whether we should agree or disagree. The function of 'true/false' consists in expressing our agreement or disagreement with the uttered sentence. Indeed, Neurath thinks that what is true for us are the sentences of an encyclopedia which is valid for us, which we accept (Neurath 1936c, p. 161). Against this background it is not unintelligible that Neurath generally tries to avoid to talk of 'the' reality. This way of speaking, which would have to refer to something existing independently of all subjects and their sentences, is in principle incapable of allowing decisions, for it presupposes (in the manner of a comparison) that we could occupy a standpoint that is, so to speak, beyond the sentence and reality.[1]

Schlick objects against the coherence theory that it did not provide a criterion of truth which would uniquely distinguish one system of sentences—every system of sentences, even that of a fairy-tale could be made coherent. But Neurath admits this ambiguity. The acceptance of certain protocol sentences always allows for an unlimited number of equally usable hypotheses which 'fit' the protocol sentences. And the contradiction of a hypothesis by a protocol sentence can be transformed into the confirmation of the whole system by adding suitable auxiliary hypotheses. Even Schlick's constatations do not allow for univocality nor do they exclude a fairy tale world, for the same constatation may be confirm that, e.g., a real object is present or that a hallucination obtains. The acceptance of the correspondence theory can similarly contribute nothing to the reduction of this multiplicity because it remains in principle uncertain as to which of the competing systems is correspondence-theoretically true.

Let me now propose the following definition of 'truth' in Neurath's spirit. A content (empirical) sentence is true if it is accepted, if it is non-tautological and free from contradiction, if it does not stand in contradiction to the other sentences of the relevant accepted system which, in their turn, are non-tautological and free from contradiction, and if it conforms with certain test sentences of the system, amongst which the protocol sentences play a distinguished role; all along, of course, the system as a whole must have empirical content.

That sentence is true therefore that can be 'integrated'. If we take Neurath's demand seriously, then we cannot attribute 'true/false' to sentences themselves, but only to the sentences *in a system of sentences* (context, encyclopedia). The result is, of course, that the same sentence can be true or false depending on the system. Falsity is thus characterised by contradictoriness in the system at issue (necessary condition). In a system every sentence is either true or false (i.e. connected with other sentences by contradictions or not).

Yet not only freedom from contradiction matters for Neurath: there is also empirical truth, the agreement of certain singular statements derived from hypotheses with protocol sentences. It seems presupposed here that the protocol sentence in question is true, for a false protocol sentence which contradicts other sentences of the context from which it is derived (e.g. a language of protocol sentences or the 'world of everyday observation') leads by purely logical means both to the affirmation and the denial of the sentence to be tested. Thus we must establish the truth of the protocol sentence, its freedom from contradiction in the system (e.g. the language of protocol sentences), its agreement with further protocol sentences—and so on, in a regress.

But I think that the coherence theory offers certain advantages in the interpretation of the regress of justification or testing. The same regress obtains as follows within the framework of the correspondence theory: (1) We assert that a content sentence is true (correspondence-theoretically); (2) the sentence is either true or false 'in itself,' *tertium non datur*; (3) the truth of the sentence is not certain. The sentence requires testing, for example by a protocol sentence. For this protocol sentence the same conditions (1-3) obtain; it too stands in need of testing etc. Thus Schlick postulated certain test sentences, constatations, which terminated the regress.

According to the coherence theory (1) reads: We assert a fact. *Ad* (2): what does it mean that a sentence is true 'in itself'? This 'in itself' could only be possible sentences in the 'universe of discourse.' Thus we can say that a sentence *p* is true or false 'in itself' if there exist sentences in the universe which contradict *p* and which do not contradict p. That is presumably one reason why one should not take the 'in itself'-position and rather confine oneself in the evaluation of truth and falsity to sentences which have actually been asserted. A sentence becomes false only once a contrary assertion has been made and appears in the system.

Any talk of *tertium non datur* similarly is senseless. For there are sentences without truth value: these are not only sentences which are considered apart from a system or context, but also within a system there are, I believe, sentences which remain indeterminate: those protocol sentences which, while positively grounding hypotheses, remain themselves untested (which only means that in the concrete system no further test sentences for them have been formulated). (This was not Neurath's intention, however.) That a sentence stands in need of justification has, according to Neurath, not epistemological, but practical and social reasons: a sentence is uncertain if there is a probability in the community of scientists (or otherwise involved epistemic subjects), that it will be revised, or confronted with denials, and in particular if such conter-assertions already are part of the system which needs to be revised.[2] A sentence is certain if such revisions and contradictory assertions are not to be expected, and in the case of protocol sentences these generally are not expected. Therefore the coherence theory alllows us at all times an unproblematical breaking off of the regress; the need for, or lack of justification, or uncertainty are not epistemological properties at all, nor do they perpetuate them. They are properties which refer to the concrete behaviour of epistemic subjects. Certainty and indeterminacy of truth value do not exclude each other.

Of course Neurath is a fallibilist, and this leads us to (3). He is of the opinion that every sentence can be contradicted and revised (so it can be 'struck off'). For every sentence we can formulate a denial; there are no epistemologically certain sentences. This fallibilism does not end up with a problem of justification. For from the fact that every sentence can be denied it does not follow that 'truth/falsity' become uncertain and need justification. 'Truth' only and always means that a concrete sentence does not contradict other concrete, asserted sentences and that it agrees with some; 'falsity' means that a concrete sentence stands in contradiction with others.

A few additional words about Neurath's empiricism which grants to protocol sentences a decisive role for testing and truth, but still is shot through with conventionalism: every protocol sentence is revisable and in a case of contradiction between a protocol sentence and the hypothesis to be tested, we are free to change the hypothesis (strike it out) or to change the protocol sentence (strike it out)—as long as we discount saving the entire system by auxiliary assumptions—in order to establish coherence. Here too then truth does not amount to the unique distinction of sentences.

But is it indeed a matter of decision whether we should strike out the protocol sentence or the hypothesis? Only such a protocol sentence will be struck out which is contradicted by other hypotheses which are supported by other protocol sentences. If one were to hand it over merely to decisions whether to strike out protocol sentences, then it is not at all clear why the control of a hypothesis by protocol sentences is demanded in the first place.

As far as I can see, Neurath's formulation is similarly unfortunate in another way, namely that in the case of the rejection of a protocol sentence the hypothesis which remains in the system becomes a true sentence, or that in the case of the rejection of the hypothesis the now integrated protocol sentence becomes a true sentence. For according to the definition given above, the contradiction-free integration of a sentence does not make for a sufficient criterion of truth, and there must also be an agreement with a class of protocol sentences. Why this empiricist insistence on such an agreement? I believe that Neurath himself also must deny that scientific systems are created purely by conventional determination—the (accepted) multiplicity of systems does not mean arbitrariness or caprice. And an observation sentence does escape our arbitrary determination to a large degree; experience is forced upon us and may run contrary to theoretical expectations. This does not mean that such observation sentences are incorrigible or untheoretical—that they are not. But the relative stability of observation sentences, which Neurath after all remarks upon (1935b, p. 129), is in my opinion more than a historical-social fact.

Karl Franzens Universität, Graz

NOTES

* First published as 'Neurath kontra Schlick. Zur Wahrheitsdiskussion im Wiener Kreis', in *Wittgenstein, der Wiener Kreis und der kritische Rationalismus. Akten des 3. Internationalen Wittgenstein-Symposiums, 1978*, Hrsg. H. Berghel, A. Hübner, E. Köhler, *Schriftenreihe der Wittgenstein-Gesellschaft Bd. 3*, © 1979, Hölder-Pichler-Tempsky, Wien. Translated with kind permission of Verlag Hölder-Pichler-Tempsky and the author by T. E. Uebel.

1 Neurath 1933a, p. 6: "There is no court of appeal outside the totality of sentences."

2 Neurath 1934, p. 104: "'Certain' is to be defined as a term within the doctrine of human 'behaviour'."

HEINER RUTTE

ON NEURATH'S EMPIRICISM AND HIS CRITIQUE OF EMPIRICISM*

In this contribution I wish to consider a few fundamental aspects of Neurath's critique of empiricism without, of course, treating it exhaustively.

To study Neurath is not easy. Most of the philosophical termini we have grown accustomed to are on his *index verborum prohibitorum*—already the terms 'perception' and 'observation' which are so important for my topic.[1] The terms 'perception sentence' and 'observation sentence', however, are not listed there—strangely enough, for obviously a perception sentence is to refer to the fact of the perception. But it seems that already this semantic relation was metaphysically suspect for Neurath. And other terms which I use are also on his *index*, like 'reality', 'truth', 'fact', 'verification', etc. So we have to proceed very carefully and use these terms as 'innocently' as possible.

I

It is one of the central theses of the Vienna Circle that it only makes sense to speak of true statements if the truth implies the verifiability of these statements, i.e. the ascertainment of their truth. With this thesis the Vienna Circle opposes the ideas of 'truth itself', of 'facts in themselves', etc.

Neurath shares this thesis in a more or less unclear fashion. Repeatedly he expresses himself as follows. To assume the existence of truth itself, i.e. of a relation of correspondence between statement and fact, means to assume a metaphysical instance of appeal beyond all statements which can decide about their correspondence (as it were an instance beyond statements which nevertheless passes judgements). Thus he reaches the view that one cannot compare statements with facts, but only statements with statements. If *establishing the truth* consists in this, then truth can only consist in freedom from contradiction and conformity of statements with each other.

As is known, Schlick thought differently.[2] He too started from the central thesis mentioned above, but thinks that verification cannot simply mean the comparison of statements, but must *ultimately* be the comparison of

T. E. Uebel (ed.), Rediscovering the Forgotten Vienna Circle, 175–190.

sentence and fact. Schlick was asking: what distinguishes *knowledge of reality* from merely arbitrary construction of sentences? Obviously the good reasons, the verifications. And if verification were only the comparison of sentences, then each and every arbitrary construction of sentences could be verified. Neurath on the other hand was not bothered by this danger of conventionalism because he believed that 'selection by life' very soon reduced the possible plurality of constructed systems.

For Schlick, the final verification consists in something like an intuitive comparison of a sentence about experience, which makes a truth claim, with the experience itself, whereby the fulfillment of the truth claim obtains. This experience (not a perception in the normal sense, but at best an immediate perception) has the character of immediate knowledge and is treated by Schlick *virtually like* an inexpressible statement—such are his 'constatations'. In them lies the point of 'contact' between statements and reality.

A person possesses immediate knowledge, apparently, only of those things which he himself experiences at the moment, like, e.g. 'pains,' 'impressions of red'. Of that which we usually call the 'external world' (including the experiences of others and our own earlier or later experiences), a person cannot have any immediate knowledge at the same time. Of course, Neurath doubts such immediate knowledge. He seems to think that knowledge always has the character of statements, and statements are never immediate but imply hypotheses and theories of all sorts. Perception too cannot serve as a bridge for this comparison, and Schlick's supposedly intuitive comparison only moves between statements.

What arguments can be given for Neurath's views? They are pretty simple arguments. If these quasi-statements, like 'red here now,' 'this now red,' 'I now have an impression of red' and similar ones, are to be *true* in the usual sense, then they must refer to facts and must be intersubjectively intelligible. There are no 'I-', 'here-', 'now-', or 'this-facts'. These ostensive terms can be understood by another, only if they refer to the speaker. The other must turn the ostensive statement into a statement about experiences of persons at certain spatio-temporal positions, if these quasi-statements are to be at all understandable. And the verification of any hypothesis depends totally on this, that a certain person at a certain place in space and time makes a certain statement, namely, that person who is at that place where the perceptual conditions are realised which allow the verification of the hypothesis.

Persons must be *identifiable* by everybody as the objects of self-ascriptions and attributions by others. If I refer to myself as a 'private' person, e.g. by introspection, and ascribe experiences to myself, then a fellow human being would not know at all who was intended as the object of this attribution, because my person would not be identifiable for another. To refer to persons as intersubjectively identifiable, however, means to attribute *bodily* features to them, to comprehend them as organisms.

Analogous considerations hold for the spatio-temporal places: they must be identifiable by everybody. If it were merely a question of a private spatio-temporal position, experienced only by one person, then this person would have to be identifiable, i.e. an *intersubjectively* perceptible, and so a bodily, person must be meant, for otherwise it would only be a person experienced through introspection or otherwise, and we would end up in a regress. Furthermore, the determination of the spatio-temporal position implies a network of comparative relations to other spatio-temporal positions, and this network cannot be immediately present, for we can only be at one spatio-temporal position at one time.

This is essentially a *physicalist* way of arguing. It makes it understandable why Neurath arrived at his protocol sentences. I refer to the relevant remarks by Carnap and the similar later remarks by Wittgenstein and his school.[3]

Therefore Neurath's protocol sentences (against which Schlick argued) treat mainly the perceptions of real things by bodily persons at certain spatio-temporal locations. And it is clear that they already, by this alone, imply an unlimited mass of hypotheses and expectations. Admittedly, Neurath formulated these as strangely involved sentences. Protocol sentences are acceptable for him not only if a perception took place and the so-or-so perceived object exists, but also if the protocolist has a hallucination or makes other perceptual mistakes, and even if the protocolist lies or otherwise does not take the protocol seriously.[4] Behind this there seems to loom the attitude of a 'behaviouristic of scientists' (more on this later), according to which every written or spoken protocol that is given must be taken seriously and can only be eliminated as unacceptable after comparison with other protocols by scientists and with other statements. But the same situation obtains concerning these other protocols, and ultimately one must rely on protocols of which one only knows who wrote them down or spoke them—nothing more. For a comparison with reality is never possible.

An observation sentence is accordingly hypothesis-laden, i.e. it implies an *indefinite* mass of expectations about the perceived objects and the perceiving persons; it implies, for example, that other persons under such-

and-such conditions will perceive the object in such-and-such a fashion, that I would perceive the same object with other sense organs under those and those conditions in such-and-such a way, and so on. These expectations, these mostly unexpressed and unconscious conditional prognoses, presuppose equally unexpressed theories, everyday theories and scientific ones. To speak behaviouristically: an observation sentence expresses an attitude of expectation which can be realised in unlimitedly many cases as a concrete expectation (an unconditional prediction). If such an expectation is frustrated, then the acceptance of the perception sentence in question becomes problematic.

Neurath's standpoint then, which I just formulated in a more familiar, somewhat Popperian terminology, appears very modern. It says[5]: Like all other sentences, protocol sentences stand in need of justification and can be 'struck off'. They are not simple or non-corrigible sentences, but very complicated hypotheses, full of theoretical 'elaborations'. They are embedded in a mass of sentences of different provenances, and therefore constitute the 'rich' but semantically fairly indeterminate sentences which in a way represent the whole of science, whereas the scientific statements of laws are the more simple sentences. But observation sentences remain in fact relatively stable in the republic of scientists (they are relatively widely accepted) without there being an epistemological explanation of their stability. Therefore no contrast of theory and observation could obtain; a pure protocol beside or before the theory is impossible, and one already must possess a rough theory in order to put questions to the observation.

As will be noted, these are thoughts with which we have become very familiar due to Kuhn, Feyerabend, Popper and others, thoughts which we did not know were expressed so early, so often, and so clearly by Neurath.

II

My problem in what follows is that of the *acceptance* of protocol sentences: does Neurath's conventionalism constitute a justification through belief, conviction, consensus? Can there be such a justification at all?

Empiricist foundationism presents us with the following problem: If in the course of the justification of a statement, I predict an experience and have this experience, then I may regard the statement in question as justified (e.g. as something 'confirmed'), provided yet further conditions are fulfilled which need not be discussed here. But then the question of justification can be asked again: "Is it *really* true that I have the experience?" Simply to ask this would not, of course, be meaningful since I assert exactly that it is true that I have this experience. Obviously it means: "Do I have a *right* to assert (believe etc.) that I have this experience?" Or even: "Is it certain, or do I know immediately, that I have this experience?" A question of this sort shall be called the '*theoretical*', the 'classical' question of justification.

Now, when it is stated in Neurath's conventionalism that a statement *p* is justified if it is reducible to a statement which I believe, of which I am convinced or of which the majority is convinced, or that *p* is justified if *p* is integrated in the system or into the encyclopedia of accepted statements—then we may ask: "But are people really convinced of these statements? Do I know that people are convinced of them?"—and again a regress reappears.

But Neurath seems to have intended to reformulate the entire problem of justification. The theoretical question after the right to believe in the truth of a statement, to accept a statement, cannot any longer be meaningful for him. Presumably he would ask what 'rights and duties in themselves' would be which have not been postulated by anybody. The question of justification can only have a *practical* sense for him: shall I accept a certain statement in order to act successfully? And this practical question is already meaningless if I have decided to accept a statement, or if I accept it, that is, expect success. The question has meaning only for situations of indecision. Thus Neurath can emphasise that all statements of science rest on a decision. And he gives advice to the undecided scientist which he must consider as useful in a practical sense: Try to integrate your statement into the system of the statements which you accept, make the system coherent! If the statement can be integrated without contradiction into the system, or if the statement itself or one of its consequences agrees explicitly or implicitly with the statements of the system, then it has been made coherent or can function as the statement of a coherent system. (Later Neurath grants that the 'total mass' of statements may possess also logical faults: freedom from contradiction does not constitute an ultimate criterion; he emphasises the logical and linguistic unstructuredness of the mass of sentences, the imprecise, unanalysed terms ('*Ballungen*') more often than before and speaks instead of the 'system' of the 'encyclopedia' and even exclaims: "*'The' system is the great scientific lie.*" (Neurath 1935a, p. 116.) But these are questions which need not interest us now.)

Essential for our purposes is that the classical, theoretical question of justification can also be addressed to the total mass of sentences: Is this total mass really free of contradiction or coherent, do you have the right to assert such a thing? Neurath would have to reject the question here as well. In the spirit of his universal fallibilism, he asserts even of logical, analytical and mathematical sentences that freedom from contradiction, coherence etc. cannot be established without doubt. (One could use Gödel's results to sharpen this thesis, of course.) Establishment of truth in the strict sense is therefore impossible in this domain as in that of empirical sentences. Concerning empirical sentences Neurath rejected the assertion and especially the discussion of their truth with the argument that such a truth was impossible to establish; it cannot be verified in the sense of a reduction to immediate knowledge because we cannot go 'back behind language'. For this reason it is a pointless battle cry to demand a justification of the truth of statements. What Neurath seems to have in mind as a discussion of justification is not any longer a discussion about whether certain statements are true or false, but instead a discussion about whether is *useful* to accept certain statements in the republic of scholars or in a similar community.

Since such beauties like freedom from contradiction cannot be established in the sense of providing a strict proof, there can be no rational discussion about truth or falsity that would be independent of factual acceptance in this area either. What is the point then of demanding that a statement be made to cohere or be integrated into the mass of sentences? That seems to be inconsequential. Neurath might as well demand the establishment of the truth of empirical statements by their verification as a way of justification.

Neurath accordingly would have to renounce the argument "because truth cannot be stablished there is no point in justifying by verification", and must rely solely on the rules of the game that hold in the republic of scholars. Indeed, he expressed himself in this way when he demands a "*behaviouristic of scholars*", in place of epistemology which operates with such concepts as 'true', 'justified by experience', 'free from contradiction'. And in this behaviouristic he may state that scholars are not at all interested in truth and falsity and knowledge of reality (although they act as if they were), that they proceed conventionalistically, i.e. try to integrate new statements without contradiction into the accepted mass of sentences and in doing so *decide* (without consideration of truth or falsity) to exclude other statements from the mass of sentences. They are obviously most interested in freedom from contradiction and coherence, but these terms ought to be metaphysical for Neurath too, because in analogy to 'truth' the existence of 'freedom from contradiction in itself' also cannot be verified. In this respect, scholars apparently are metaphysicians, and a behaviouristic of scholars cannot change this, because it could fulfill no critical or enlightening function without the use of epistemological standards. If the scholars were oriented towards 'truth' and 'knowledge' in the classical sense, then the behaviouristic of scholars would have to accept this and state that the scholars orient themselves towards such metaphysical ideals, that these ideals are practically effective and accepted. 'In themselves' they are metaphysical ideals; but the expression 'in themselves' should already make one suspicious: that the use of the expression 'true' is meaningless if truth cannot be established constitutes itself an epistemological position and cannot be derived from the behaviouristic of scholars.

Commonly, 'truth' and 'freedom from contradiction' are held to be *concepts of validity*—that makes for their genuine epistemological character. This means roughly: even if we do not know or assert that a sentence is true, it nevertheless holds of it that it is true 'in itself', independently of human attitudes (if it is true, of course). The Vienna Circle wanted to oppose this postulated realm of validity by coupling truth with verification. Verification was to consist in the reduction to a true sentence which is known immediately. With this definition (Schlick!) one remains naturally in the realm of epistemology as a theory of validity. But Neurath wants to leave this realm and implicitly defines 'verification' as the 'acceptance' of sentences, but for this he requires the concepts of freedom from contradiction and coherence.

The crux of this is the following: one wanted to criticise the validity-theoretical explication of 'truth in itself' by introducing the certainly intuitively plausible demand for verification, but already for the explication of the concept of 'verification' and 'verifiability' one needs an explication of the concept of truth which must differ from the validity-theoretical one. In short, when the Vienna Circle criticises the concept of truth of the theory of validity, it must presuppose its own concept of truth (whatever it may be—I do not claim to know it), and reference to verification is totally unable to decide the dispute between the Vienna Circle and the theory of validity: the theoretician of validity must have a different concept of verification than a follower of the Vienna Circle.

This is a problem both for Schlick and Neurath. With Schlick, matters get particularly confusing because he wants to use the concept of truth of the correspondence theory and despite this has to oppose the conception of a truth in itself (due to his meaning criterion). Neurath, on the other hand, certainly wanted to give an explication of the concept of truth during one phase of thinking: people simply think that when they call their assertions 'true', these assertions can be integrated free from contradiction in the totality of statements they accept. It is just that the concept of verification (or 'confirmation'—Neurath prefers this non-absolutist term) is totally changed in its meaning thereby and becomes a concept of the reduction of sentences to observation sentences which we accept. For theorists of validity the concept of verification cannot have this meaning; how then could Neurath disagree with him about 'truth in itself' if each has to presuppose his own conception of truth?

A theorist of validity could of course simply respond: "Just as there exists truth in itself, so there exist the respective facts in themselves—that there exist trees and tables is a fact in itself, i.e. independently of a subject's assertions and perceptions. And the fact of accepting sentences exists similarly in itself." So here too Neurath cannot refer to a certain concept of 'acceptance' that would be, as it were, neutral (like: "Truth may exist in itself or not, but we all agree that we accept sentences!"). Of course, Neurath could also reply that, certainly, we speak of trees and tables, but that we do not mean by this that they exist 'in themselves,' nor that they exist 'for us', and that we tend to make an entirely innocent use of terms like 'fact', 'reality', 'truth'. It is this use which Neurath wishes to reconstruct, though in a not totally innocent way, in his coherence and consensus theory. But how this dispute could be decided I do not know.

III

What then remains of *empiricism*? Neurath always wanted to be an empiricist; 'logical empiricism' was one of his pet labels. To be sure, Neurath always demanded the empirical testing of theories, their confirmation and 'shaking'. But this demand can now only claim to have a practical usefulness: it speaks for observation sentences that they are in general accepted by the observer himself without problems and without difficult decisions, and that other people easily tend to agree. The question of justification—"should I accept these observation sentences?"—poses itself rarely, and these observation sentences are rarely at issue. In principle, of course—so Neurath says—we are free to strike off the observation sentence in the case of a conflict between hypothesis and observation sentence, but nature mostly saves us this conflict. We mostly accept observation sentences naturally, it is a biologically and culturally useful habit of consent; after all, we have plenty of possibilities to save the threatened hypothesis by revision of additional hypotheses, even if we have accepted without question the supposedly contradictory observation sentence into the mass of sentences.

But it is still only *behavioural habits*, habits of consent, which distinguish the observation sentences and give them a certain stability, and the habits may change, and certainly once were different. If the acceptance of metaphysical, magical, or even only theoretical hypotheses is more important to us than the acceptance of the observation sentences which contradict them, then a Neurath-empiricist cannot object. Were he to argue in the manner of common sense that "you will have a hard time getting around in life if you don't bring your predictions in line with experience", then he would not say anything of substance. For what does 'getting around' mean? It only means 'gaining satisfaction from predictive success', and then the argument goes: "You will not gain satisfaction from predictive success if you don't have predictive success"—nothing more. With Neurath, 'predictive success' can only mean that a prediction agrees with an accepted observation sentence; it is a process of transferring the attitude of consent from the observation sentence to the prediction. And if at some places of our encyclopedia we are no longer interested in predictive success because it appears more important to us to uphold certain non-empirical sentences, then we cannot object—it remains, as Neurath says, a matter of decision.

Nevertheless there is *learning by experience*, be it by association, expectation, habituation, be it through trial and error, and this learning by experience seems indispensible for the survival of the individual as of the species. Isn't that enough justification of empiricism, especially for the naturalistic 'epistemologist', the behaviourist Neurath? It is notable that he avoids speaking of experience, perception itself, something which comes easy to the biologist. Instead he musters a most artificial way of speaking which I have re-presented and enriched here ('predictive success' as 'conformity of the prediction with accepted observation sentence' etc.), and which was intended to keep one squarely in the realm of sentences. The biologist and the psychologist, of course, are not at all inclined to speak of 'observation sentences' but instead of 'observations'.

It is a *philosophical* way of speaking which is oriented mostly towards the philosophical, theoretical model of justification. The philosopher notes that there is something that is not at issue in the justification dispute over which sentences are true and which ones are not true, namely, that we construct sentences, that we accept them, reject them etc. We are not arguing over whether we construct sentences, accept them, reject them, etc., because we do not want to, but because we want to investigate whether our assertions are true and what makes them so, and not whether we make these assertions. We simply make assertions, these are acts on our part which are not at issue, because we de not want to discuss concerning acts whether we make them. If we are discussing something, it generally makes no sense to discuss whether we are discussing something (except, as we know, in philosophical discussions). From these trivialities, the philosophical conclusion is then drawn that it does *not require justification* that we make assertions, accept them, etc.

So Neurath's way of speaking seems only to formulate what is not in question in the problem of justification, and so remains in the area of the 'immanence' of assertions or speech acts. If predictive success is explicated in the fashion outlined above, then one speaks as a philosopher who asks himself: "Are we *really* having the experiences of which we speak?", and if one thinks the question undecidable one prefers to use the careful way of speaking.

The biologist, psychologist, etc. simply accepts that we have experiences, just as he accepts that we construct sentences and adopt attitudes of consent, and he accepts the life-preserving function of learning by experience; of course he does not believe that it is more certain that we construct sentences and accept sentences than that we have experiences.

On the one hand, Neurath seems to furtively assume that we have no *right* to believe in reality, etc., that we only have the right to believe that we make assertions, accept them etc. On the other hand, Neurath seems to strike off the very question of justification, and in this sense we are not more justified to believe in our assertions than in reality. If one has struck off this question, then it is *in principle* equally permissable to believe in assertions, in experiences, observations, and in the existence of material objects—there are no more epistemological prohibitions or epistemological preferences.

If Neurath's question of justification consists in asking "Shall I accept *p*?", then this means "Is it *useful* for me to accept *p*?", and the answer to this question—if it is a naturalistic one and not validity-theoretical—must take account of the whole situation of the questioner, his aims, the pre-given aims of the community and institution to which he belongs, as well as his knowledge and his experiences. Here it is useful for the individual to be oriented towards experience, i.e. to be guided by the empiricist principle

"Adjust your views to experience if you want to suffer as few misfortunes as possible". If we want to consider justification as a practical question and for this reason take seriously an empiricist principle like "If you want to avoid experiences which bring pain, then make your assertions and predictive behaviour conform with experience", then questions like "Did you really have this and that experience?", "Am I just imagining I have experienced pain?" become meaningless. For the practical attitude *presupposes* just that we had and will have good and bad experiences, that we want to avoid bad experiences etc. Of course, one can discuss whether all this really exists or whether it is an illusion, but he who poses a practical question does not *want* to discuss whether he really has aims, really has had or will have experiences, really accepted certain theories, etc. That's not his problem. For a naturalistic or behaviouristic theorist, the question "Shall I accept *p*?" can only mean "Is it useful to exhibit a certain attitude of belief", and this attitude of belief consists in, amongst others, doing certain things and not doing others, bringing about certain experiences and circumventing others. This makes questions like "Am I really doing this?", "Do I really have this experience?" meaningless.

If then a naturalistic or behaviouristic theorist asks "Shall I accept *p*?", then certainly this implies the demand that "*p* must be empirically tested", given the acceptance of utilitarian rationality. We can and should test and experiment empirically, as always has been common, but we need not ask whether this test is ultimately a comparison of statement and reality or only a comparison of statements among themselves, and we need not decide in favour of the latter. This alternative only comes into view when the question of justification is formulated in a philosophical manner: if we confront an, as it were, immanent domain of language and statements and wonder how we may in this domain gain knowledge of a transcendent reality beyond language. And it is the *philosopher*, not the naturalistic thinker Neurath, who answers: We cannot escape language, we can only compare statements with statements, and hence justification and testing only amount to a comparison of this sort.

Of course, the standpoint of coherentism can be very appropriate and realistic, but not as a philosophical standpoint, only in Neurath's behaviouristic of scholars. It would be *philosophical* to say of the behaviouristic of scholars: "Ultimately we can only say that scholars make statements, discuss and accept them, bring them into mutual agreement etc., but we cannot say that they have experiences." From the philosophical perspective, however, it is just as questionable a fact for an empirical discipline like the behaviouristic of scholars that scholars accept sentences, as that they have experiences. And if we don't take the philosophical perspective, then we may simply assume both. Nonetheless, as an empiricist the theorist of the behaviouristic of scholars may assume that in their discussions scholars are mainly concerned to establish wide-spread coherence amongst the mass of accepted sentences, and to strike off what disturbs the coherence of what has been accepted, and that in doing so they often discount observations, even sometimes the principle of non-contradiction, because they consider certain predictive successes more important. But an empiricist cannot say that scholars are *justified* only to establish coherence between accepted sentences.

I believe I have characterised the empiricism of the '*natural world-conception*' which proceeds from problems of action. This empiricism is different from that of philosophy (epistemology) which proceeds from the question: What can be doubted in principle, what cannot be doubted in principle? Similarly, it is a problem of action whether I should doubt something, whether doubting would be useful.

IV

How then does this empiricism relate to Neurath's *conventionalism*, the Neurath principle (cf. Haller 1982a [ch. 9]) which teaches that the conformity of theory and observation can be established in arbitrarily many different ways? It seems to me that this conventionalism is limited—as Neurath himself emphasised—by *practice* (the natural selection of life). It is practice which in general limits conventionalism to the superstructure of the explanatory theory and does not extend it to practical, empirical, experimental effects and to observations themselves, in short, to all that which is to be explained. For if we have problems of action, then we want to bring about certain empirical effects, prevent others, and do not wish to discuss whether these empirical effects exist, whether they remained 'the same' or whether they also changed their character once we altered the explanatory theory.

Of course, conventionalism can in the end be extended so that every perceptual experience may be declared a perception of reality or an hallucination. But he who believes himself to be hallucinating sometimes or permanently has problems of action of a totally different order than the usual ones (if it still makes sense to call them problems of action). Similarly, this naturalistic empiricism does not change Neurath's view that observations are laden with hypotheses and expectations. It is just that problems of action are generally such that they do not question whether the expectations inherent in observations will be met or not. Only *after* certain expectations have been disappointed will a discussion begin over why this happened.

But is this all not contradicted by Neurath's *holistic* fallibilism, his strongly articulated view that it is always the *whole* of science, the entire mass of sentences that is at issue? I believe that we do not have to think so. Perhaps Neurath only wanted to express the following. Since the encyclopedia forms a more or less coherent mass of sentences, it is *possible* in the course of the discussion of a certain hypothesis that other hypotheses will also be questioned, in so far as these hypotheses form a mostly implicit logical context, that is, in so far as that hypothesis presupposes an undetermined number of other hypotheses for its own 'validity'; there are no logical or epistemological reasons for interdicting such a discussion, there is no '*noli me tangere*' for statements. It seems to me to go too far to think that in the total system of science a certain statement presupposes every other statement, i.e. the whole system, and that the discussion or alteration of a statement necessarily *must* involve the discussion or alteration of every other statement, i.e. the whole system. As Neurath's follower Quine makes clear (e.g. 1951, sect. 6), within the totality of science there are peripheral and central locations, and an alteration at a peripheral location need bring with it neither alterations in the centre nor alterations at other peripheral locations (but may do so under certain circumstances).

For there is, on the other hand, the fact which I just articulated in Neurath's sense, namely, that, when they discuss a problem, scientists and epistemic subjects generally *want* to call into question as few statements of the total mass as possible. Simply by formulating the problems as they do and by using turns of phrase like 'Presupposed that ...', they intend to leave intact as much as possible of the accepted mass of sentences. Already the formulation of a problem makes many assumptions which must remain unquestioned as such, for otherwise the problem becomes a different one. (For example, if I formulate the initial position of a problem of action by "I have this-and-that goal, had this-and-that experience, ...", then the problem is completely changed if I doubt whether I have these goals and experiences.) So if the conservative, anti-revolutionary attitude of conventionalism is to be meaningful in science and daily life, then it must be realisable, and it would not be realisable if holism were to be formulated in the radical fashion just rejected. For *this* holism means a theory of permanent revolution in cognition: Every alteration of an individual statement would cause alteration of the total system of statements and generally a meaning change of all these statements. It speaks against this conception that a change of meaning occurs only once alterations have been made to the center of the system, those 'analytical' or 'empirical' statements which we permanently presuppose with our individual peripheral statements. That is to say, such alterations are always possible, but the practice of life rarely forces these upon us. (Like Quine, Neurath envisages the revisability of logical, analytical and similarly fundamental statements.)

I do not think therefore that the further holistic principle of Neurath, according to which the observation sentences are the rich sentences which contain, at bottom, all of science, is intended to mean that the acceptance of an observation sentence presupposes the acceptance of arbitrary scientific theories, even the wole of science, and so also presupposes the acceptance of that theory which was to be tested by the observation sentence in question. For if the observation sentence would not be semantically *independent* of the competing (i.e. mutually exclusive) theories which it is supposed to test, then it would become unintelligible why we even test competing hypotheses by observation sentences. Similarly, it would become impossible to uphold the conventionalistic principle that indefinitely many competing hypotheses may be built upon one and the same observational datum, for this datum could not be considered as *the same* when linked to the different hypotheses: its character would be completely changed by the competing hypotheses which it would semantically presuppose. If this were so, we could not even interpret the process of empirical testing conventionalistically (and also not relativistically).

Something far more simple seems to be intended, namely, that not only everyday theories, but also indefnitely many scientific theories can be called into question when we are discussing whether certain observation sentences should be accepted. When we discuss whether a sensory illusion occurs, e.g. concerning a broken stick in water, we can refer to our sense of touch—thereby presupposing an "everyday theory" like: "If something is broken it must appear broken to our vision and to our independent sense of touch." We tend to presuppose such everyday theories when we discuss the reliability of observations, for on their basis we form expectations about which we can say whether they have been fulfilled or disappointed. But we can also explain the matter by reference to an optical theory: "Given this optical theory, we are dealing with a sensory illusion, not a perception of reality." In this case we call into question the optical theory and the whole of physics into which it is embedded—better: we *may* call it into question if we are prepared to do so (and, in the extreme case, consider the broken stick a perception of reality and try to rebuild natural science accordingly).

At which point, however, does the fundamental question of the acceptance of an observation sentence even arise? I said that this is a practical question, that it is the question "What disadvantage would I suffer if I were to accept this observation sentence?" Suppose someone were to formulate an observation sentence from which I come to understand that so and so has happened. Now I may ask: Is he lying or did he make a perceptual mistake? If he is lying or made a mistake, then my belief in his observation sentence will have disadvantageous consequences; I will derive expectations which will be disappointed. The only way to investigate the matter is by observing myself whether so-and-so has happened, or by recognising through observation whether he lied or made a mistake. In doing so I myself construct observation sentences which I believe in. And if I accept them, then the question becomes meaningless whether I should accept these observation sentences—as I said already above. Every process of justification finally ends in the best Humean spirit in this belief which does not constitute unmediated knowledge. Yet this no longer expresses an epistemological proposition, e.g. that we have no right to doubt this belief, or that this belief may not come into question; it *may* come into question because it implies indefinitely many expectations which at one time may not be fulfilled. It is at this point that it becomes *useful* for my actions to ask the question "Shall I accept this and that observation sentence?", and to answer it.

Karl Franzens Universität, Graz

NOTES

* First published as 'Über Neurath's Empirismus und seine Kritk am Empirismus', in *Schlick und Neurath. Ein Symposion*, Hrsg. R. Haller, *Grazer Philosophische Studien* **16/7** (1982) 365-384, © 1982, Rodopi, Amsterdam. Translated with kind permission of Editions Rodopi and the author by T. E. Uebel.

1 All points of reference for the views I here attribute to Neurath are contained in Neurath 1981. For my topic in particular the following essays are relevant: Neurath 1931b, 1931c, 1932a, 1932d, 1933a, 1934, 1935a, 1935b, 1936f, 1936c, 1936d, 1938b, 1941a, 1944, 1946a.

Neurath's holism and conventionalism which I discuss later on can already be found in his earlier works, e.g. 1916a, 1921a.

2 Schlick's relevant views are expressed in his 1934, 1935a, 1935c, 1935d.

3 Cf. Carnap 1932b, 1932c, 1932d.

4 In order to avoid the appearance of giving a correspondence-theoretical formulation one should really say that a protocol sentence for a person *P* is acceptable not only of *P* accepts that the protocolist perceived, or that the perceived object exists, but also if *P* accepts that the protcolist was hallucinating or lying.

5 See, apart from Neurath's writings listed in fn. 1, also my summary characterisation in Rutte 1982a [ch. 7].

RUDOLF HALLER

TWO WAYS OF EXPERIENTIAL JUSTIFICATION*

The reasons which we do or could furnish for the existence or non-existence of a matter of fact are often nothing else but the outcome of attempts to vindicate ourselves, answers to the question of the right of a particular epistemological attitude towards belief. These reasons may concern facts (*quid facti*) or the justification of attitudes (*quid iuris*). They may carry their power of proof within themselves, or they may derive it from other reasons. If they carry their power of proof within themselves, if they are self-evident, then they can carry the burden of proof of a deductive chain which rests on them. Of these reasons it is also said that they are incapable of proof and do not need any.

All the reasons which we may command to justify an opinion, or to support a fact, are either derived or non-derived ones, and only the latter are held to be self-evident. Aristotle's principle of contradiction, that the logical product of *p & -p* is false, may serve as the classical example of a self-evident reason. Here we are dealing with a presupposition which could not be proved without a *petitio principii*, insofar as our proof must presuppose what is to be proved. Epistemological questions present us with an analogous problem already at the start, namely, when the possibility of the theory is disputed on the grounds that it is impossible to prove it without using what it presupposes, namely, some knowledge. Despite this objection we may easily distinguish two kinds of answers to the question of the justification of experience: those which for whatever reasons assume non-derived premisses as the starting and end points of a justificatory chain, and those which deny their possibility and existence. In this way one can distinguish between those who accept a 'justification model' of knowledge, and those who reject it.

In this lecture I do not want to explain the content of these two mutually exclusive models, but to trace their philosophical background in the Vienna Circle—naturally, given the breadth of the topic, in a somewhat simplified fashion, and with a certain generosity regarding the neglect of subtle and not so subtle investigations and arguments.

Aristotle claims that all proof consists in the derivation of what is conditional from judgements which represent the reasons for, or causes of, matters of fact as necessary conditions. He defines 'knowledge' accordingly: "We know something in the fullest sense if we know not only that it is the

T. E. Uebel (ed.), Rediscovering the Forgotten Vienna Circle, 191–201.

case, but also that it could not be otherwise." For this reason he held that from empirical, merely contingent statements one can derive only contingent ones, whereas from necessary statements only necessary ones followed. The traditional interpretation of Aristotle therefore has often held aprioricity to be the distinctive property of judgements of principles and particularly of judgements about causal contexts, because aprioricity was defined by necessity, that is, both terms were used interchangeably. But that is not necessarily correct, as Kripke showed. A statement can be necessarily true without being true *a priori*, for it is possible that the truth of a statement which is necessary may not be determinable *a priori*.

The picture of the history of the Vienna Circle is beginning to change. The undeniable new interest in what was essential to its ideas soon found that there did not exist the uniformity of basic conceptions which is suggested by various descriptions, especially those of its opponents. Most immediately there were two trends in the Circle which pulled further apart, with increasing tension. It is not very difficult to characterise the grouping of main representatives, but it is much more difficult to present the roots of their philosophical differences.

The main representatives of both camps were on the one side Schlick and Waismann and on the other Carnap, Hahn and Neurath. Of course, one could characterise the contrast also—and with good reasons—by the names of Wittgenstein and Neurath, although the former never participated in a meeting of the Vienna Circle. For a long time now, it has not been a secret that Wittgenstein had an overwhelming and dominant influence on the members and the themes of the Vienna Circle in its high phase (cf. McGuinness 1979, Baker 1979, Mulder 1968).

It is known very widely to what degree Wittgenstein influenced the development of Waismann's 'Theses', and of many chapters of his *Logic, Language, Philosophy* (posthumuously published), and it also known how significantly the thinking of Schlick after the publication of the second edition of his *General Theory of Knowledge* was influenced by the writer of the *Tractatus*. This was so not only because Schlick was present during many of the discussions of philosophical questions which Wittgenstein engaged in during his visits to Austria, but also because he took dictations from the philosopher from Cambridge—just as Moore did during his visits to Wittgenstein in Norway. (A few of these can be found in his *Nachlass*.) Then there is also Wittgenstein's unquestionable influence on Carnap which I do not wish to explore here.

But somewhat less well known perhaps, or less well regarded, is the degree to which Neurath acted as the real opponent of Wittgensteinian theses in the Vienna Circle, constantly striving to hold the unified science aspect of rational empiricism against the metaphysics of the *Tractatus*, and the conception of philosophical activity as elucidation and of philosophical sentences as elucidatory sentences of a peculiar character. Repeatedly, Neurath complained in letters to Carnap that Schlick and his followers slid into ambiguous metaphysics, and that "Carnap, Frank, Hahn, Neurath" would "be the eternal quartet".[1] In an earlier letter he had already said of the latter that they'd be the ones who would have to represent the Vienna Circle just as the third estate once "declared itself representative of the nation".[2] Such distancings were not only preceded by personal disappointments—e.g. Schlick rejected on grounds of *content* a manuscript by Otto Neurath, co-founder of the Vienna Circle, for the series *Schriften zur wissenschaftlichen Weltauffassung* which he edited with Frank—but this antagonism had still deeper causes.

One of these causes—and in my view the most important—is the completely different conception of what *philosophy* itself is or should be, and, correspondingly, also a completely different understanding of the *relation between science and philosophy*. Wittgenstein sketches a definite delimitation between philosophy and science which is then more or less followed by Schlick and Waismann, whereas epistemological naturalism consists in the denial of just this thesis. According to the latter, there is no strict separation between scientific and philosophical statements, there is only *one* science as a totality, and its parts, the individual sciences. Neurath was the first in the Vienna Circle who advocated this position in all its sharpness.[3] And Carnap and Quine followed him in this respect in totally different ways.

Given the constraint of brevity, I do not imagine I could take account of all important nuances which it is necessary to keep in view if one wants to establish what the commonalities and differences amongst the members of the Vienna Circle were. But this does not matter, at least not presently. On the other hand, however, there is no hope for reaching a proper understanding of the role of the justification of experience in the Vienna Circle without taking account of this fundamental difference of their conceptions. Whether there is a deeper connection still between the two issues I do not say nor wish to investigate here.

Wittgenstein's definition of philosophy is: "All philosophy is 'Critique of language'." (1921, #4.0031) This criticism of language should not result in *philosophical* sentences; instead, the true method of philosophy should consist in *saying nothing*, in making no factual assertions. If there were no *philosophical* sentences, then what kind of sense was possessed by those sentences which constituted *elucidations*? That is one of the questions which are philosophically relevant and whose character is also revealed as philosophical. For, if one were to divide all sets into two classes, those that contain *themselves* as member and those that do not, the philosophically interesting question is one of self-reference, resulting in the case mentioned in Russell's antinomy.

If then all sentences *are either* meaningful *or* meaningless, and if only those are meaningful for which the circumstances can be specified under which we would call them 'true,' then the question of *self-reference* is just as relevant. And quite consistently, Wittgenstein decided in favour of the answer which was to characterise philosophical sentences as senseless.

Naturally he does not deny their purpose, to serve the sole aim of clarity and to delimit that which can*not* be said. "Philosophy limits the disputable sphere of natural science." (1921, #4.113) But it is not itself part of natural science, where 'natural science' stands, according to the author of the *Tractatus*, for the totality of true sentences. In other words, the seemingly philosophical sentences are neither true nor false, but meaningless, yet they have a function: they serve the purpose of *criticism*. In a certain respect they are like the *aqua fortis* which separates silver from gold, but is itself only an acid one disposes of when it has done its work. Similarly the means of criticism—the elucidations—become superfluous once that which can be said, i.e. the truth-valuable sentences of science, has been separated from that which cannot be said.

Now, at first glance Neurath and the other members of the Circle would seem to agree without reservations with the thesis that "there is no 'philosophy' existing side by side with the sciences as a discipline with its own special statements; all meaningful statements are contained in the sciences." (Neurath 1932a, p. 58.) And Neurath claims in the same passage that *all* members of the Circle agreed in this.

But Neurath by no means draws the consequence that the activity of the philosopher should be excluded from unified science. He rather asserts that there was no need, nor was it possible, to separate the 'clarification of concepts' from the business of the sciences.

"More and more members of the Vienna Circle"—so Neurath in an essay of 1932—"accept the *radical formulation*. It knows only *one* scientific language, it uses 'philosophical elucidations' not even for propaedeutic purposes, it rejects everything that is not controllable. The most radical formulation subjects all sentences, even the protocol sentences, to scientific criticism." (Neurath 1932c, p. 568.)

The *radical formulation* is that of physicalism, and for Neurath physicalism is, as we know, not a reductive program. Rather, the physicalist language is essentially the everyday language of the 'natural conception of the world' (Avenarius) or, as Neurath puts it literally: "The language of physicalism is nothing new as it were; it is the language familiar to certain 'naive' children and peoples."[4] Thus it is a learnable language. Yet metaphysical expressions could be accepted *into* it—expressions whose meaning does not permit their integration into the total context of statements—so a decisive role falls to the question of the transition from the everyday language to the physicalist everyday language. I believe that the only criterion which Neurath finally accepts for the latter is that of 'predictability'. When it comes to acceptance into the system of scientific statements, the mechanism of integration remains always the same: that which cannot be integrated into the totality of already harmonising statements either is rejected as incorrect, *or* it leads to the reform and revision of the system. (I shall not enlarge on the Neurath principle here (cf. Haller 1982a [ch. 9] and Rutte 1977, 1982a [ch. 6]).)

More important in our present context is Neurath's rejection of philosophy as a self-subsistent discipline or task: he rejects not only certain other philosophical standpoints which differ from unified science and its *unified language*, but *all* forms of philosphy. And since from early youth onwards Neurath always returned to his *idée fixe* of an *index verborum prohibitorum*, he also placed the termini 'philosophy' and 'epistemology' on his list of forbidden words, next to the expressions 'truth' and 'falsity', 'verification' and 'falsification', 'fact', 'world', 'I', 'reality'. Nevertheless, those philosophical problems which can be 'transformed' into empirical questions are accepted as legitimate problems of unified science. Thus Neurath admits that at least one part of epistemology, namely that which can be treated meaningfully, must be investigated by scientific research. And with this he formulates very clearly the program of a naturalised epistemology which Quine presented in 1968, perhaps not totally accidentally, in Vienna. But Neurath still is not as naive as Russell or we might think he is (cf. Russell 1940). He sees in the avoidance of the use of certain words only a preventive measure, but no guarantee of delimitation. Neurath repeatedly uses the argument that there is no authority in the republic of the sciences, that everybody possessed the right to occupy the place of the head of tribunal, and decide with the court of justice that a statement, and even if it is a protocol statement, will not be accepted (Neurath 1944, p. 13).

I need not stress how imprecise Neurath remains when it comes to determining the empirical character of this radical formulation. But let us turn to his successors, of whom I have mentioned on the one side Quine, and on the other side those who used his arguments for relativism and pluralism like Feyerabend and Kuhn. It is not difficult to note that the justification of experience still ends at that rock which Neurath did neither wish to, nor could, move, the *observation sentence.* — Please do not advance as criticism the multi-purpose argument that every sentence is theory-laden, or to a certain *degree* theoretical, for *this* thesis is one Neurath *himself* advocated since 1913. (I have long found it strange that those who defend this thesis and have read Neurath do not refer to him.) So even with the qualification that no observation sentence is ever written onto a *tabula rasa*, that we can *never* begin completely anew with the construction of a scientific language, the observation sentence remains the cornerstone of empiricism as of semantics, to employ a term of Quine's (cf. Neurath 1941a, p. 221; Quine 1969). Expressed differently, if we want to do science, we must begin with the *uncleansed* language as we use it in its raw condition *every day*, and if we reformulate our results we return to it whether we speak of the *rectangularity* of a square table or of the *red lines* which we see in a *microscope*. Like Mach and Avenarius, whom he mentions in this connection several times, Neurath proposes to proceed from the *natural conception of the world* and to return to it. The key to this conception of the world is the *observation sentence*.

I cannot enter into the clarification of this point at the present time, although nothing seems more important than enlightenment as to why the role of starting and end point of knowledge falls to the observation sentences, which precise semantic status they occupy, and how a criterion would have to be constituted that could distinguish them from other experiential sentences.

These questions, which are neither purely logical nor purely factual, were considered by Wittgenstein to have an *elucidatory* character, and their discussion was placed either *below* or *above* that of natural scientific questions. And already in his programmatic essay on 'The Turning Point of Philosophy' Schlick set out that the giving of the meaning of statements could not in the end be done by statements, for every such elucidation would lead into infinity, unless they found a self-evident terminus in an actual showing. Such showings, however, are *actions* and not sentences.

This process cannot proceed endlessly. It always comes to an end in actual pointings, in exhibiting what is meant, thus in real acts; only these acts are no longer capable of, or in need of, further explanation. The final giving of meaning always takes place therefore, through *deeds*. It is these deeds or acts which constitute philosophical activity. (Schlick 1930a, (1959, p. 57).)

Neurath reveals this attempt as metaphysical by declaring: "*We do not need a metaphysical ladder of elucidation.* On this point we cannot follow Wittgenstein..." (Neurath 1932a, p. 60.) Entry into unified science, i.e. the totality of a mass of sentences from both the everyday and the scientific language can*not* be effected by "meaningless sequences of words". In 1930 he only held that such elucidations were not needed; by the time of the Prague congress of 1934 he declared: "We strictly reject this conception, it rests on the error that 'sentences about sentences' are meaningless, that is the entire logic of science, which is part of unified science." (Neurath 1936g, p. 763.) In his *The Logical Syntax of Language* Carnap objected against Wittgenstein's 'mythology of the unsayable' not only by claiming that the distinction between saying and showing is inapplicable to syntax—that it is false for there are, as his own syntax showed, correctly formulated sentences about sentences—but he also objected, like Neurath, against the pseudo-sentences represented by elucidations. And Carnap stresses that this difference of opinion is not merely a theoretical one, but that it has practical consequences. This became apparent in Wittgenstein's inability to demonstrate a significant difference between the sentences of the logic of science and of metaphysics. Thus a number of statements—to be precise, of pseudo-statements—of the *Tractatus* fell into the domain of metaphysics.

At this time Carnap still thought that the *pseudo-object-sentences* considered—sentences "which assert something about the meaning, content or sense of sentences or linguistic expressions of any domain" (1934/7, p. 285.)—could be eliminated by a syntactic manoeuvre. But later on he seems to have given up this belief. Be that as it may, the terminology of whole and part was used here again, and the legitimate task of philosophy was integrated as a part into the whole of science. Thus logical-philosophical analysis proceeds by means of genuine sentences, whose objects are also sentences. In short, the 'logic of science' (an expression for the theory of science used in this period by Carnap and accepted by Neurath) is understood as a metascience; being part of the entire system of the sciences, it does not and cannot claim a separate status. For this reason, there is no more talk of 'philosophy' in the traditional sense: 'philosophy', the name of a venerable science, belongs in the *index verborum prohibitorum*, and does not designate a separate task.

How then does unified science as the successor of philosophy explain the fundamental relation between what is experienced and the objects of experience? Like Schlick, Wittgenstein had explained the representing relation which exists between the elements of a picture and the elements of a state of affairs by the *co-ordination* of elements, and repeatedly used an image which Schlick also uses to elucidate the nature of his '*constatations*': *both* speak of the *points of contact* between the objects of knowledge and the objects of reality (cf. Wittgenstein 1921, #2.1-2.17). If we develop such a model of reality—and a picture is such a model—then the coordination of the elements, which after all *connects* the picture with reality and which grounds the identity of what is shared, the form, may be interpreted as a *touching*. The "coordinations", so Wittgenstein, "are as it were the feelers of its elements by which the picture touches reality." (Ibid., #2.1515.)

For Schlick the image of touching was prefigured already in his *General Theory of Knowledge*. There too, empirical sentences are models of univocal designation, i.e. co-ordination, and just *these* are tested by verification or hypotheses. A hypothesis is a true sentence *if* a univocal designation of a fact has been achieved, for a judgement is true if and only if it "uniquely designates" a fact (1918/25, p. 60). Verification consists in the identity of a perceptual judgement and an observational judgement such that the former does not say more than that something is experienced at a specific time and place and under certain circumstances. The identity of both judgements can only depend on the existence of that which is the case if the judgement is true, and it is this identity of what is asserted by the judgement that is to be established by verification. "The act of verification", so Schlick in his 'The Turning Point in Philosophy,' "is always of the same sort. It is the occurrence of a definite fact that is confirmed by observation, by means of immediate experience." (1930a (1959, p. 56).) But with the constatations we leave the domain of legitimation by means of sentences, for constatations are not sentences which express a judgement but are momentary endpoints of experience, points of contact, as it were, between the realm of words and experiences and the realm of objects and facts.

The real reason for the fact that the meaning of sentences points beyond themselves lies in that it makes a difference *whether* they are true or false. To state this is the ultimate function of that which was designated immediately: it is in *this* sense that Schlick called it 'final'. That which makes an endpoint of a self-justifying sentence like "I am appeared to yellowly" has no doubt to do with the construction of concepts, since it contributes to the determination of the experiential concepts and to the mode under which we comprehend what is given intuitively (cf. Wittgenstein 1956, p. 248, 237). But the grounds which speak for the truth of that which is being observed, are not determined by these modes and concepts.

The pseudo-sentences with whose help we lay out the grounds are such that they depend on the decision of a more or less homogeneous group of people which determines, first, whether a certain expression possesses this exact meaning, and which decides, second, whether a sentence may figure as a candidate of that mass of sentences into which it will ultimately be integrated.

> When a new statement is made we must see how it is to be integrated, whether as a reality-statement, a hallucination-statement, an untruth. A statement which is incompatible with the whole system is excluded as an untruth unless we are resolved to change the whole system so that this statement can find its place therein. (Neurath 1931d, p. 408.)

There is no need to emphasise specifically how much of what is acceptable is mixed here with what must be rejected. It is certainly natural to interpret the agreement of the opinions of a group, a community, after the fashion of how an individual subject judges a series of offered judgements for their compatibility with judgements already accepted. In this respect, the group or community is no different from an individual subject. What gives it epistemological dignity is only the pragmatic argument that more eyes—say, a hundred—see more than two or four do. But even this is not as matter of principle, but only of degrees or of time.

Yet this is still not yet the weakest point which I want to emphasise. This weakest point rather is the again and again proffered assurance that the *compatibility*, ultimately the truth of statements, depends on actions, and intentions to act, which we integrate, eliminate, decide etc.. In order to be practicable (if I may use this expression for the realisation of types of actions or schemas), these actions require definite reasons for their justification just as much as epistemological justifications do. Whether I *hold true* that a line will appear redly, or whether I *integrate* or *eliminate* the statement "A red line will appear" into a mass of sentences, depends on the reasons which support my holding true or the decision which I make. In this way decisions are affected by regress arguments just as much as epistemological justifications.

Of course, as we know, one of the reasons for switching the level of discourse was the complication involved in the idea of the coordination and correspondence of sentence and fact. In the attempt to eliminate the *metaphysics* of coordination and to preserve the unity that the mere differentiation of levels of sentences bestows, it was decided that decisions and acts should carry just that weight for the acceptance of sentences, which was not entrusted to reasons. But by *resolving* to accept a sentence as an observation sentence, or to integrate a statements into the totality of our attitudes, we show that reasons guide the decision, at least insofar as the existence of other attitudes is presupposed and taken into account. And for the longest time Neurath did advocate freedom from contradiction as a criterion for this decision, until it too falls before the tolerance of 'anything goes'. "The question which contradictions can just be tolerated, which not, how one behaves altogether in the development of the whole of science, is a question of behaviouristics, of history of science", so Neurath wrote in 1936 (1936c, p. 169).

Here it finally becomes clear, I hope, that the problem of justification has only been displaced onto another level, but has not been overcome. For the problems repeat themselves for a *historical statement*— i.e. that a certain group of people decided in favour of a certain theory, say that of continental drift—in a manner I need not specify further. Without wishing to limit the authority of history, we nevertheless must not concede from the start that it can decide whether contradictions be *allowed* to appear in the system of the attitudes of an individual subject or of a group. For if we set out to explain why they appear, then even the worst explanation will be one which appeals to a possible reason and does not prefer, without any reasons, *one* interpretation to the totality of possible ones.

If we ask ourselves now what is the motivation for this softening of the initial position, then it seems to me that one good reason has already been noted, the wish to treat all questions uniformly. *Every* sentence is accorded the same initial justification for acceptance into the convolution of unified science, and *no* sentence is accorded a particular dignity. There is no movement in this context until it is to be decided *which* sentence *can* be eliminated. If one were to take the standpoint that this is determined solely by the development of *future* science, then indeed the history of tomorrow would be the judge of today, and the history of the day after tomorrow the judge of the history of tomorrow.

This also shows clearly that the *desire* for uniformity does not suffice to bring it about: "As empiricists we will always start from our everyday formulations, we will always test our theories and hypotheses with their help.

These rough sentences with their multiple indeterminacies are the starting and end point of all of our science." (Neurath 1938b, p. 871.) And indeed, this sufficiency does not obtain. The path taken by Neurath already leads him to the abolition of the postulate of precision: its place was to be taken by the flexibility of concepts and of epistemological attitudes. For more could not be asserted—to express a complex matter quite simply—than that one person or group of persons accepts *these* sentences at one time, and *those* sentences at another. That they are accepted does not make them true, and that they have not been accepted does not make them false. For on Neurath's boat 'true' and 'false' go overboard, like 'verification' and 'falsification'. Every confrontation of name and object, sentence and fact, method of projection and what is projected, was rejected as a form of absolutism. There is no struggle between error and truth; what the physicalist unitarian scientist describes rather, are historically emergent groups of thinkers who advocate the plausibility of their program. And since all possess equal rights, even the timing of a claim does not matter. I think I do not need to continue this sketch. Everyone who knows the discussion of the standpoints of Kuhn and Feyerabend concerning the supercession of programs or paradigms will have noted the parallel. History has already caught up with Neurath's explanation that the rejection of absolutism and the acceptance of a democratic model leads "to [the] history and the sociology of the sciences and to a stressing of the social implications of language." (Neurath 1941a, p. 229.)

It would be interesting to specify in detail how the so-called anti-positivistic and anti-foundationalist standpoint is connected with the teachings of the positivists. But that is the theme of another lecture.

Karl Franzens Universität, Graz

NOTES

* First published as 'Zwei Arten der Erfahrungsbegründung', in *Schlick und Neurath. Ein Symposion*, Hrsg. R. Haller, *Grazer Philosophische Studien* **16/7** (1982) 19-33, © 1982, Rodopi, Amsterdam. Translated with kind permission of Editions Rodopi and the author by T. E. Uebel.

1 Letter Neurath to Carnap, 18 June 1933.

2 Letter Neurath to Carnap, 13 March 1933.

3 Cf. letter Schlick to Carnap, 7 August 1930.

4 Neurath 1932a, p. 66; see also his 1941b, p. 923, where Neurath emphasises explicitly that he does not propose to reduce all sentences to so-called physicalistic ones.

LOLA FLECK

OTTO NEURATH'S CONTRIBUTION TO THE THEORY OF THE SOCIAL SCIENCES*

In the Vienna Circle, Neurath was the representative of sociology. His education had well-prepared him for this role in the discussion group around the philosophy professor Moritz Schlick. In 1902 Neurath began his university studies in Vienna with mathematics and physics, but soon changed to political economy, history and philosophy, and received his doctorate in these fields in Berlin, 1906. Since at the beginning of this century, sociology did not exist as an academic discipline, this combination of studies may well be considered a substitute for the study of sociology.

The following remarks focus on Neurath's theoretical contribution to the social sciences. But this should not leave the mistaken impression that Neurath's work would be exhaustively characterised thereby. I also must mention his efforts in applied sociology and his socio-political practical work.

As the director of the Social and Economic Museum in Vienna he made a decisive contribution to the development of pictorial statistics. The first attempts at the graphic representation of social processes Neurath had undertaken already in 1916 as director of the Museum for War Economics in Leipzig. With his picture statistics Neurath developed not only the tools of social enlightenment, but also a non-metaphysical language in order to render visible the complex, many-dimensional social relations (he later called them "aggregations").

Neurath's path-breaking though little-known thoughts about the theory of life situations—employing predecessors of today's 'socio-economic indicators'—can also be counted as applied sociology. His rejection of one-dimensional valuations of wealth lead him to construct a 'relief [map] of total life situations' (*Lebenslagenrelief*) which combines the factors "housing, food, clothing, education, pleasures, work time, and discomfort." (Neurath 1920c, p. 11.)

Finally, I may be permitted to mention, and count towards his work in applied sociology, Neurath's conception of economy-in-kind and the calculi for it which preoccupied him throughout his life. Neurath envisaged a

T. E. Uebel (ed.), Rediscovering the Forgotten Vienna Circle, 203–208.

comprehensive calculation for a national economy on a non-monetary basis. Money and the theory of money, which after all is only a tool for better economic organisation, had come to dominate the development of theory in economics and had prevented certain questions from being asked in the first place. It was accepted without question that "monetary calculations correctly reflect the relations of wealth anyway" (Neurath 1910a, p. 37). Against this Neurath pointedly summarised his own position: "However, if one wants to characterise the state of a human being, then one must name the goods and the evils in the same breath." (Ibid., p. 39.)

Which other scientist who is interested in the socio-technical utility of his scientific results can offer such a manifold of political activity? This activity of course brought with it inconveniences: as the director of the Central Economic Commission in Bavaria Neurath was arrested after the suppression of the Munich Soviet Republic in 1919. The decision which he reached "after much reflection", namely, "to conclude [his] life of contemplation and begin one of action, to help to introduce an administrative economy that will bring happiness" (1919a, p. 124), ended with being sentenced to 10 months in prison and the loss of his lectureship in Heidelberg.[1]

Neurath's engagement in the socialisation movement is wholly consistent with the totality of his philosophical efforts. For him, science is a recipe for practical activity. Social scientists cannot provide univocal recipes, of course, they do not stand on a 'neutral' position outside from which such a thing would be possible, but "[w]e argue differently and act differently, when we know the material provided by the social sciences." (Neurath 1944, p. 46.)

Let me now consider how, according to Neurath, a social science would look which was to achieve this.

Already in 1910, in his essay on the theory of the social sciences (1910a), the then 28 year old Neurath rejects the possibility of a systematic distinction between the social and intellectual sciences (*Geisteswissenschaften*) and the natural sciences. It was neither the case that the areas to be explored could be univocally be assigned to one rather than the other, nor that their methods differed to such a degree that a separation between the sciences would be justified.

Instead, we must conceive "all statements and all sciences as coordinated" (Neurath 1944, p. 8). Neurath envisages a type of research which proceeds according to common principles, which overcomes the mere coexistence of the sciences; in his perspective, partial inquiries contribute to the common work of many, to a unified science. Of course, one will continue to group scientific sentences "according to purposes of one's orientation" (ibid.), but the clear delimitation between fields of inquiry, or even the hierarchical organisation of the sciences, had to be given up.

Neurath's conception of a universal science or of unified science is often misunderstood, sometimes even misused.[2] Unified science is misinterpreted as a unified system of science, itself non-contradictory, which allows only those research techniques which are used in physics. Accordingly, the aim of the scientist would be the search for progressively better confirmation and elaboration of this system, for the aim in the far future would be the completed edifice of science which clearly reflects our world and allows for unproblematical predictions.

That is not what Otto Neurath said. In one of his last essays we read instead:

> My 'pluralist' attitude ... immediately objected to all absolutist attempts to speak of one comprehensive 'world picture' as the best, or as other absolutists would say, the 'relatively best', even if one were to look at it as an 'ideal' only. I would not think of 'the system' as a 'model' and stressed that one had to remember the 'gaps' and 'gulfs' from the beginning. (Neurath 1946a, p. 231)

Because of the many misinterpretations of his idea of a unified science, Neurath in his later years speaks more often of encyclopedism, or of a model-encyclopedia. "Encyclopedism does not pretend to present a thoroughly consistent system of statements, though our program tries to extend the area of consistency." (Neurath 1944, p. 12.)

In order to clear up the misunderstanding which the concept of unified science brought with it, it would be possible to replace 'unified science' by 'interdisciplinarity'. 'Interdisciplinarity' not only as the fashionable name for the willingness to discuss research across scientific boundaries, but as the demand for the integration of the research techniques of the individual disciplines. "We stressed the point that actual cooperation in fruitful discussion should demonstrate how much unity of action can result, without any kind of authoritative integration." (Neurath 1946a, p. 230.)

At the very beginning of scientific research stands the language of everyday, with all its impreciseness and a manifold of indeterminacies. It is the task of the scientist to test whether the formulations of a hypothesis under discussion are empirical ones, that is, whether they are "made in a language which allows comparison with observation-statements." (Neurath 1944, p.

23.) Still independently of that, the scientist must consider how the hypothesis in question may in future be scientifically tested and supported.

Let me repeat that I do not suggest as our position at the start ... "simple basic assertions", "atomic ideas", "sense data", or their phraseological substitutes, i.e. something elementary, primitive and poor. On the contrary, I suggest our starting with a full lump of irregularities and indistinctness, as our daily speech offers it. *Afterward* we may find some regularities in it ... (Ibid., p.18; italics L.F..)

The demand that all scientific sentences have to be comparable with observation sentences—a demand placed by Neurath at the beginning of all scientific activity—amounts to a demand for an empirical language. Scientists must develop the language of science from daily speech by elimination of all unempirical and metaphysical expressions. This unified language, or 'universal jargon', is intended to make it possible for us to speak in a similar fashion about different objects without introducing *prejudicial distinctions*. Although observation or protocol sentences are always somehow indeterminate—names of observers and terms of observation are indeterminate—they are nevertheless, if we commit ourselves to empiricism, the only thing with which we can compare the sentences expressing a hypothesis. A comparison with facts is impossible, for one can only compare sentences with sentences. We only have the possibility of taking terminological care and assuming a critical attitude towards the acceptance of sentences which can hardly be tested with reference to observation sentences.

Logical empiricism does not provide us with a wizard's sieve which holds all this uncritical stuff, as logic does not prevent us from amking huge buildings of metaphysical speculations, consistent in themselves, but unempiricist. Books on racialism may be written in plain empiricist language, but nevertheless lack a critical attitude. (Ibid., p. 19.)

Once Neurath's demand for a terminological empiricism is fulfilled, then we can accept, or seek to support, or reject an assertion or hypothesis, or postpone our decision. If the social scientist faces two or more 'comprehensive sets of sentences' which are in competition, then he must *decide* whether he has enough time and energy to 'try out' more than one of these, or he must decide to work only with one of these systems of hypotheses. One cannot scale these hypotheses and so choose the best. It is also not fruitful to introduce dichotomies and construct an asymmetry pf theory/anti-theory. Neurath argues that the main elements of contradictory hypotheses are to be isolated and then to be newly combined, so that a whole matrix of conceivable hypotheses is obtained (cf. Neurath 1915a, 1916a).

Neurath welcomes this coexistence of hypotheses and calls it 'pluralism'. Especially in the social sciences, where it is difficult to consider individual relations in isolation and where we find aggregations— "with ... human beings, animals, houses, streets, swamps, and fields"[3]—there it is the scientists who combine the same scientific material in different ways. Thus a whole network of hypotheses comes to be. The scientist considers how he could test and support these arguments and hypotheses—but:

> Assaying [testing] one hypothesis implies, as it were, assaying all hypotheses together. And, in starting with assaying some hypothesis and going on successfully then with the argument, we may be hampered by findings in different parts of the cosmic history which alter the whole attitude of our arguing. (Neurath 1944, p. 19.)

Scientists will try to integrate their hypotheses into a set of other hypotheses, observation sentences and other accepted sentences. Contradictions do not always mean the rejection of an hypothesis, for the hypothesis in question may be a useful one (as long as there is none more attractive). Neurath is very suspicious of all attempts that claim to rule out a hypothesis by means of a crucial experiment; he fights against talk of truth; rarely does he speak of the possibility of verification or falsification, rather of the 'corroboration' or 'shaking' of hypotheses; and particularly "[i]n the field of aggregational analysis we always have to deal with more than one possibility" (ibid., p. 24). As he already put it in 1910, it is the scientists who decide, and they must always seek to know "how long it is useful to maintain the entire edifice of the theory and to explain certain facts by auxiliary hypotheses, and when it is more useful to reconstruct the entire system anew." (Neurath 1910a, p. 44f..)

Neurath is not very optimistic with regard to the potential for scientific predictions. Not only must the formulation of a prediction be accounted for as part of social life, but "scientists are no better off than the man in the street as far as their predictions are concerned. They only collect more material dealing with past uniformities." (Neurath 1944, p. 22.) The social sciences do not provide mankind with recipes: "we have to expect gulfs and gaps everywhere, together with unpredictability, incompleteness, and one-sidedness of our arguing." (Ibid., p. 27.)

Even if certain techniques have been employed in one science more often than in another, the procedures of all empirical sciences are the same; for logical empiricists, the protocol sentences are always the basis for further discussion. Otto Neurath shared with the logical empiricists of his day the optimistic attitude toward the possibility of science serving the human life world: science serves the conscious fashioning of lives.[4] At the same time, Neurath did not overestimate the reputation of science:

> One gets the impression that through the ages people who have to make sudden decisions frequently, such as generals, statesmen, industrialists, heroes, adventurers, pirates, are more devoted to the sayings of astrologers, soothsayers and other guides of this type than are the rest of their contemporaries. (Neurath 1944, p. 44.)

Redaktion Wissenschaft, Österreichischer Rundfunk

NOTES

* First published as 'Otto Neurath's Beitrag zur Theorie der Sozialwissenschaften', in *Arbeiterbildung in der Zwischenkriegszeit*, Hrsg. F. Stadler, © 1982, Österreichisches Gesellschafts- und Wirtschaftsmuseum, Wien, pp. 100-103. Translated with kind permission of Österreichisches Gesellschafts- und Wirtschaftsmuseum and the author by T. E. Uebel.

1 For further biographical details, see Fleck 1979.

2 A great many misinterpretations seem to rest in the fact that only Neurath's *Empirical Sociology* is taken account of. By contrast, the later contributions by Neurath are much more differentiated and constitute in certain ways a return to the level of argumentation of 1910. For this reason I rely on these later contributions. An example of misuse is presented by the Moonies and their 'Unity-of-Science' congresses.

3 Neurath 1944, p. 20. Neurath stresses explicitly that his 'aggregational program' must not be confused with a metaphysical holism.

4 In 1929 Neurath wrote, for example: "Anti-metaphysicians strengthen the power of the proletariat!" (Neurath 1929a, p. 338.)

PAUL NEURATH

SOCIOLOGICAL THOUGHT WITH OTTO NEURATH*

That I call my paper 'Sociological Thought with...' rather than 'The Sociology of Otto Neurath' is, because—in spite of his having written, in 1931, a whole book called *Empirical Sociology* and in 1944 a monograph *Foundations of the Social Sciences* and already in 1910 a major paper '*Zur Theorie der Sozialwissenschaften*' ('On the Theory of the Social Sciences')—he never wrote anything that could be presented as *the sociology* of Otto Neurath.

What he wrote instead was largely two things: on the one hand a comprehensive critique of what was then taught in German universities as 'sociology', usually called '*Gesellschaftslehre*', that, instead of dealing with the real problems of society and its structure and man's social actions in it, was rather concerned with discussions about the intrinsic nature of man and his society as such, much of this in metaphysical terms; and on the other hand prolegomena towards a future sociology, that would have to be an empirical science, dealing with human beings as active and passive participants of and in society, influenced by it and in turn influencing it. For Otto Neurath, sociology was to be a science based on empirical observations, with conclusions and hypotheses derived from them or based upon them, that could then be shown as being or as not being in agreement with empirical evidence, and either be accepted or, if need be, be revised or be rejected, just as in any other science.

In all of this it will be useful to remember that what today we understand as 'sociology' was at the time, when Neurath wrote his *Empirical Sociology* some sixty years ago, practically non-existent as a separate discipline in German universities. In fact, the very name 'sociology' was then only 90 years old, having been coined by Auguste Comte (1798-1857) only in 1838 in his *Course de philosophie positive*, whereby he meant a kind of superscience at the apex of a pyramid of sciences of man, there replacing philosophy, which was still too much overshadowed by the kind of theological thinking that should have gone out with the French revolution. Small wonder, then, that what was being taught as 'sociology' in the late 19th and early 20th century was still largely concerned with the construction of comprehensive systems of society, still heavily permeated with metaphysics. With that in mind one may better understand the subtitle Neurath gave to his book: 'The Scientific Content of

T. E. Uebel (ed.), Rediscovering the Forgotten Vienna Circle, 209–222.

History and Political Economy', as a kind of program still somewhat in the tradition of Comte, trying to come to a unified science of man and his society.

True enough, the first beginnings of a true empirical sociology were already there. Durkheim (1858-1917), e.g., had already shown, in his *The Causes of Suicide* (1897) the social nature of the ups and downs and differences of suicide rates as linked to variations in the degree of social integration—and done so on the basis of thoroughly empirical data, so that one could check and re-check, and then either agree or disagree with his interpretation of the data; but the data themselves were empirical, not speculation.

There were already the great studies of poverty in England around the turn of the century, all thoroughly empirical—but those who made them were not the sociology professors of their day but amateurs, interested in social reform. Chales Booth (1840-1916), who conducted the biggest of them all, the 14-volume *Life and Labor of the People of London* (1889-91, 1902-03), was the successful owner of a steamship company and B. Seebohm Rowntree (1871-1954), who did the study of poverty in York (1901), was the director of a local chocolate factory.

There were already the beginnings of the Chicago ecological school of sociology under men like W. I. Thomas (1863-1947) and Robert Ezra Park (1864-1944) with its heavy concentration upon the spatial distribution of various forms of social pathology linked to the spatial distribution of social living conditions—all of it based upon empirical observation and empirical data collection. But when Neurath wrote his *Empirical Sociology* in the late 1920's, most of this work still lay in the future. Most of the famous Chicago Series of Sociology was published between 1929 and 1940.

There was also already before World War I the work of the *Verein für Sozialpolitik* (Society for Social Politics) in Germany, in which Max Weber and Ferdinand Tönnies participated—the latter, by the way, a kind of revered father-*ersatz* for Neurath, whom, when he was studying natural science and mathematics in his third semester in Vienna in 1903, Tönnies persuaded to transfer to Berlin, to study history and economics under Eduard Meyer and Gustav Schmoller. The *Verein für Sozialpolitik* had e.g. conducted studies of the occupational choices of industrial workers and problems of Polish seasonal workers in the sugar industry of Saxony. But for the participating professors this was more in the nature of a sideline, not the essence of their work as sociologists. In that latter capacity they were still very much preoccupied with the construction of overarching grand theories of human society. Whereby most of them made it a point that theirs was not a

Naturwissenschaft, but a *Geisteswissenschaft*—a distinction that in that form never existed in the Anglo-Saxon universities or literature. Which is why, although it is easy enough to translate the former term with 'natural science,' it proves quite impossible to translate the latter—the best one could do would be a long circumscription and explanation of what may be meant with it. There simply does not exist a single term in the English language, that could convey the emotional loading and the intellectual pretension that the term *Geisteswissenschaft* implies, when it is used in juxtaposition to, and in contradistinction from, *Naturwissenschaft*.

Much of what Neurath had to say in his *Empirical Sociology* was, and of course still is today, a polemic against this kind of a distinction, that tries to treat, e.g. sociology as something on a higher level than lowly natural science, rejecting with disdain the suggestion that the usual empirical methods of the latter could be applied to the former. The attack begins already with the opening sentences of Neurath's Preface:

> It is customary in present-day German literature to presume that there exists besides the kind of cognition or perception that exists in the natural sciences [*naturwissenschaftliche Erkenntnis*] a quite different one, that is usually called *geistewissenschaftliche Erkenntnis* [the kind of cognition or perception specific for the *Geisteswissenschaften*]. The thesis, upon which this distinction rests, can, within the framework of the scientific world-conception, not be reduced to facts that could be formulated in an empirical manner—it is devoid of meaning, it is metaphysics. The distinction is usually seen in that the natural sciences have 'only' the task to bring some 'order' into emprirical facts, while the *Geisteswissenschaften* are beyond this based on *verstehen* [understanding; P. N.]. If then one analyses this concept of *verstehen* in any scientific manner, then everything about it that can be transmitted with the help of words, proves to be statements about order, exactly as in the natural sciences.[1]

Neurath takes issue with this kind of 'sociology' primarily through presenting—in a chapter called 'Metaphysical Countercurrents'—pages of quotations from four of the then most popular and most respected German University sociologists—Werner Sombart, Othmar Spann, Max Scheler, Max Weber—that in their wording can speak so much for themselves and for the metaphysics in them, that there is hardly any need for much additional comment from Neurath to point it out. I can of course not bring you the quotations in their full length, each of which comes to about one to two pages in the original, but must restrict myself to single sentences here or there, just to give you the flavor.

Here are some sentences from Werner Sombart:

> Spirit cannot be dissolved into soul [*Geist läßt sich nicht in Seele auflösen*]... we give to the spirit what is the spirit's and thus we cannot be satisfied in any of the cultural sciences

[*Kulturwissenschaften*] least of all in sociology [*Gesellschaftslehre*] with the mode of perception that is characteristic of the natural sciences...

Empathy [*Verstehen*] is a central concept of our science. All endeavor of cultural science is directed at *verstehen*, i.e. at cognition from innermost to the surface [*Erkenntnis von innen nach außen*], in contrast to natural science, that can only perceive from the surface towards the innermost [*von außen nach innen*]...

If the great problem of history is how the spirit becomes soul, then it will be the task of *geistewisenschaftliche* sociology to present this this process as a social, i.e. as a relational phenomenon in its typical form. It is necessary, however, to insert a warning: not to oversimplify that contrast that I have been at pains to elaborate, between psychological (Western) and noological (German) sociology, as though this were only a juxtaposition of the material to the formal.

To empathize psychologically means to empathize with the soul of others [*Psychologisches verstehen heißt Seele verstehen*] and rests upon immersing oneself into the soul processes of others. To comprehend noölogically means to comprehend with the spirit [*Noölogisch verstehen heißt Geistverstehen*] and is synonymous with understanding sense and meaning [*Sinn- und Bedeutungsverstehen*]...[2]

From Othmar Spann we have these sentences:

If there is to be a *Gesellschaftslehre* then ... it will be necessary to present proof of the epistemologico-logical possibility of a whole that does not come about through the mutual impact of its parts...

... he who has not recognized the methodological difference between the path from the whole to its members as distinct from that from the independent part of an agglomeration to the pseudo-whole, has not yet passed the threshold to the true *Gesellschaftslehre*.[3]

To which I may add a gem that I have been carrying with me ever since I heard it myself from Spann when I was his student in Vienna well over half a century ago: "*Der innere Geistesgrund ist das seinstragende Übersein.*" If this can be translated at all, it might be rendered as "The inner spiritual foundation is the super-being that is carrying the being"—which, just to make sure, has nothing to do with Sigmund Freud's concept of 'ego' and 'super-ego' for Freud was, of course, anathema to Spann. As it happened, at first I wrote in my notes "*Der innere Geistesgrund ist das seiltragende Überseil*" (which might be rendered as "The inner spiritual foundation is the super-rope that is holding the rope"), but learnt that Spann had indeed said "*das seintragende Übersein.*"

Here are some quotes from Max Scheler:

The more clearly a group is merely a class (i.e. a unit that is classified i terms of possessing or not possessing property), the more it is merely a sum of individual economic subjects without any reality as a totality [*Ganzheit*], without a continual existence beyond the individual subjects, without a special spirit [*Geist*], will, soul [*Seele*], as it is already possessed by the life communities [*Lebensgemeinschaften*] (family, home town, tribe, gens, a people) and even more so by the genuine total persons [*Gesamtpersonen*] (Church, nations, states, cultural circles)...

The positivistic doctrine of the three stages in any of its forms, be it those given to it by Comte, Mill or Spencer, or in Germany by Mach and Avenarius, is fundamentally erroneous. Religious-ideological cognition and thinking, metaphysical cognition and thinking and positive cognition and thinking are not historical phases of the development of knowledge, but are essential permanent spiritual attitudes and forms of cognition that are already given with the essence of the human spirit as such.[4]

It is only with regret that Neurath finds it necessary to have to include Max Weber in this series, whom otherwise he regards highly. He introduces the appropriate quotation with this remark:

Most earnestly concerned with an empirical treatment of sociology, Max Weber nevertheless always maintained a principally anti-behavioristic attitude; he who time and again ascribed to the spirit [*Geist*] of various epochs a moving force in the social realm—an attitude that still brings him into the neighborhood of Sombart, Spann and Scheler.[5]

The two-page quotation from Weber is actually a composite of passages from the famous opening pages of *Wirtschaft and Gesellschaft* (*Economy and Society*) where Weber lays the foundations of his 'interpretative sociology' (*verstehende Soziologie*). Other than with the authors quoted earlier, I find it impossible in this case to take out just a few short sentences to give you the flavor of it. Because here it would really mean tearing sentences out of their context and thereby giving a false impression of what Weber wanted to say. Perhaps it helps if I mention that Talcot Parsons in his translation of these opening pages needs a footnote of 18 lines in small print to disentangle the many different meanings that Weber gives to this one word '*verstehen*' in different places (and, by the way, another half page of small print to disentangle Weber's various uses of such terms as '*Sinn*', '*Handeln*', '*Verhalten*', terms that the unwary might feel tempted simply to translate as 'meaning', 'action', and 'behavior').

At any rate, Weber tries to make clear that to explain human action requires a form of *verstehen* that distinguishes the sciences of human action, specifically sociology and history, from other sciences, among which he specifically enumerates law, logic, ethics and mathematics. To all of which Neurath remarks:

With Weber, time and again, empathic immersion appears in place of science. Over this poetic activity there is no test, no control, nor does it belong in a scientific account... If someone maintains that he can only predict historical processes 'empathically', we would have to ask him on what he based his empathy. If he tell us that, if he can point to verifiable

> data from which he reaches his empathy, from which in turn he presses forward to his predictions, then one can as well derive this prediction of a process *directly from the data mentioned.* (Neurath 1931c, p. 357.)

For a better understanding of the social context of this whole discussion about metaphysical vs. non-metaphysical sociology it may be useful to remember that all of this did not take place in a social or political vacuum. Most of those who, like Alfred Vierkandt, whom Neurath also briefly quotes, represented a 'new idealism' in sharp opposition to positivistic philosophy and science, were either, like Vierkandt himself, politically conservative or, like e.g. Othmar Spann and Hans Freyer (both of whom Vierkandt specifically mentions in the citation chosen by Neurath) aggressively anti-democratic. Whereas those on the positivistic side were for the most apart either non-political liberals or, like Neurath himself, socialists, or in many cases at least in sympathy with the latter.

Today, almost sixty years later, Neurath's comment on the passage he chose from Vierkandt sounds almost prophetic:

> The struggle between metaphysical and scientific sociologists is unavoidable; it takes place against a background of much greater social struggles, of which these discussions are only a mild reflection.[6]

Within but a scant few years this somber prophecy that still carried an undertone of slight hope as though perhaps not all were fully lost as yet, became a full reality that can be read off the lists of those whom fascism expelled from their positions, incarcerating or even killing some and driving into exile most of the others—either into inner exile or into emigration. Vierkandt himself, politically conservative, but not a National Socialist, was allowed to remain in Germany, but not allowed to continue teaching, while Neurath was among those who were driven into emigration.

In a sequence of three chapters, 'History', 'Political Economy' and 'Uniting History with Political Economy', Neurath elaborates on the theme announced in the subtitle of the book, emphasising that they both deal with man's social actions in their social settings, and that therefore these are not two sciences, but only two interlocking parts of one—the distinction can only be one of convenience. In this context he distinguishes between theories of history that can, and those that cannot, be subjected to scientific examination. The latter he relegates to the realm of metapysics. Among the former he mentions especially the then widely discussed *Phasen der Kultur* (*Phases of Culture*, 1908) by Franz Carl Müller-Lyer (1857-1916) and among the latter particularly Hegel's *Philosophy of History*. Concerning Müller-Lyer's *Phases of Culture* he emphasises that, regardless of whether one agrees with him or not, and even regardless whether Müller-Lyer's system is internally consistent or not (in full agreement or not with what we know about the course of history), it can always be checked against actual historical events and their sequences and, if need be, either be corrected or even be rejected as not being born out by empirical evidence.

Empiricism tries above all to understand existing peoples in their concrete ways and to deduce the present state from the past in all its concreteness. It may well happen that a people which has already experienced many transformations still lacks the wheel and the cart (ancient America) while another, which on the whole is rather poor in its inventions, possesses both of these. (Neurath 1931c, p. 332.)

But what can one say from this point of view to statements like this one from Hegel:

The German spirit is the spirit of the new world, whose purpose is the realization of absolute truth as the infinite selfdetermination of freedom, of that freedom whose content is its absolute form. It is the destiny of the Germanic peoples to be the carriers of the principles of Christianity.[7]

Whatever the meaning of statements like these, they are certainly not capable of being verified on the basis of empirical evidence as being or not being in agreement with historical sequences as we actually know them. About statements of this kind Neurath remarks in his *Foundations of the Social Sciences*:

Some arguments dealing with social problems, particularly with history, become equivocal because writers sometimes prefer an involved and crooked language where plain empiricist and univocal expressions might sound unpleasant to readers. Instead of saying bluntly: "One human group killed another and destroyed their buildings and books", certain historians prefer to say "Forced by its historical mission, the nation started spreading its civilization over the earth." (Neurath 1944, p. 8.)

Within the same context he cites, in *Empirical Sociology*, Robert Malthus (1766-1834) for one of those general assertions about the course of history that cannot possibly be verified by facts:

The human institutions are only light and superficial compared with the deeper causes of evil which lie in the laws of nature and in the passions of men. It follows that some human beings are exposed to want, according to unavoidable laws. They are those unfortunates who in the great lottery of life have drawn a blank.[8]

To which Neurath remarks: "We can see here a kind of regularity being invoked that, standing outside of the social order, makes its input always felt." In contrast to Malthus he cites the latter's contemporary Sismondi (1773-1842) who

as critic... does not explain poverty as due to some biological laws; he sees poverty as the product of the social machine. He shows that the income of the poor cannot buy the products that do or might come on to the market... he... has the merit of having shown, with special emphasis, that crisies are essential features of the new economic order. (Neurath 1931c, p. 344.)

(As in other cases cited above, what is important is not whether one does or does not agree with Sismondi, but that he, and others too, saw poverty as a product of social institutions, an assertion or interpretation that is at least capable of being confronted with empirical evidence.) In the same context Neurath emphasises that, according to Marx and Engels, the very same social conditions that produce the mechanism of capitalist order with its crises and poverty, also produce the conditions for the change of this order (ibid., p. 345.).

Although a socialist, Neurath was never an orthodox Marxist. In fact, in one memorable replique to Helene Bauer, who in criticising his ideas about the role of money under socialism had quoted Marx, he remarked that "What is correct does not become even more correct because Marx teaches it too, but neither does it become incorrect because Marx thinks otherwise."[9] Nevertheless, he considered Marx's and Engels' materialistic concept of history, whatever its occasional weaknesses and inconsistencies, as the best available form of historic and economic analysis, both because it has man's social actions within, and influenced by, the social setting within which he acts, as the center of its discussion; and because its statements and, with the passage of time, its predictions too, can in principle be subjected to scientific scrutiny and then can either be shown to be right or wrong in view of what we do know that actually happened.

Later chapters of *Empirical Sociology* contain numerous socio-historical examples, of which I can only cite three. Commenting on the well-known fact of high illegitimate birthrates in certain heavily Catholic agrarian areas, and remarking that this is obviously not due to a weakening of the influence of the Catholic religion in those areas but only a reflection of agrarian laws of inheritance (Neurath 1931c, p. 381), he says:

It is more promising if we look at the social conditions of birth statistics and population growth as parts of the whole social and economic order. One cannot find a uniform population law for all times and peoples as Malthus tried. We see, on the contrary, that each social type has its own movement of population. The habits of sex and family life cannot always be related to the contemporary mode of production, but often to the previous one. (Ibid., p. 384/5.)

Which, written about 1929 or 1930 sounds like a remarkably clear anticipation of one of the main causes of the population explosion in the developing countries after World War II: that in in many of these countries, in spite of a dramatic reduction of infant mortality rates—and of death rates in general, largely through the eradication of many killer diseases—the social habits governing, e.g. people's ideas about the right number of children to have, are still rooted in an as yet not distant past, in which generally half of all children born died before they were yet five years old.

At one point, where he talks at length about the fact that Luther and the Reformation played essentially a subsidiary role in the struggle of territorial princes for independence from Emperor and Pope, he remarks:

The terms 'Lutheranism' and 'Calvinism' describe people with certain habits and their cults, with distinct ways of speaking and writing. People who want to fight against their monarchs are inclined to spread a doctrine that monarchs may be dethroned; territorial rulers who want to get rid of their monarch like to hear that he is the representative of the Anti-Christ. (Ibid., p. 384.)

Which is consonant with Neurath's repeated criticism of the main thesis of Max Weber in his *Protestant Ethic* that people first had to become Calvinists before, under the encouragement of their new religion, they could become capitalists. Neurath sees it the other way around: first they developed out of the social and production conditions of their time their striving for wealth, and then they twisted their Christian religion into the special form of Calvinism under which this striving could appear as pleasing to the Lord, and the actual attainment of wealth as a proof of the Lord's blessing.[10]

As an illustration of the need to analyse social behaviour in context (he speaks of 'coherences') with the process of production, he points out that "Those villages which first overcame blood feuds were better suited to cooperate in dam building than those with blood feuds. Overcoming of blood feuds and large scale water economy are technically coherent." (Neurath 1931c, p. 382.) But disparaging any tendency of becoming too rigid about the need for this kind of connection, he also sounds a warning that, after all, one can every so often envisage, and even encounter in history, different combinations, e.g. very different social habits associated with the same kind of general production conditions. (Witness his birth-rate example.)

After having brought up the topic of sociological prognoses occasionally here and there, emphasising that sociologist are not prophets—all they can do is, like any other scientists, try to combine observations of past sequences of events into as plausible as possible surmises of what the future may hold—Neurath devotes the whole last chapter of his *Empirical Sociology* to this topic, and does so again a few years later in a separate paper (1936e). Already in his book he stated:

> Where prediction is possible, empirical science can succeed. Sometimes a change in a sociological position may be predictable though the direction of the change cannot be known... but this happens in physics too: a cone spinning on its top will fall, but which way cannot be predicted. (Neurath 1931c, p. 362.)

As an example of such an unstable situation, in which change could be expected, but where hardly anybody could anticipate when it would come and what direction it would take, he mentions the time shortly before the French revolution. All sorts of minor changes had already taken place here or there, the rising bourgeoisie had already occupied a number of new positions, but all of this had taken place slowly, more or less step by step, without any sudden changes. Thus, when the *États Genereaux* were called into session in May 1789, hardly anybody anticipated more than the usual bickering of parties and groups jockeying for positions or, to stay with the picture above: that the cone would keep spinning as before. Yet it turned out that this was the decisive event that made the cone topple over. Suddenly the situation changed completely. And what was perhaps most surprising: that one element of it, that had been considered as the most stable of them all, the holding of land by the aristocracy, was one of the first to collapse—peasants suddenly began to occupy and divide the land among themselves.[11]

There are, however, as Neurath points out, important limits to sociological prognoses, primarily because the sociologist is himself part of the situation concerning which he is trying to make his predictions—and he may, by making his predictions, influence its coming true.

> Sociological predictions ... are, as products of an era, co-determinants of what they assert. Predicting an eclipse of the sun does not influence the eclipse; but predicting the position of the market does influence the stock exchange, and asserting that a revolution will occur does influence the coming of a revolution. The asserting and denying prophets have become agents through their predictions, which does not mean that their forecasts must always have a reinforcing character; it may happen that their influence is positively paralysing. To reckon adequately, one must promptly take this into account. (Neurath 1931c, p. 405/6.)

Haller and Rutte, the editors of Neurath's collected philosophical and methodological writings in German, point out that this theme of the self-fulfilling prophesy, which Robert Merton, in his well-known article of this title of 1948, ascribes to W. I. Thomas, occurs in Neurath's writings twice before in much earlier years: once in his *Anti-Spengler*, a polemic against the latter's *Decline of the West*:

One who as a prophet or as the disciple of a prophet steadily supports a certain prohecy, may prove the correctness of his prophecy through influencing reality. The prophecy becomes the cause of its own fulfillment. (Neurath 1921a, p. 160.)

and once already in 1911 in an article '*Nationalökonomie und Wertlehre*' ('Political Economy and the Theory of Value'):

Prophecies in the social realm do have influence upon development. In that way they are different e.g. from astronomical prohesies that have no influence upon the stars. But in the social realm a certain prophecy is frequently a co-determinant for it own fulfillment. (Neurath 1911, p. 113.)

Neurath perhaps comes closest to today's mainstream sociology with his suggestions concerning the proper presentation of standards of living or, widening the range of the problem, the presentation of total life situations. Already in 1910 he had criticised the prevailing tendency of the economists of the day to compare life situations of various categories of people in terms of simple uni-dimensional scales, taking e.g. differences in average incomes per capita as expressions of corresponding differences in standards of living (1910a). In his later writings he warned sociologists not to fall into the same trap. Instead he called for a multi-dimensional approach, taking into account many different aspects of total life situations—an approach which in retrospect may sound like an early anticipation of today's development of 'social indicators'.

True enough, Frederic Le Play (1806-82) in his *Ouvriers Européens* (1855) had already made a beginning with his detailed household statistics of workers' families in various countries. But that could not yet suffice to describe the differences in total life situations of, say, a group of workers and a group of rentiers even if both happened to have similar average incomes; or perhaps the differences between two groups of workers with similar incomes but perhaps considerable differnces in their average life expectancies or in their rates of accidents and illnesses on account of differences in their working conditions; or, in the same direction: a group of perhaps comparatively well-paid industrial workers and a group of lower civil servants with similar or even somewhat lower incomes, but with greater differences in job security, pensions, education, leisure time activities, etc.

In his last monograph, *Foundations of the Social Sciences*, he made this suggestion:

One may compare two social patterns as far as death rate, suicide rate, illiteracy, use of radio, and other items may be concerned. Let us speak of various 'silhouettes', composed of the items in question, in analogy to individual 'profiles'. (Neurath 1944, p. 32)

Actually, he had done just that five years earlier in his book *Modern Man in the Making* where he had indeed, with appropriate pictorial statistical symbols, presented such 'silhouettes' for ten different countries (USA, Great Britain, France, Germany, India, etc.), each one showing for the respective country (1) the average length of life of the female population; (2) the literacy rate; (3) the number of suicides per million; (4) the number of radio sets per 100 people—and all ten 'silhouettes' on one single page, so that they could be compared at a glance (1939, p. 61). This form of presentation could obviously be extended to include a variety of other items as well.

In his book *Wirtschaftsplan und Naturalrechnung* (Economic Planning and Economy in Kind, 1925) and in an article 'Inventory of the Standard of Living' (1937c), Neurath elaborated on this idea in a more theoretical manner. A file of many different kinds of total life situations should be established, each one described in the multi-dimensional manner indicated above, but taking into account also in some ways the amount of enjoyment and the amount of suffering that generally goes with it. Then, in a kind of three-dimensional presentation for which he suggested the name of '*Lebenslagenrelief*' (relief of total life situations) the frequency distributions of these various types of social life situations within various categories of people could give a picture of the variety of total life situations within, say, a whole nation or within various segments of it. That would then also enable a better comparison between various groups or categories, both as to their general level and also as to the amount of variation within each group or category. Neurath did not fully work out this proposal, but bequeathed it as a task for future sociologists to develop the full theory and the appropriate scales and measuring devices—all of it, as said above, as an early anticipation of today's 'social indicators'.

He had come to this problem while, several years before the First World War, he developed a whole theory of war economy, based on his studies of economies in previous wars, especially the Napoleonic wars and the

American Civil War, but also some economic aspects of war in antiquity, that he had presented in his history of economics in antiquity (1909a). It had struck him, that war apparently did not necessarily always mean a general lowering of the standard of living. It seemed, that the economic planning of the provision of raw materials and their distribution in war time, together with the attendant regulation of production and consumption led to so much elimination of the usual waste and wasteful production, that the greater efficiencies achieved in this manner could *per saldo* lead to a better standard of living. All of this led him eventually to an interest in, and then to an advocacy of, planned economy in peace times as well, which eventually led to his involvement, as a 'social engineer' as he called himself, in the planning of the socialisation of the Bavarian economy in 1919.[12]

Let me close with one more biographical item. After the end of World War II, in late 1945, he was invited, because of his city planning experience in Vienna, as an advisor on the social side of the slum clearing project in the little town of Bilston in England. They called him 'Consultant on Human Happiness', because his task was, as it might be that of a any sociological consultant in such a project today, to see to it that in the zeal for mechanical efficiency in the reconstruction of that part of the town the human needs of the people who would live in it would not be overlooked. One of his many suggestions was that, contrary to earlier intent, older people not be segregated in a separate part of town which might have led to greater efficiency in health care etc., but that they should be enabled to live as long as possible within the general community. All that was necessary for this was to reserve for them a sufficient number of small ground floor appartments with easy access and a few extra conveniences and a buzzer with which to call for help in an emergency. It was the last thing that this 'social technician' could do for others, and it was quite in line with a word that I learned from him long ago in my youth: "The sum total of human happiness in this world is too small—it must be made larger." In the midst of writing, lecturing, heading his Institute and advising on human happiness he suddenly died, practically in mid-sentence, of a stroke, in December 1945, a few days after his 63rd birthday.

Queens College, City University of New York and Universität Wien

NOTES

* Lecture given to the Boston Colloquium for the Philosophy of Science, Boston University, March 1, 1988, and at Columbia University, April 6, 1988. Not previously published.

1 Neurath 1931c (1981, p. 424; this passage is not translated in 1973); insertions P.N..

2 Ibid. (1981, p. 458/9; this passage is not translated in 1973); insertions P.N..

3 Ibid. (1981, p. 459; this passage is not translated in 1973).

4 Ibid. (1981, p. 460/1; this passage is not translated in 1973); insertions P.N..

5 Ibid. (1981, p. 461; only the first part of this passage is translated in 1973, p. 356/7); insertion P.N..

6 Ibid., p.361, translation amended.

7 Hegel as quoted ibid., p. 336/7.

8 Malthus as quoted in Neurath 1931c, p. 343/4.

9 Neurath 1923a, Bauer 1923, Neurath 1923b; the quoted sentence is on p. 288.

10 Ibid., p. 384. Into this context fits also his very approving review of I. B. Kraus, S.J., *Scholastik, Puritanismus und Kapitalismus*, to whom he attests, under the fitting title of '*Marxismus eines Jesuiten*' ('The Marxism of a Jesuit') that he is also contradicting Max Weber, essentially on the same grounds as Neurath himself, and based on much concrete historical material; see Neurath 1931e.

11 See ibid., (1981, p. 480; this passage is not translated in 1973.)

12 This involvement began first under the legitimate social democratic government under which he was appointed as President of the Bavarian Central Economic Office (*Zentralwirtschaftsamt*), then, after the communists had taken power, also under the communists. After the latter had been defeated, he was sentenced by a court martial to 18 months imprisonment for aiding and abetting high treason for remaining in office under the communists. Incidentally, Max Weber served as character witness on his behalf, as did several other well-known persons, testifying to the fact that he had done nothing to enrich himself, that he was only obsessed with his idea of a planned socialised economy. Actually, he did not have to serve this penalty. The Austrian government, with Otto Bauer, one of the leading socialists, as Secretary of Foreign Affairs, intervened on his behalf and got him back to Vienna, which, however, cost him his position as *Privatdozent* in Heidelberg on account of abrogation by prolonged non-teaching, since he was not allowed to reenter Germany until many years later.

KARL H. MÜLLER

NEURATH'S THEORY OF PICTORIAL-STATISTICAL REPRESENTATION*

The present study seeks to achieve three goals:

(1) highlight the contours of the Vienna Method of Picture Statistics (*Wiener Methode der Bildstatistik*), invented primarily by Otto Neurath;

(2) explore the *linguistic* dimension of its symbolical arrangements of statistical facts; and

(3) evaluate the importance of this specifically Neurathian pursuit in the context of the 'linguistic turn' taken by the Vienna Circle, arriving at a perhaps surprising conclusion.

This declaration of intent already makes clear that many topics within the horizon of the Vienna Method of Picture Statistics will not be considered, such as the significance of this form of information transmission in the context of the highly innovative educational work of 'Red Vienna', the possibility of a reconstruction of this mostly forgotten picture-statistical know-how in the context of modern day computers, etc. The reconstruction of the more *philosophical* questions that is undertaken here should nevertheless be of interest, in particular since Neurath's varied, but highly repetitive writings on the theory of pictorial-statistical representation (Neurath 1991) have not yet received the attention they deserve.

1. The Principles of the Language of Picture Statistics

The history of Neurath's method of pictorial statistics—developed primarily in his Social and Economic Museum (*Gesellschafts- und Wirtschaftsmuseum*) with the designer and graphic artist Gerd Arntz—and its political-pragmatic point is dealt with elsewhere in this volume (Stadler 1989a [ch. 19])[1], so let me begin at once with the systematic analysis of this method. Since Neurath's picture statistics is to be analysed as a symbolic or pictorial language, I shall experiment by characterising it from a *semiotic* point of view.[2] And in order to acquaint readers unfamiliar with Neurath's typical ways of expression and procedure in this area which are relative inexplicit—they are more in the

T. E. Uebel (ed.), Rediscovering the Forgotten Vienna Circle, 223–251.

nature of *aperçus*—I shall reconstruct his theory by way of a montage: aside from providing the framework, I shall primarily let him speak for himself.

The rather inexplicit character of his theory of pictorial-statistical representation stands in considerable contrast to the richness of his mature method, an impression of which can be gleaned from illustrations 1 and 2 (Neurath 1939, pp. 53, 65). To judge from his writings, Neurath conceived of the language of picture statistics as constructed along the lines of three types of principles.

1.1. The Semiotics of ISOTYPE: The Sign

Pictorial statistics, as is apparent from illustrations 1 and 2, must have available a large number of symbols. Indeed, from the late 1920's up to the time of his death, his team accumulated a fund of symbols numbering between several hundred and several thousand elements, depending on how one counts them. These elements mainly served to represent the fields of economics, technology, and the social and life worlds, and offered relatively few symbols for the fields of politics and science (Müller 1987).

Neurath developed the following postulates for this sizable set of basic symbols on which the Vienna Method of Picture Statistics was built.

First, *the basic symbols have to speak for themselves*:

> For our picture language one general list of a limited number of signs is needed for international use ... The signs have to be clear in themselves, without the help of words as far as possible—that is, 'living signs.' They have to be different from one another, so that there will be no doubt of their right name, when they are seen again. They have to be so simple that they may be put in lines like letters. The signs have to be of such a form that the onlooker will not get tired of seeing lines of the same sign. The sign 'man' has not to give the idea of a special person with the name XY, but to be a representative of the animal 'man'. (Neurath 1936d, pp. 32/3)

Second, *the signs have to be independent of color*:

> One of the important rules for the designer is that it is necessary to be able to see the sense of an ISOTYPE picture from its camera copy in black and white. (Ibid., p. 46.)

Third, *it holds for the use of color in general that, apart from a general demand for consistency, there are no obligatory rules*:

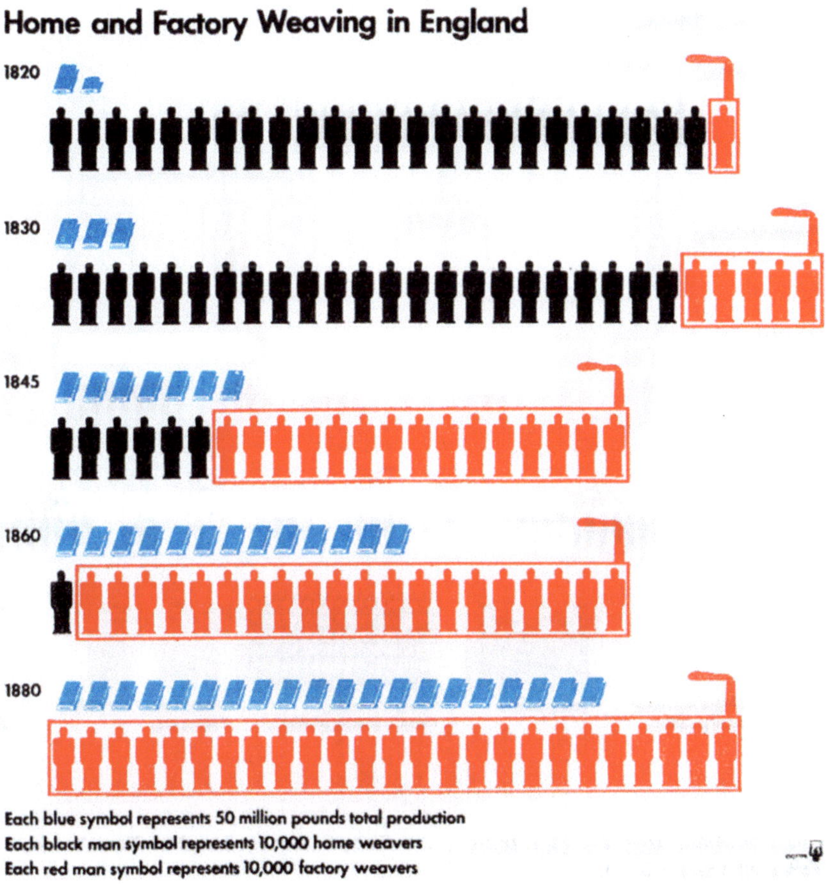

From *Modern Man in the Making*, A. Knopf Inc., London-New York, 1939, ©1967, p. 65

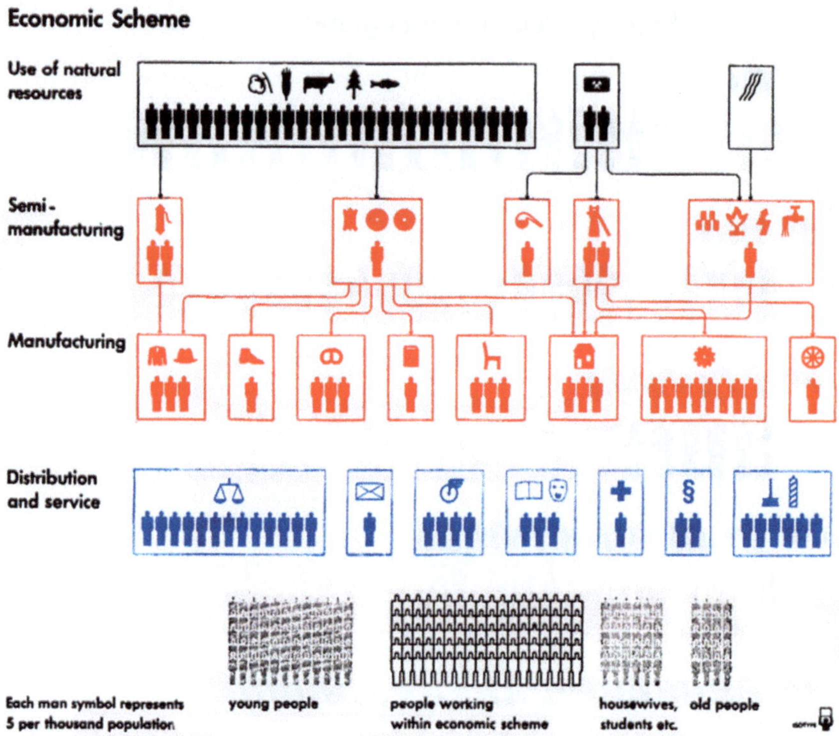

From *Modern Man in the Making*, A. Knopf Inc., London-New York, 1939 ©1967, p. 53

The rules for colours are not so fixed as the rules for forms. There are only seven colours for use in ISOTYPE pictures: white, blue, green, yellow, red, brown, black; and only some of them may have a further division ... or be mixed with others. (Ibid., p. 42.)

Fourth—and this is essential—*signs must be drawn without perspective*:

Perspective should not be used. Perspective involves making objects of the same size smaller or larger according to their distance from the viewer, which means that they cannot easily be quantified. (Twyman 1975, p. 10.)

Fifth, *signs must be combinable*; this is one of the more interesting rules, for it guarantees the existence of highly informative symbols:

There is a sign for 'shoe' ... and there is a sign for a [factory] ... By joining these two words to make the new word 'shoe-[factory]' we get the word for a [factory] in which shoes are produced. The picture of such a [factory] in the ISOTYPE picture language is the building with the great smoke-outlet, with a shoe [superimposed] on it. (Neurath 1936d, p. 50.)

Such combinations could be expanded considerably, e.g. workers or employees with symbols of their trades or professions, national flags, etc. Finally, *reduced to their essence, the signs must nevertheless leave a lasting impression on the mind*:

The order of signs seen by the eye has to be in relation to the best order for keeping in memory marks in the mind. ... The purpose of teaching-pictures is to have an effect on the mind. (Ibid., pp. 64, 62.)

1.2. The Semiotics of ISOTYPE: Combinations of Signs

Neurath gives three rules which concern the arrangement of basic signs. The *differentia specifica* seems to be Neurath's *rule of quantities*, which can be quoted en bloc since it is self-explanatory.

The first rule for quantitative pictures ("number fact pictures" (ibid., p. 73)):

A sign is representative of a certain amount of things; a greater number of signs is representative of a greater amount of things. Everyone is conscious of the value of this system in comparison with the old system of making the signs themselves greater or smaller. ... These old pictures are not able to make it clear that all greater amounts are made by the addition of smaller amounts, and they give no idea what will come out if we take a smaller amount from a greater one. These old pictures are not able to give a clear idea of the effect of a change in in the number of births and deaths. The ISOTYPE system of picture making is able to do that. (Ibid., pp. 73ff.; italics K.H.M..)

Pictorial statistics are to be read like a book, that is, from top left to bottom right:

Generally, the direction of the eye has to be from the top left side to the foot of the right side, as with books. (Ibid., p. 79.)

Finally, *spaces must be arranged in the manner of maps*:

If the comparison is not between the amounts at different times, but between the amounts in different countries, there will sometimes be a distribution of signs on a geographic map. Even if it is better to have no map, the memory of a map has to be respected. The positions of the countries have to be the same as on the map in general use. (Ibid., pp. 81/2.)

1.3. The Semiotics of ISOTYPE: Picture Stories

When arranged in sequence, single pictures constructed in this fashion should in turn follow principles similar to the ones so far enumerated, e.g. concerning the consistency of symbols and colors.

Complex facts are transformed into several single pictures which—combined against their picture-statistical background—must in their totality yield a coherent and relevant "story" (Ibid., p. 65). An over-arching theme must emerge, each of whose separate episodes is 'told' by one of the single diagrams.

All pictures together make one unit, and it is important for the reader not to be troubled in any way, if he is conscious of all the marks which teaching pictures have made on his memory. (Ibid., p. 64.)

Neurath aims for nothing less than isomorphisms, as far as the interaction between his language and its readers, or rather, between his pictures and their viewers' everyday perception, is concerned:

Reading a picture language is like making observations with the eye in everyday experience: what we may say about a language picture is very like what we may say about other things seen by the eye. For example: a man has two legs; the picture-sign has two legs ... Our experience is that the effect of pictures is frequently greater than the effect of words, specially at the first stage of getting new knowledge. ... pictures whose details are clear to everybody are free from the limits of language: they are international. Words make division, pictures make connection. (Ibid., pp. 20, 22, 18.)

This concludes the montage of the basic principles of Neurath's picture statistics; even an intensive search did not unearth significant additional ones.

1.4. Pictorial Statistics and Encyclopedia

That there exist only these few normative pronouncements by Neurath on a professional activity which kept him busy over decades does not imply that it lacked *substantive* relations to his philosophical or social scientific work, as though it were a hobby horse with only the *loosest* of connections to the rest of a person's endeavors. I would be remiss, then, if I failed to point out, at least briefly, the *direct* and *systematic* connections between Neurath's symbolic designs and the encyclopedic and sociological aspects of his research program. Pictorial statistics, after all, constitutes the *true projection* of elements of his *encyclopedia* and his widely neglected *sociology* into the realm of *non-discursive representation.*

First of all, we can affirm this connection on the level of *language*. The *protocol sentences* of the universal language of the encyclopedia and the arrangements of symbols of the Vienna Method of Picture Statistics share one fundamental feature. The matters of fact that can be stated by protocol sentences, as well as the theoretical terms and statements they support via reductions or chains of reductions, mark out the *possible domain* within which a symbolic pictorial statistics *may* operate. Both the visualisation of more or less *simple* observable facts and the pictorial transformation of higher-level concepts or statements is possible *only insofar* as Neurath's encyclopedic program can be implemented.

I need only point to the specific *logical* relation between encyclopedia and symbolic design. It may be the case that for a not insignificant number of protocol statements no 'hieroglyphics' can be found. The opposite, however, is never the case: *whenever* a matter of fact can be symbolically represented, there *exists* (potentially) a corresponding encyclopedic statement, either at the level of spatio-temporal protocols or on a level that is reductively related.

The second point is brought home by the fact that both symbolism and physicalistic thing-language apply exclusively to extensional relations. If conceptions like 'physicalism', 'reduction', 'reduction chains,' etc., have the pronounced effect of *deflating* most contemporaneous problems such as those of the *Geisteswissenschaften*, and discarding them as pointless, then the same must also hold for pictorial statistics. Neurath repeatedly stresses this *therapeutic* impetus, indeed the latter's point as a *therapeutic symbolism*:

What a triumph it was when one had freed oneself of the limits of picture language, what a triumph when language became supple and allowed itself to be applied to all demands of scientific work after one had learnt the command of logical tools. But, of course, this liberation from pictures also lead astray, lead into realm of the meaningless. Nominalization gave rise to ever new problems. (Neurath 1933b, p. 6.)

The Vienna Method of Picture Statistics was thus conceived as a continuation of the 'elimination of metaphysics' (Carnap) with other and even sharper weapons, namely those of symbolism.

The third point is that the general pragmatic conception of Neurath's orchestration of science also finds its continuation in his picture statistics. In analogy to his encyclopedia—which admittedly possesses a (heterogenous) set of norms, but still remains strongly inconclusive since their application requires a *separate* 'decision' on each occasion—Neurath's picture statistics also exhibits this characteristic of non-coercive norms and directives for its use. Though Neurath did not *expressiv verbis* declare it a goal to develop a comprehensive semiotics for ISOTYPE, it would be possible in principle and would even have been desirable; nevertheless, *no* such semiotic system could have established the *redundancy* of 'decisions' concerning how many and which of the rules would have to be applied.

It is impossible to apply all conventions automatically at the same time. It is a matter of pedagogical discretion, of a special *decision* therefore, which conventions are to be taken as more important in a particular case. (Ibid., p. 7; italics K.H.M..)

We meet yet another familiar feature of Neurath's encyclopedism in the field of picture statistics, the *Neurath principle* (Haller 1982a [ch. 9]). Its holistic insight holds here as well: it is never the case that individual ISOTYPE rules are at issue in isolation or in sequence. Rather, what is at issue is always their system as a whole—and, interestingly enough, not only the present one, but also future ones that are likely to be used.

We have to view every display as a part of powerful system, of a new *Orbis pictus*; we must not only consider how the new display can be brought into harmony with the existing ones, but, beyond that, how it will fit with displays that will be created later on. (Neurath 1933b, pp. 20/1.)

This hypercomplex decision rule—it requires us to confront the problem that it is *not* possible to logically predict *future* knowledge—may conclude my survey of the rather *close* connections of encyclopedism, sociology and picture statistics in Neurath's research program.

2. *The Dimensions of Neurath's Picture Language*

The close 'family resemblances' of picture and word languages are not at all anathema for Neurath, as is shown by this selection from his *International Picture Language*:

> This book ... will give some facts about the relation of ISOTYPE to languages generally, about some language rules of ISOTYPE, and about its international value. ... In building up such a language great care has to be taken to see that as much as possible may be said with the help of the smallest number of words. That is to say, for our picture language one general list of a limited number of signs is needed for international use. ... In the Far East we see one language for writing, but a great number of languages for talking. We have made one international language... We have given it the name ISOTYPE. (Neurath 1936d, pp. 7, 32, 17.)

What *kind* of language is at issue Neurath never makes explicit, however. I shall now reconstruct the answer with respect to the three levels of linguistic analysis this question could involve.

2.1. *The Symbolic Limits of Neurath's Picture Language*

To begin with, a comparison that is virtually forced upon us yields a revealing evaluative point of view. Neurath himself attempted to approximate the goal of a true picture language by trying to fit ISOTYPE to Ogden's BASIC English.[3] The latter's basic stock of the vocabulary consists of 800 words for 'things' (nouns), 150 words for 'qualities' (adjectives) and 100 words for 'operations' (a mixed bag of elementary verbs, prepositions, pronouns, articles, adverbs, etc.). Neurath himself sought out this point of reference by supplementing the BASIC vocabulary with an ISOTYPE fund of symbols. Already their comparison shows that in the 1930's and early 1940's

(1) only a *small* number, to be precise a third, of BASIC *nouns* have a symbolic ISOTYPE representation; accordingly, *two thirds* of all nouns *cannot* be (properly) symbolised;
(2) there are numerous other BASIC terms and types of terms—in essence *all* 'operations' and 'qualities'—for which there are apparently no symbolisations at all.

The following arbitrary set of words, alphabetised only later, could at best be symbolised in a *most cumbersome* fashion (and then only through a fortunate 'twist of fate'):

{across, and, because, certain, clear, come, complete, decision, dependent, development, early, free, get, give, good, great, I, important, invention, knowledge, late, like, loud, must, only, organization, other, out, please, possible, quite, reaction, reason, same, say, stretch, such, tendency, together, view, where, with, who, yesterday, you}

This shortcoming of concrete realisations of the desideratum of a picture *language* already at the level of basic symbols seems remarkable. *Prima facie* Neurath appears to have thought of full symbolic languages more as something desirable than as something indeed practicable.

2.2. The Syntactic and Semantic Limits of Neurath's Picture Language

Not only the fund of symbols showed characteristic limitations already in Neurath's time; the syntactic and semantic relations also were only very weakly developed. Neurath and and his team made little or no effort to develop a simple grammar of picture arrangements wherein the basic relations between individual symbols—atomic pictures, connections, molecular pictograms, etc.—would find their specific characterisation and definition.

By way of creative approximation I can offer the following impressive disanalogy between the living word language and a Neurathian symbolic language (derived from the work of Stephen M. Kosslyn (1980, p. 31)). The example is the sentence

A ball is on a box.

A corresponding picture and an overview of the disanalogies between them is given in table 1. Its heuristic import lies in allowing us to *clearly* delimit the missing *connections* and *termini* of a picture language.

This juxtaposition of propositional and pictorial representation does not, however, point to a 'gay apocalypse' of any form of pictorial language, for the following reasons. Points 1 through 4 could after all be turned into points in favor of pictorial languages: why not go for languages which operate with surface isomorphisms, non-arbitary signs, necessary part-whole relations, etc.? If this standpoint is taken, points 1 through 3, which identify grammatical incapabilities of pictorial configurations (no distinct relation and arguments, no clear syntax), remain undecisive.

Propositional Representation (Description)	*Quasi-pictorial Representation* (Depiction)
On (Ball, Box)	
1. Relation	1. No distinct relation
2. Argument(s)	2. No distinct arguments
3. Syntax	3. No clear syntax
4. Truth value	4. Truth value only under a particular description
5. Abstract	5. Concrete
6. Does not occur in a spatial medium	6. Occurs in spatial medium
7. No abstract spatial isomorphism	7. Abstract spatial isomorphism
a) No necessary part/whole relations	a) Necessary part/whole relations
b) Size and orientation optional	b) Size and orientation necessary
c) Arbitrary marks	c) Non-arbitrary marks
8. No abstract surface-property isomorphism	8. Abstract surface-property isomorphism
a) No necessary part/whole relations	a) Necessary part/whole relations
b) Shape not necessary	b) Shape necessary
c) Arbitrary marks	c) Non-arbitrary marks

Table 1: The Disanalogy of Word and Picture Language

Nevertheless, concerning points 1 through 3 we must indeed note that Neurath did not make any statements that could be construed as answering their claims. Nor did he mention what would be admissable relations in ISOTYPE or what would be necessary and sufficient conditions of picture sentences. Least of all, finally, did he provide definitions that might serve as a rudimentary grammar of picture languages. We must conclude then that the most fascinating and innovative aspect of his Vienna Method—rule-governed inferences from diagrams and inductive as well as deductive reasoning with pictures—has been left *virtually undeveloped* by Neurath.

2.3. The Pragmatic Limits of the Neurathian Picture Language

How better to continue the investigation of Neurath's picture language than by exploring its pragmatic dimension, asking with reference to ordinary language analysis: *Which* 'language games', *if any*, could ISOTYPE be applied to? For answering this question, Neurath's following statement provides an obvious starting point:

> But the uses of a picture language are much more limited than those of normal languages. It has no qualities for the purpose of exchanging views, of giving signs of feelings, orders, etc. ... it is a help inside its narrow limits. (Neurath 1936d, p. 20.)

By his own testimony, Neurath's picture language can only be used in highly select special areas. We can name the following concrete restrictions in comparison to ordinary language.

What can be represented in the picture language are only language games of assertion and communication, but nothing from the rich spectrum of questions, orders, wishes, appeals, inclinations, promptings, etc. This alone shows that an adaptation of Austin's classic as *How To Do Things With Pictures* would end up a rather small volume dealing solely with *locutionary* activities.

However, even within this class of language games the limitations are still more severe. Only diachronic processes and synchronic structures can be reproduced in picture statistics, but not individual events. This condition makes for a very restricted view of everyday communication in picture-statistical terms, as the following example makes clear. A complete sentence concerning a temporally as well as spatially sufficiently delimited process, such as

A child played for two hours with two others in a sandbox.

presents a multitude of difficulties for representation by symbolic pictures. The resolution of these difficulties would be quite unsatisfactory, at best extremely *ad hoc*. For besides the lack of symbols for 'play' as noun or verb, there remains the problem of visualising an interaction with the structure {one child: two children}; there are also problems in adequately capturing the designation 'two hours' in pictures; in reflecting the temporal dimension of the sentences (its past tense); not to mention still other, less obvious matters, e.g. the necessary transformation of the indefinite article in 'a child,' etc.

Not without interest for a philosopher of language is a further limitation that springs from the relation between qualitative and quantitative concepts. According to his own testimony, picture languages can be applied only to quantitative relations, but not, or only inadequately, to qualitative relations. A basically simple sentence of the form

> The French imports increased *more heavily* between 1770 and 1780 than in the period from 1760 to 1770.

cannot be represented according to the semiotic criteria which Neurath offers. For qualitative statements of this type Neurath's quantitative rules would have to be expanded by a set of qualitative conventions guiding the symbolisation of asymmetric relations like 'bigger than', 'longer than', 'more than', 'warmer than', etc.

A last restriction follows from the nature of the symbolic construction itself whose ambience of abstraction and *Neue Sachlichkeit* (New Mattter-of-Factness) only allows for the visualisation of general processes and structures, but not of individual phenomena. Neurath's symbolic repertoire would hardly allow room for micro-logical processes. An example would be a typical diagram from Neurath's *Modern Man in the Making*, say the number of strike days in the USA between 1927 and 1936 (Neurath 1939, p. 99). Though it would be possible to represent this type of fact also for smaller regions (West Virginia) or cities (Chicago, Pittsburgh etc.), this would only change the ordinary language diagram headings, not the general type of symbolic arrangement, the ISOTYPE ensemble.

From these restrictions vis-a-vis the possibilities for using ordinary language it can be easily seen that Neurath's statistical picture language is a very weak language with a necessarily narrow range of application.

I should add that these restrictions do not imply that a picture language exhausts itself in assertions without much theoretical content and thus could not be theory-laden. The possibilities for iconic languages to represent theoretical and less observable facts are as rich or poor as the possibilities of the given scientific language to relate theoretical terms or relations to observable and testable processes and events—an undertaking whose success is to be evaluated independently.[4]

This brief and incomplete survey of possible pictorial-statistical language games raises the question: are there cognitive reasons for Neurath's insistence on calling ISOTYPE a genuine language *sui generis*, given the strong limitation of possible symbols, the very weak syntactic-semantic relations and the restricted forms of application? Indeed—and this is the apparent paradox—can there be a *theory* for Neurath's profound silence?

3. *Neurath's Picture Language and the Vienna Circles*

The strange status of Neurath's picture language as described above will probably become clearer in light of the attitudes taken toward it by the protagonists of the Vienna Circles.[5] If they dealt with it at all[6], they thought of ISOTYPE as an interesting popularisation attempt in the context of the workers movement, but placed it thematically at considerable distance from their own efforts. In contradistinction to his project for the *International Encyclopedia of Unified Science*—but similar to his socio-economic work—Neurath's work on ISOTYPE did not call forth cooperation or a symbiotic division of labour. But this only answers a small part of the question, for Carnap's volumnious analyses of 'Testability and Meaning' (1936/7) and his Introduction to Semantics (1942) had already existed and could have been employed for a concise picture language. Why did Neurath not do this?

3.1. *Neurath's Picture Language and its Distance from Carnap's Language Systems*

The answer is: nothing would have been further from his mind than to recourse to Carnap's proposals. And why? Neurath's radical pragmatics prevented him not only from agreeing to a concept of truth that was divorced from the concept of acceptance. It also kept him from accepting the research field of semantics, with its references and propositions, its segmentation of levels and languages, as a genuinely encyclopedic one.

The main reason for this rejection lies in a principle essential for Neurath, recently dubbed by Frank Hofmann-Grüneberg as the "principle of avoiding artificiality": "A theory of empirical truth must be applicable to the 'universal slang' and must not presuppose an artificial language system." (Hofmann-Grüneberg 1988, p. 152.) That the break between Carnap and

Tarski on the one hand and Neurath on the other ran along the lines of their attitudes towards semantics is shown by the following montage of quotes.

Rudolf Carnap's great vision is powered by two things. The first is the vision of achieving of a semantic theory of truth:

In recent years many philosophers and scientists interested in the logical analysis of science have become aware that we need, in addition to a purely formal analysis of language, an analysis of the signifying functions of language - in other words, a theory of meaning and interpretation... It will be seen that this theory, if sufficiently developed, contains not only a theory of designation, i.e. the relation between expressions and their meaning, but also a theory of truth. (Carnap 1942, p. v.)

The second is an appreciation of the high relevance of this new field of research, and therefore of the need for its logical clarification and precise specification:

Semantical concepts are often used, not only in science but also in everyday life. When, for example, a person says that a certain word is used by him in a different sense than by somebody else, or that a certain assertion is true or that it is false, that a particular statement is analytic, i.e. true for purely logical reasons, that another statement follows from the first, or is compatible with it or contradicts it, then in all these cases he applies semantical concepts. Thus some of the problems of semantics deal with familiar concepts and are by no means new. The task of a systematic construction of semantics is to find adequate, exact definitions for the customary semantical concepts and for new concepts related to them, and to supply a theory based on these definitions. (Ibid..)

It goes without saying that such an agenda would have been of highest relevance for the orderly construction of a picture language.

Yet Neurath is unable to see, first of all, its *usefulness*, the *encyclopedic* point of separating truth, semantics and ontological neutrality on the one hand and acceptance, epistemology and empiricism on the other. Instead he tries for years to convince Carnap to transform semantics into a decision-theoretical calculus for criteria of the acceptance of statements:

It would, for example be perfectly expedient to use the term 'true' for all statements that are 'valid' for us in the sense given above, that is, are either part of our encyclopedia or can be deduced from it. 'False' would then be the qualification of the statements that are in contradiction with the encyclopedia. In addition, there would be statements of which we cannot decide whether they are 'true' or 'false'... (Neurath 1936c , p. 161.)

Secondly, Neurath does not find the systematisations offered by Carnap desirable. For this I need only cite the presumably final verdict of Neurath, the theorist of 'decisions': "I do not accept semantics."[7] It must be mentioned, however, that Neurath was not simply following here a misplaced anti-metaphysical sentiment. ("Can I see propositions? No."[8]) In fact, there existed principled cognitive reasons to doubt that semantical analysis was so overwhelmingly fruitful after all.[9] For this I need not cite Neurath, but can turn to an analogy by Ferdinand de Saussure, the pioneer of early linguistics.

> Take a knight, for instance. By itself is it an element in the game? Certainly not, for by its material make-up—outside its square and the other conditions of the game—it means nothing to the player; it becomes a real, concrete element only when endowed with value and wedded to it. Suppose that a piece happens to be destroyed during a game. Can it be replaced by an equivalent? Certainly. Not another knight but even a figure shorn of any resemblance to a knight can be declared identical provided the same value is attributed to it. (Saussure 1915, p. 110)

Here Saussure articulates a general objection against all attempts to recur to extralinguistic facts instead of intralinguistic ones:

> For ... Saussure ... the fundamental error of nomenclaturism is rather like supposing that appealing to something outside of the game of chess is necessary in order to explain the significance and function of the chessmen. Not only is such appeal unnecessary, but it would betray a profound failure to grasp what chess is. Likewise, the appeal to what lies 'outside' language in order to explain the significance and function of linguistic elements betrays a profound failure to grasp what language is. (Harris 1988, p. 25.)

I hope this rough sketch makes clear why Neurath thought there was little his picture language could gain from Carnap's semantical systems. Due to their metaphysical trend, Carnap's systems seemed pointless or detrimental to the encyclopedic and picture-statistical *praxis*.

3.2. Neurath's Picture Language and its Proximity to Wittgenstein's Language Games

So far we have learned that one possible theoretical elaboration of ISOTYPE language, namely the Carnapian one, was not explored by Neurath. Accordingly, we might conclude that, besides Neurath's philosophical and sociological work, there existed yet another, related field of his activity that was also left relatively undeveloped theoretically. Such a bare conclusion, however, would only continue to leave unrecognised one of the most *hidden* intellectual affinities within the Vienna Circles, the *high degree* of 'family resemblances' which exists between central aspects of the research programs of Otto Neurath and—Ludwig Wittgenstein.

Once the possibility of this connection is raised, it may be thought most plausible to start the investigation with certain passages from the *Tractatus*, e.g.: "In order to understand the essence of the proposition, consider hieroglyphic writing, which pictures the facts it describes. And from it came the alphabet without the essence of the representation being lost." (Wittgenstein 1921, #4.016.) And via the picture theory of the sentence we'd end up somewhere in the ontology of the *Tractatus*.[10] Yet I need mention only one reason for not exploring such a parallel at all: the background assumptions of the early Wittgenstein and Neurath differ radically. A revealing documentation by Rose Rand of the Vienna Circle's response to certain passages of the *Tractatus* before, during and after its reading shows Neurath as the virtual antipode of Wittgenstein, one who is left remarkably unaffected by most of the latter's theses.[11] Indeed, the underlying ethical dimension of Wittgenstein's work, which Alan Janik and Stephen Toulmin made the central point of their *Tractatus* interpretation[12], seems to make for the most essential 'family *un*resemblance'. Wittgenstein places the field of the problems of life *beyond* that which can meaningfully be said.

> We feel that even if all possible scientific questions be answered, the problems of life have still not been touched at all. Of course, there is then no question left, and just this is the answer. (Wittgenstein 1921, #6.52.)

For Neurath, the discursive realm (what can be spoken about) remains in a strong sense *doubly inconclusive*: once in that decisions play the main role for the acceptance of meaningful sentences, and once in that the problems of life *can* be talked about *ad libitum*, but will in general *not* be resolved thereby.

> It was a fundamental error of Descartes that he believed that only in the practical field could he not dispense with provisional rules. Thinking, too, needs preliminary rules in more than respect. (Neurath 1913b, p. 3.)

Because of this difference in fundamental convictions the analogy between the *Tractatus* and ISOTYPE, though potentially fruitful, will not be pursued here.

Instead, I shall shall propose the following general hypothesis:

(1) There exists a near congruence between the foundations and basic heuristics of Otto Neurath's socio-economic research program and the social foundations and essential heuristics of the philosophical research tradition of the later Wittgenstein;

(2) valuable additional insights concerning Neurath's theory of pictorial-statistical representation can be gained by considering relevant remarks by Wittgenstein.

These two leading theses concerning Neurath and Wittgenstein may appear uncommon, perhaps even tending to the absurd, for given the protagonists' style, their interests, their political engagements and particularly the reception each has received, we might be led to expect no overlap between their characteristic ideas whatsoever. It is by no means my intention to cover up their differences, but these differences must not lead us, as is common, to dismiss the points of convergence and agreement between them, points I shall now lay out.

3.3. Agreement on Social Foundations and Heuristics

The main reason why the parallelism between Neurath and the later Wittgenstein might be thought to be the result of a mere 'glass bead game' in the history of ideas is that far too little is known about Neurath's rather original foundation of the social sciences. One concept stands at its center, *custom* or *habit*. Let me briefly delimit its meaning.

One of the possible unities in the social world is marked for Neurath by customs or habits, for "[a] system of entangled habits characterises our existence." (Neurath 1931c, p. 374.) Neurath only treats examples of customs: they can consist in not eating horsemeat, or preferring a quack to a doctor; customs also manifest themselves in dress rules and fashions, and they determine "the books one writes, the fairy tales one tells" (ibid.[13]), or the tendency "of building large temples for [a] god, or for the dead." (Ibid., p. 375.)

Since Neurath's concept of custom can be applied so very widely, we may draw the first parallel: the field of application for *customs* is the same as that of Wittgenstein's *language games*. Regarding language games Wittgenstein, too, says that they are "countless", that they are "not someting fixed, given once for all" (Wittgenstein 1953, §23.) and that their entirety forms *clusters*, *forms of life*. According to Neurath—and also Wittgenstein[14]—such habits, such language games to a large part represent forms of established learning, stabilised form of behaviour[15], which expand and change over the course of the natural history of human beings. This is the first point of convergence or agreement then: their use of *language games* and *customs* establishes the identity of the praxeological foundations of the social world in Wittgenstein and Neurath.

But the similarities between the two protagonists reach much further still—and deeper. The key concept for effecting their *rapprochement* is that of *morphological investigations*. This form of investigations can be found, basically since the time of Goethe[16], in two equally powerful but distinct ways of proceeding. In the first sense, morphology is to be understood as a static enrichment, in that a *certain* event, a *specific* structure or a *particular* process is placed in the context of *possible* events, structures and processes. In the second sense, morphology is to be understood as a dynamic description, in that changes in the *gestalt* of events, structures, processes or more complex systems are to be represented. The first procedure maps, if you like, the space of abstract possibilities, whereas the second maps actual concrete metamorphoses. What is to be shown is that both types of morphology are in fact employed as heuristic devices by both of our protagonists. I begin with morphology in the first sense.

It is well known that Wittgenstein is characteristically prone to transfer existing facts, accustomed ways of behaviour and rooted meanings into a context of interesting possibilities.

A main source of our failure to understand is that we do not command a clear view of the use of our words. - Our grammar is lacking in this sort of perspicuity. A perspicuous representation produces just that undestanding which consists in 'seeing connexions'. Hence the importance of finding and inventing intermediate cases. (Wittgenstein 1953, §122.)

This method of Wittgenstein's, first adopted in the 1930's, can be summarised in a nutshell. The spectrum of philosophical concepts is confronted with a great number of possible applications, such that concepts are torn from their exemplary contexts and placed in new and surprising ones; this shock is to make the bruises disappear which reason got from running against the limits of language. "Nothing is more important for teaching us to understand the concepts we have than constructing fictitious ones." (Wittgenstein 1982, #19.)

Otto Neurath, too, again and again stresses the tremendous usefulness of the first morphological heuristic, though for reasons of an improved systematisation of the domain of the empirical sciences. One example is the aim of an improved order in the history of science:

In order to obtain a scientifically satisfactory systematisation, one must first, willy-nilly, try to give a complete survey of combinations of the elementary notions; by the application of certain principles a selection from the logically possible combinations could already be created. After surveying this totality, one could investigate which of these combinations are realised in 'nature'. As not all chemical compounds are represented in the minerals, so in the world of real combinations of ideas as well, not all possible theories, which can be derived from certainelementary notions, are represented. ... If we have certain elementary components of a view as given, we can provide a survey of the possible combinations in the simplest way by admitting each characteristic as well as its negation as an element of a combination, for example, a, b, c, and their negations a_1, b_1, c_1. ... For example, from three elementary notions and their negations we get eight combinations in all. (Neurath 1916a, p. 15.)

Another example is given by procedures in economics:

Comparative economics will be able to use very abstract models and schemata. ... The facts so established can often be summarised in a table; this offers the advantage that simultaneous changes can be surveyed without one change being seen as dependent upon another, and that missing entries show missing facts... The creation of such models is unlimited. One can create models whose components—humans or things—have properties which do not happen in reality. But it is also possible that these components have real properties whose combinations and changes are not yet known in reality. (Neurath 1917a, p. 110.)

The second morphological heuristic also finds application by both Neurath and Wittgenstein, though with different accents. Wittgenstein develops it in order to describe in great detail existing forms, *gestalts*, their changes or metamorphoses, and finally their new appearances. To give but one of many typical examples:

"What is 'fear'?" - "Well, the manifestations and occasions of fear are as follows:—" - "What does 'to be afraid' mean?" - "The expression 'to be afraid' is used in this way:—" - "Is 'I am afraid...' therefore a description of my state?" - It can be used in such a connection and with such an intention. But if, for example, I simply want to tell someone about my apprehension, then it is not that kind of description. (Wittgenstein 1982, #20.)

Neurath, by contrast, developed an original conceptual apparatus for dealing with such changes under the headings of 'Extrapolation' and 'Coherence'.[17] For Neurath, habits (or language games) constitute the units for 'extrapolations', they are stable routines which nevertheless expand in all possible directions. What is to be understood by this in detail may be presented by the following example, which is probably still topical today.

If a child is put off a white mouse by means of a sudden noise which sounds everytime the child touches this mouse, then it will withdraw once this mouse appears. But not only if this familiar mouse appears, but when any mouse appears. The fright could be caused by a rabbit or a fur; even threads of wool could already cause disquiet. Non-living angular pieces of wood have a rather pleasing effect; supposedly, children particularly like to hold them in their hands after such fright experiments (compensation). What has happened? The child "extrapolated" its habit, extended it beyond its original domain. It transferred the fear of one white mouse to other mice, to rabbits, to furs. (Neurath 1931c.[18])

'Extrapolation' designates what is otherwise known as 'analogy formation', 'inductive reasoning', etc., namely, the expansion of established routines, of practical problem solutions into new, hitherto untouched domains (cf. Holland *et al* 1989 or Holland 1989). 'Extrapolation' therefore comprehends unlimited cases, reaching through gradually more complex behaviour and actions, still *without* too much admixture of conscious theory, all the way to thoroughly cognitive examples of the *development of theories* themselves, the expansion of a previous set of 'intended applications' by the addition of *new* ones. According to Neurath, a sociology dealing in 'extrapolations' would be well advised to minimise its search for motives, but maximise the relevance of *non-intended* appearances.

The appearance of new customs need not be derived from a particular sensitivity of humans, even less from a growing inventiveness. We can derive everything much easier if we think of extrapolations as having come into existence unnoticed, until they are favoured by selection, thereby favouring related habits. (Neurath 1931c.[19])

How to *evaluate* these extrapolations? Neurath assumes a genuinely *evolutionary* perspective: such learning processes are determined by the process of *selection*:

Clearly, certain customs give superiority to groups in which they prevail; the custom of loathing dead animals and dead men often helps to avoid infection. To differentiate loathing and fear makes modern man superior to primitive peoples. We vary our carefulness towards machines and adjust to success. Certain habits which have neither positive nor negative effects can persist for a long time... (Ibid., p. 377.[20])

However, according to Neurath this process of selection does not automatically favour comparatively more complex combinations of habits; even thresholds of complexity themselves and robustness as such, i.e. the possibility of adapting to change in environmental conditions, are part of the development of extrapolations (ibid., p. 378).

If extrapolations represent forms of social learning, and thus dynamic processes, then 'coherences' represent synchronic connections between such customs in given spatio-temporal domains. To introduce 'coherences' by an example:

Those village communities which first overcame blood feuds were better suited to cooperation in dam building than those with blood feuds. Overcoming blood feuds and large scale water economy are technically coherent. (Ibid., p. 382.)

Coherences then answer to the question of the *harmony* of customs, be they the ones of an individual, a group, a nation, or the whole world.

The analysis of coherences has a predominantly heuristic point, despite—in fact, particularly because of—its synchronic character. It seeks to structure socio-economic change in an original and stimulating way:

> Here are the basic findings which can help us to initiate changes. ... The question which statistical magnitudes are dependent on each other and in which way, permeates the whole of sociology. Here it becomes obvious that separation into comparative study of history, comparative politics, political economy, and anthropology, etc., is impossible. (Ibid., pp. 380, 387.)

Accordingly, the central area of research into coherences includes the splitting of the spectrum of customs into clusters of coherent, indifferent and contradictory forms; questions of the fit of new customs with the old established ensemble; problems of the fit between newly initiated customs and the forms of speech and behaviour that can be expected to follow; and, finally, analyses of the simultaneity or dissimultaneity of customs (a point of relevance for the self-understanding of social groups that can hardly be exaggarated). "The habits of sex and family life cannot always be related to the contemporary mode of production, but often only the previous one." (Ibid., p. 385.)

We can see another piece of evidence for the view that the parallel between the heuristics of Wittgenstein and Neurath is not artificial, but genuine, in the fact that both of them distance themselves from the great morphologist of their day, Oswald Spengler. Interestingly enough, both employ the same type of argument: they oppose the supposed element of necessity and inevitability which pervades his entire work.

> Spengler could be better understood if he said: I am comparing different cultural epochs with the lives of families; within a family there is a family resemblance, though you will also find a resemblance between members of different families; family resemblance differs from the other sort of resemblance in such and such ways, etc. What I mean is: we have to be told the object of comparison, the object from which this way of viewing things is derived , otherwise the discussion will constantly be affected by distortions. Because willy-nilly we shall ascribe the properties of the prototype to the object we are viewing in its light; and we claim "it must always be..." This is because we want to give the prototype's characteristics a purchase on our way of representing things. ... Similarly the question always to ask when exaggerated, dogmatic assertions are made is: What is actually true in this? Or again: In what case is that actually true? (Wittgenstein 1977, p. 14e.)

Neurath, who repeatedly stresses the *fruitfulness* of the two kinds of morphological approaches and considers *legitimate* their extrapolation into sociology, takes Spengler to task in a way that bears a family resemblance to Wittgenstein's critique: he accuses him of "pseudo-rationalism".

It is not the individual results, the wrong facts, the wrong proofs, that make Spengler's book so dangerous, but above all his method of conducting proofs, and his reflections on proof as such. Against this one must defend oneself. Anyone who wants to shape a happier future with hope and striving should know that none of Spengler's 'proofs' is enough to prevent him; and whoever wishes to come to terms with the idea of 'decline' should know that he does so on the basis of a resolution, not a proof! ... The book satisfies a strong contemporary yearning for a complete world view. ... Many feel liberated by Spengler. We want to liberate from Spengler, from that kind of mind which, tempting and violating, destroys clarity of judgement and precision of reasoning, and distorts feelings and observations. (Neurath 1921a, pp. 161ff.)

I have claimed that significant agreement concerning social foundations and analytic heuristics holds between Otto Neurath and Wittgenstein. Whoever *still* views this absence of difference between them as accidental should take note of the following.

Note first that in the late 1920's there existed within the Vienna Circles a certain problem whose solution Carnap attempted at a certain distance, Wittgenstein and Neurath head-on (though *strictly* independent of each other): the private language problem. To show the *similarity* between the position of Neurath and the later, though much more elaborated position of Wittgenstein, I can cite a recent, very detailed reconstruction of Neurath's standpoint.

Neurath held that any language use was socially mediated. That one person understood another, of course, presupposed that the extension of the other's expressions was a public one. But Neurath's point did not stop there. A speaker's sense of the meaning of the expressions of his/her language was shaped in and by the process of communication: it did not precede it. For Neurath, the public accessibility of the extension (of at least some if not most) of the expressions of a language was also a precondition of the constancy of the use of a given language by an individual. How could that constancy be assured? By being corrected by other speakers! What this meant was that a speaker's sense of the meaning of his/her words depended on the public accessibility of their extension. (Uebel 1989, p. 409.)

Secondly, I can serve notice of a great number of further agreements between Wittgenstein and Neurath, besides the similarities already mentioned. Table 2 gives an initial, still incomplete and expandable overview of these agreements.

Later Wittgenstein	*Neurath*
Language games	Customs, habits
Primacy of ways of action	Primacy of decisions
Knowledge justifiable as satisficing but revisable, sometimes in need of revison	Knowledge justifiable as satisficing but revisable, sometimes in need of revision
Truth also viewed as a phenomenon of coherence and consensus	Truth also viewed as a phenomenon of coherence and consensus
Against a trilemma of justification (against infinite regress)	Against a trilemma of justification (against infinite regress)
Against private languages	Against private languages
Linguistic behaviorism (problem of self and other minds)	Linguistic behaviorism (problem of self and other minds)
Radical constructivism	Radical constructivism
Radical holism	Radical holism
Empirical magic	Empirical magic

Table 2: Agreements between the Later Wittgenstein and Neurath

4. *The Developmental Potential of Neurath's Pictorial-Statistical Representations*

The results reached above provide the surprising framework—for it is shared by the later Wittgenstein—for the proper Neurathian theory of pictorial-statistical representation: no longer does Neurath's silence on many weighty matters and aspects of his picture language signal disregard, rather it opens quite different perspectives.

A *radical* pragmatist like the holist Neurath had no need for explicit rules—which in any case were unable to precede or rule over the relevant 'customs'—as long as there existed a consensus in his team about how to apply the implicit rules they did use. Neurath's *pragmatic* refusal to develop an elaborated network of rules—due to the multiplicity and ambiguity of 'language games' and 'contexts'—goes as follows.

> The 'words' and 'statements' of the picture language—signs and their order—are not the same in pictures in a book as they are in pictures on walls. A special organisation is necessary to make clear certain relations between them... Such adjustments to special purposes have to be made in any language. (Neurath 1936d, p. 15.)

Or in another variation:

> To be good at a language more is needed than a knowledge of the words and of the rules: the sense of a word and of a group of words is different under different conditions. It is an art to get the full effect from the words. In the same way it is an art to get the full effect from pictures. (Ibid..)

Neurath's *holistic* resistence against all too strong systematisations is occasionally expressed as follows: "Every ISOTYPE picture is like part of a great picture book or encyclopedia, because all of them must go together." (Ibid..) This is a goal that can even be attempted only under extremely weak conditions of consistency.

Still further, a third reason could be given against an elaborated network of rules, namely, the one-many relation between picture language and ordinary language:

> The ISOTYPE picture language is not a sign-for-sign parallel of a word language. It is a language which may be put into words in very different ways. ... The parallel in a normal language of a complete 'language picture' is a complex group of statement; and an account in words of what is in a group of language pictures would make a book. (Ibid., pp. 18f.)

Still, so as not to conclude my evaluation of Neurath's theory of pictorial-statistical representation with 'satisficing' relations to human practice, I shall conclude with a more theoretical point. Neurath's ISOTYPE allows the exploration of two large and separate areas of investigation. These cannot serve as parts of the—in Neurath's slim rulebook non-existent—theory of symbolic representation. Yet they can serve to indicate in what direction this particular form of an iconic communication 'custom', these picture-centred 'language games', *could have* developed.

4.1. A Multiplicity of New Codes

The first of these areas of investigation is marked by the *conventional* character of the relation between symbols and their meanings. On this point

Neurath, Wittgenstein, Nelson Goodman and Umberto Eco all agree. In the words of Neurath:

Every language is complex—even the simplest. Its store of words and its rules make it possible to put a statement into other words or one group of statements in the place of the other. ... All this is true not only of our normal language, and the special instruments of business and science, but, with special changes, of picture languages... (Neurath 1936d, pp. 14/5.)

Especially when viewed against the background of manifold customs and minimal codes, the discrepancy between BASIC word types and ISOTYPE elements need not necessarily indicate the deficiency of the latter. Why not go for genuine languages which typically do *not* possess a basic verb phrase-noun phrase structure, but are built up from a single type of symbol and gain their different types of expressive means by the *combinatorial system of the composition of elements*? The first great developmental potential of ISOTYPE designs lies exactly in this: the exploration of many different kinds of visually guided applications. Besides their *didactic*, educational uses, they invite employment for danger signals, traffic direction, and international, simple, but intuitively intelligible information systems; all of these are applications that were only tentatively explored in their day[21] and did not gain comprehensive usage.

4.2. Analog Surplus Value

In addition, an ISOTYPE picture language possesses a general advantage over the established BASIC language or even the more complex normal languages: potential analog surplus value. I turn to Fred I. Dretske for the definition of this concept and its opposite, 'digital' (Dretske 1981, p. 137).

Information with the content F is encoded *digitally* "if and only if the signal carries no additional information about *s*, no information that is not already nested in *s*'s being *F*." An *analog* transfer of information presupposes the existence of informational *surplus value*: "When a signal carries the information that s is F in analog form, the signal always carries more specific, more determinate, information about s than that it is F." Take the example of the *digitally coded* information, the sentence "The cup has coffee in it":

No more specific information is supplied about the cup (or the coffee) than that there is some coffe in the cup. You are not told how much coffee there is in the cup, how large the cup is, how dark the coffee is, what the shape and the orientation of the cup are, and so on.

Now take the analog form of information, a picture:

> the picture tells you that there is some coffee in the cup by telling you, roughly, how much coffee is in the cup, the shape, the size, and colour of the cup, and so on.

The occasional *superiority* of analog representation becomes clear: an ISOTYPE language carries a *potential manifold of information* that also concerns figure, colour, proportions, relations, etc. Such analog surplus value can be realised on three levels, the level of symbols, the level of symbol arrangements and the level of background-foreground partitioning.

The main avenue to gaining analog surplus value is the *symbol* itself. This becomes evident if we search for symbolisations which "giv[e] an idea of the different times and of the facts in connection with the different living conditions" (Neurath 1936d, p. 54) and which, in addition, possess a particular attribute, namely, a great distance from the traditional form of life (either diachronically or synchronically). Already, e.g., regarding the type of cup pictured, the simple 'sentence' "The cup has coffee in it" may show a great variety of symbolisations providing interesting information.

The second way of generating analog surplus value consists in the particular *arrangement of symbols* with respect to each other. Especially with the highly abstract ISOTYPE language nothing would prevent the composition of visual elements such that new idiosyncratic figures come into being, as it were, symbols of symbols, or words from symbols. The third and last way of generating analog surplus value consists in the *partitioning of foreground and background*, e.g., designing the background areas by means of figures. Admittedly, this is difficult since the information carrying foreground figures would have to be so placed that they are grouped around a background figure. Yet this is by no means impossible.

To summarise. A multiplicity of ISOTYPE languages could exploit the transfer of analog information, something which traditional word languages are in general able to do only rarely. What would be the advantage of this? It would consist in achieving an increasingly dense information transfer within a *constant* area of display by means of the form of symbols, their arrangement vis-a-vis each other and their placement in the fore- or background. Possessing such a perspective on the development of information transmission would and should propel the respective 'language community' into highly complex achievements. Concerning this communicative 'drift', we may cite another point already noted by Neurath:

The writing and talking language is only of 'one expansion'—the sounds come one after the other in time, the word-signs generally come one after the other on paper, as for example the telegram signs on a long, narrow band of paper. The same is true in books—one word over another in the line under it has no effect on the sense. ... Some of the picture languages in existence are only of one expansion, they are made up of long lines of small signs, put together in rolls. Another sort of picture language, such as the ISOTYPE system, makes use of the connection of parts not in one direction only, but in two, and the effect is a language picture. (Ibid., pp. 60f.)

The freedom of dimensionality gained by the analog codification of iconic languages has the notable evolutionary advantage of initiating a variety of processes of compressing information without loss of retrievability.[22]

Institut für Höhere Studien, Wien

NOTES

* Specially written for this volume; translated by T. E. Uebel.

1 See also Neurath 1942a, Arntz 1982, 1988, Stadler 1989b and Dvorak 1990. It may be remarked, however, that the principles assembled below already help to make plausible the practical results Neurath's method had already achieved in Vienna: "The experience in people's education show us that the Vienna Method of Picture Statistics can transmit knowledge of and insight into complexities without engendering the feelings of inferiority which a learner is subject to in the confrontation of masses of words which he does not understand. For this reason pictorial statistics is so desirable for less intelligent or less educationally prepared children. They are encouraged to feel that they can handle social structures like they handled the structures of toy building blocks. Hopefully, we shall soon have more material on this..." (Neurath 1931a, p. 578.) As far as its use in Vienna was concerned, however, this hope of Neurath's was to be largely disappointed by the adverse political developments.

2 A brief remark on the terminology used: concepts like 'picture language', 'symbolic language', 'pictorial language' and 'iconic language' will be used as equivalents, and the existence of their referents in Neurath's ISOTYPE system is presupposed.

3 On the relation between BASIC English and ISOTYPE see Neurath 1937b.

4 Compare by way of overview Balzer et al 1984.

5 This pluralisation of 'Vienna Circle' has primarily substantive reasons: it has proven advisable for the treatment of different topics to introduce either a division of 'the' Vienna Circle into temporal phases (Müller 1985) or a division into interest-related reference group with at least two or more such 'Circles'. Compare also Uebel 1989.

6 No exchanges over theoretical questions concerning ISOTYPE took place between Neurath and the other members of the Vienna Circles, even informally; apart from sympathetic acceptance, this work of Neurath's found no echo within his own Vienna Circle reference group. And within the international encyclopedia movement there existed only one person for whom ISOTYPE possesed direct relevance, namely, Charles W. Morris.

7 Letter Neurath to Carnap, April 1, 1944; Vienna Circle Archive, Amsterdam.
8 Letter Neurath to Carnap, September 25, 1943; Vienna Circle Archive, Amsterdam.
9 Compare e.g. Carnap's often repeated assertion "Pragmatics is the basis for all of linguistics." (Carnap 1942, p. 13.). On a philosophical parallel between the Carnap of the Aufbau and Neurath's ISOTYPE see Rungaldier 1979.
10 For the framework for such an interpretation see e.g. Schulte 1989, p. 74-93.
11 This documentation can be found in the Carnap Collection in the Archives for Scientific Philosophy at the University of Pittburgh.
12 Compare, e.g., Wittgenstein's own remarks in a letter to Ludwig von Ficker (in Wittgenstein 1969a, pp. 94/5 and Janik and Toulmin 1973, p. 192) and Janik and Toulmin's general exegesis.
13 In 1981, p. 493; this passage is not translated in 1973, p. 383.
14 How else would it be possible to fruitfully understand as sentence a sentence of the type "We are satisfied that the earth is round" (Wittgenstein 1969b, §299) if not in the context of self-stabilising learning processes?
15 On the importance of this specific point for more contemporary system-theoretical conceptions of action see Müller 1991b.
16 See his natural-scientific writings, e.g. Goethe 1977, pp. 133f. or 415.
17 Compare on the following for greater systematicity Müller 1989, pp. 115-196.
18 In Neurath 1981, p. 480/1; this passage is not translated, only summarised in Neurath 1973, p. 373.
19 In Neurath 1981, p. 486; this passage is not translated in Neurath 1973, p. 377.
20 The third sentence is not translated in Neurath 1973; see Neurath 1981, p. 487.
21 See the diagrams nos. 1, 2, 37, 8, 10, 11, 13 and 19 in Neurath 1936d.
22 For greater systematicity on the above see Müller 1991a.

PART III

NEURATH AS POLITICIAN OF KNOWLEDGE: THE PARTISANSHIP OF ENLIGHTENMENT

FRIEDRICH STADLER

OTTO NEURATH: ENCYCLOPEDIST, ADULT EDUCATIONALIST AND SCHOOL REFORMER*

1. Biographical Remarks

Born 1882 in Vienna as the son of the Viennese social reformer and political economist Wilhelm Neurath and Gertrud Kämpfert, Otto Neurath gained his university qualification at the *Wiener Staatsgymnasium* in 1902 and started his many-sided studies with mathematics and physics, but finally settled on political economy, history and philosophy. During this time he became acquainted with Philipp Frank, Hans Hahn and his sister Olga Hahn, and his later first wife Anna Schapire who died in 1911 after the birth of their son Paul. The multi-lingual and socially critical feminist Anna Schapire conducted international studies about philosophy, literature and political economy and earned her doctorate in 1906 with a thesis *Arbeitsschutz und die politischen Parteien in Deutschland* ('The Protection of Labour and the Political Parties in Germany').[1] It is more than likely that she stimulated Neurath's socio-economic and political engagement. She was influential also in his move to Berlin where he continued his studies in political economy under Eduard Meyer and Gustav Schmoller, and already in 1904 published his article '*Geldzins im Altertum*' ('Monetary Interest in Antiquity'). As a sideline he tried to prove himself as a literary historian with an edition of, and introduction to, Marlow's *Faust* (1906c), until he gained his doctorate *summa cum laude* under Eduard Meyer with his published dissertation *Zur Anschauung der Antike über Handel, Gewerbe und Landwirtschaft* ('On the Conception of Antiquity concerning Commerce, Trade and Agriculture', 1906a). Three years later he published his *Antike Wirtschaftsgeschichte* ('Economic History of Antiquity', 1909a), which he had submitted as an alternative dissertation. Returning to Austria he absolved his military service 1906-7, married Anna Schapire in 1907 and taught economics and history until the beginning of the First World War as an (assistant) teacher at the *Neue Wiener Handels Akademie*. In 1910 Neurath published a *Lehrbuch der Volkswirtschaftslehre* ('Textbook of Economics', 1910c), which was officially approved for use in business schools, and also edited, with his wife Anna, a two-part *Lesebuch der Volkswirtschaftslehre* ('Reader in Economic Theory',

T. E. Uebel (ed.), Rediscovering the Forgotten Vienna Circle, 255–264.

1910). In 1912, after Anna's death, Neurath married the mathematician Olga Hahn, who had lost her sight at the age of 22 in 1904. With her, and alone, he published a number of mathematical works which, together with his contributions to ethics and social science, determined his later position in the Vienna Circle: holism in the theory of science, anti-metaphysical unified science, logical empiricism and epistemological relativism.

These comprehensive studies allowed him to explore and shape the problematic domain 'science-philosophy-metaphysics' and to prepare the foundations for the physicalistic unified science and empiricist encyclopedia he later was to promote. All the while, his interest in visual methods of enlightenment had been stimulated in the scholarly atmosphere of his Jewish-liberal parental home, an interest which throughout his life made him a political man in a practical sense, as the inventor of the *Wiener Methode der Bildstatistik* (Vienna Method of Picture Statistics) and ISOTYPE (International System of Typographic Picture Education).

Until the end of World War I, however, Neurath mainly published research works in political economy. He became a member of the *Verein für Sozialpolitik* and, assisted by a travel stipend from the Carnegie Foundation for International Peace, undertook research on the socio-economic conditions in the strife-torn Balkans. This allowed him for the first time to develop his views on war economy as a model of socialist planned economy on the basis of an economy-in-kind (*Naturalwirtschaft*). Yet he still found time to lecture, like Hans Hahn and Philipp Frank, at the Philosophical Society of the University of Vienna, and to work in the tradition of Ernst Mach and Popper-Lynkeus on the systematic connection between political economy, value theory, decision theory and a theory of life conditions ('basic needs'). (The aim of a theory of social science was taken up again in his *Empirical Sociology* of 1931.)

The theory and practice of war economy were also necessary preconditions for the total socialisation schemes which Neurath developed towards the end of the war and attempted to realise in Bavaria in the revolutionary post-war period. There he was to fail, together with the entire movement, as a centralist economic planner in the role of the President of the Central Economic Commission of the Munich *Räterepublik* (Soviet Republic). Earlier during the war he had accepted the leadership of the department for war economy in the Austro-Hungarian War Ministry, and in 1916 he was appointed director of the Leipzig Museum of War Economy. In 1917 he gained *Habilitation* in political economy in Heidelberg, but lost his lectureship due to the political prison sentence he received in Munich in 1919.

These biographical marginalia deserve to be mentioned at once for a number of reasons. Of all the theorists of the Vienna Circle, it was Neurath who, next to Zilsel, was most concerned to integrate the social sciences within the empiricist concept of the 'scientific world-conception'; it was he who undertook the most manifold tasks for this scientific circle as its organiser, propagandist and programmatic thinker. Moreover, Neurath also sought to take account of the political preconditions and implications of the project of social reform on the one hand, and of the *International Encyclopedia of Unified Science* on the other: on the one hand as a functionary in various movements, in the party and adult education of 'Red Vienna,' and as the director of his museum until 1934, then in exile as the founder of the International Foundation for Visual Education, and, on the other, as author of the program of the Vienna Circle, as activist in the *Verein Ernst Mach*, and as promoter of the international Unity-of-Science movement from 1934 until his death in 1945.

During his entire life then, Otto Neurath fought in theory and practice for the democratisation and popularisation of knowledge, and for a widened concept of culture. His was the concept of a materialistic culture of everyday, which was empirically and sociologically determinable in terms of 'standards of living' (*Lebenslagen*). His efforts towards socio-economic enlightenment for all classes in Austria and abroad corresponded to these intentions.

After World War I and Munich a convinced socialist, Neurath propounded in theory an idiosyncratic variant of Austro-marxism ('Epicurean Marxism', or 'Epicurean Materialism'); in practice, he worked, alongside his engagement in adult education, in the Viennese city politics of the 1920's for the Communal Economy Movement (*Gemeinwirtschaftsbewegung*), the Worker Ownership Movement (*Genossenschaftsbewegung*), and the Housing Movement (*Siedlungsbewegung*). His most outstanding and effective achievement, however, was the foundation and directorship of the GWM (*Gesellschafts- und Wirtschaftsmuseum*, Social and Economic Museum) in Vienna. This unique institution of adult education carried out valuable work in Austria and abroad from 1924/5 until 1934. Here the Vienna Method of Picture Statistics was invented and put to use.

The impressive number of Neurath's articles, contributions and independent publications comprehends the following, partially overlapping fields of inquiry (which stand in a methodological and substantive inner connection), in chronological order:

- economics, political economy, statistics, history of economics, war economics, planned economy;
- settlement-, worker ownership-, communal economy-movements, city planning;
- Marxism, socialism, social democracy (Austro-marxism);
- scientific philosophy and world-outlook, unified science in the Vienna Circle;
- GWM, pictorial statistics and ISOTYPE, visual pedagogy and museology.

Neurath's theoretical and philosophical work in the Vienna Circle ran in parallel to all these practical enterprises. With the help of the principles of empiricism, and of modern logic and mathematics, Neurath strove to turn what so far had been speculative-idealistic philosophy into scientific discipline, and to turn science 'empirically rational' by giving it a non-metaphysical orientation that was informed by the analysis of language. Neurath attempted to spread and popularise these ideas also in the *Verein Ernst Mach* which existed from 1928 to 1934. In the person of Neurath we can see how scientific thinking is connected with activity for social reform in a comprehensive attempt at enlightenment on a materialist basis. The humanisation of life and the democratisation of knowledge emerge as the constitutive aims in his various fields of activity, from political economy and the scientific world-conception to his cosmopolitan visual pedagogy. The remarkable continuity of the principles and methods in his life and work shows Neurath as a concrete Utopian of a socialist society. His theory and practice could be summarily characterised by the two slogans 'encyclopedism' and 'social reform'.

2. *The Social and Economic Museum in Vienna 1925-1934*

On the initiative of Neurath, the Social and Economic Museum was founded in 1925 with the support of the City of Vienna. The members of the *Verein GWM* in Vienna were: the City of Vienna, the Free Trade Unions (*Freie Gewerkschaften*), the Workers Council (*Arbeiterkammer*), the Consumers Union (*Konsumgenossenschaft*), the social insurance institutions and the Workers Bank (*Arbeiterbank*).

This novel institution produced continual exhibitions in the public hall of the city hall of Vienna and in the 1st and 12th district. Until 1933, the GWM organised or sent works to some 36 national and international exhibitions. The themes presented give a strong impression of systematic encyclopedism: health, woman and child, social politics, housing, city planning, education for peace, schools, the workers movement, art, social insurance, architecture, etc. In addition, the GWM produced independent national and international touring exhibitions, as well as its own publications and contributions to various journals and books.

In 1927 Josef Frank (brother of the Vienna Circle philosopher-physicist, Philipp) joined as architect, and a year later they were joined by the designer and artist Gerd Arntz, who invented symbols and systematised the technology of production. The scientific department under Aloys Fischer, the department of 'Transformation' under Marie Reidemeister (later Marie Neurath), and the technical co-workers completed the team. Neurath and his associates acted for several years in close cooperation with Otto Glöckel's social democratic movement for school reform, since pictorial statistics corresponded particularly well to Glöckel's pedagogical principles. The GWM achieved a remarkable organisational and educational effect: in order to intensify international cooperation, the Mundaneum Institute was founded in 1931; related groups were established or associated in Berlin, Amsterdam, Prague, New York, London and in the Soviet Union.

After this phase of productive work in Austria and abroad, the GWM found its demise in the political consequences of the 12th February 1934. A number of its functionaries were arrested, valuable installations were confiscated. Renamed '*Österreichisches Institut für Bildstatistik*' (Austrian Institute for Picture Statistics), it came under the control of the Austro-fascist *Ständestaat* (étatist state), until the national socialists took it over for their propaganda. Despite the initial confiscations, Neurath was able to transfer a large part of the collection to Holland and England, where he died in 1945.

3. *The Vienna Method of Picture Statistics and ISOTYPE*

It was the aim of the Vienna Method to represent socio-economic facts, connections and contexts in their historical development by simple, easily comprehended symbolic figures. Numerous real things and complex matters of fact were to be represented by a certain number of signs and symbols, such that the same sign was always used for the same object. This method of visualisation results from bringing together the ideas of content and magnitude; the comparison of sets is to be effected in such a way that a larger set of objects is represented by a larger set of signs (without perspective). In this way, social facts could be reflected quantitatively. Improved and further developed in exile abroad, this method was appropriately renamed 'International System of Typographic Picture Education', abbreviated ISOTYPE (also in Greek 'always the same sign').

Already the phase of the founding of the GWM in 1925 demonstrated Neurath's talent for organisation, but also the trust and support he received from the City of Vienna. In a '*Denkschrift*' (memorandum) of 1924 he developed his concrete ideas for the 'teaching museum for the present' which was to facilitate knowledge and understanding of socio-economic contexts. Pictorial statistics and ISOTYPE must be seen in the larger context of Neurath's thought: it was his main concern to make a comprehensive contribution to general and international education and culture. This cosmopolitan and egalitarian claim corresponds to the idea of a unified science, which after his emigration Neurath also tried to realise in practice in the Unity-of-Science movement. This aim wholly coheres with the framework of his emancipatory educational concept of social enlightenment through visual education as a necessary condition for the humanisation of life. For this reason, pictorial statistics is only a part of general visual pedagogy, and schools are only one of its fields of application.

The hoped-for effect on the masses was not achieved, however, although more exhibitions were organised in Holland and important publications followed. Neurath's and Marie Reidemeister's escape to England in May of 1940 ended their activity on the continent. Gerd Arntz stayed in Holland and continued his work after 1945 alone. After the death of her husband, Marie Reidemeister Neurath carried on very fruitful work under difficult post-war conditions in Oxford with the newly founded Isotype Institute and moved to London in 1948.[2] Nowadays, Isotype continues to be developed at the University of Reading.

4. *The Vienna Method of Picture Statistics and School Education*

Since the Vienna Method was particularly appropriate for schools, Neurath and his co-workers cooperated closely with the social democratic school reform

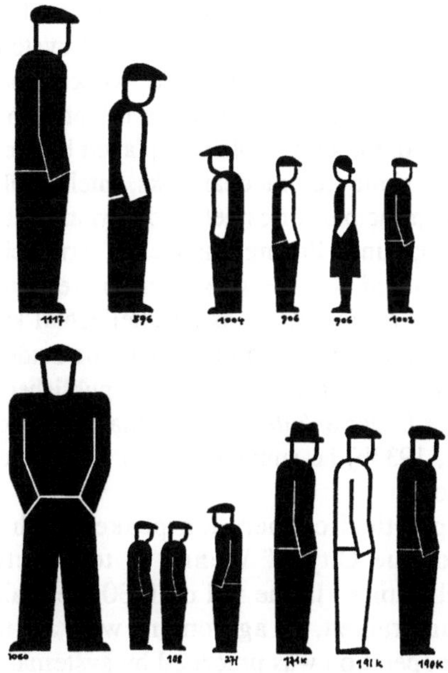

'Unemployed Persons'
From the *Lexicon of Symbols* compiled between 1928 and 1940, unpublished.

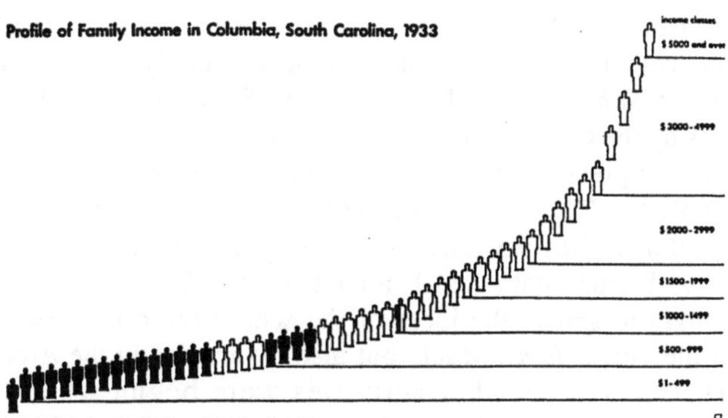

From *Modern Man in the Making*, A. Knopf Inc., London-New York, 1939, ©1967, p. 94

movement initiated by Otto Glöckel. From 1918 to 1934 it represented the attempt to radically reform the traditional school system with its state- and church-dominated 'drill schools', and to build 'work schools' whose instruction was to follow socially just methods which are oriented towards the partnership of teacher and pupil. An important role was played by the pedagogic principles of self-motivation, proximity to practical tasks, intelligibility and concentration in education, and the principles of esthetic education—all implemented against a solid scientific background. Picture statistics in turn facilitated these aims. In 1929 Neurath mentioned the Viennese school reform movement and its predecessor, the *Verein Freie Schule*, as an intellectual kin of adult education, having just stressed that education had to enable the student to comprehend the 'statistical age'. In several articles and in a book published in 1933, *Bildstatistik nach Wiener Methode in der Schule* ('Picture Statistics according to the Vienna Method in Schools', 1933b) he summarised its procedures and principles and some of its applications.

The concrete practical cooperation peaked with the decision of the Schools Council of the City of Vienna to test picture statistics in an experimental school setting. In the fall of 1930 the main boys school in the 14th district was instructed, in agreement with its teachers, to run the experiment. This cooperation was preceded by systematic preliminary efforts at the GWM: in one experiment, statistics was included in the syllabus as an independent discipline, and the GWM established the field of 'intellectual life and school education' as a continual theme for its department of Life Conditions and Culture. Various school classes visited this teaching museum; experiments in visual pedagogy were conducted in kindergartens; even publications resulted from this cooperation, especially the journal *Das Bild im Dienst der Schule* ('The Picture in the Service of the School') with supplements by the GWM from 1927 to 1930.

The most impressive applications of picture statistics are the publications *Die bunte Welt* ('The Colourful World', 1929) and *Gesellschaft und Wirtschaft. Bildstatistisches Elementarwerk des GWM in Wien* ('Society and Economy. A Picture-statistical Primer from the GWM in Vienna', 1930). Neurath sought to spread the ideas of the school reform movement through lectures in the *Verein Ernst Mach*, but also by articles in the *Arbeiterzeitung* ('Workers Times'). Further activities were begun when the school experiment was started. As is well-known, the February of 1934 brought the end of the promising school reform and at the same time the already mentioned dissolution of the GWM in its original form.

Neurath's real methodical and didactic contribution to visual pedagogy was his *Bildstatistik nach Wiener Methode in der Schule*: Neurath stressed social enlightenment through schooling of the intellect without emotional appeal, and the schooling of the pupils' creativity and graphical capacities as particular advantages. His quantitative pictures (*Mengenbilder*), for example, had been tested sucessfully at the Viennese Montessori kindergarten and the Montessori school. Its orientation towards practical applications also brought about a kinship with the then developing pedagogy of Freinet in France, who was able to give expression to the idea of work schools in its school for printing works.

The experiments begun in the school years 1930/31 and 1931/32 were held in both streams of the four classes of the school and always had as their aim the optimisation of the ways and means of concrete instruction in various disciplines. The results were discussed in numerous conferences; many interested individuals from Austria and abroad visited the classes. An exhibition of the quantitative pictures employed, and lectures about their work by the teachers involved, constituted the conclusion of these experiments. It was noted that an introduction and a certain amount of time was needed for the application of the Vienna Method, but that in all disciplines the most varied applications were possible, from the presentation of completed quantitative pictures to their production in the classroom. The drawbacks of the method were said to be, depending on the subject at issue, the required simplification of the numbers or quantities involved, its time-intensiveness, and its unsuitability for teaching German language and literature; its advantages were the possibility for active learning, the enhancement of the pupils' motivation, the sound comprehension of the subject matter at the end of working through the respective problems, the possibility to connect different subjects, and particularly its usefulness for the subject of history.

It is time to rediscover Otto Neurath as a part of the repressed Austrian-Jewish culture, and to continue his work and his vision in science and education. The first examples of this already exist.[3]

Universität Wien

NOTES

* First published as 'Otto Neurath (1882-1945): Enzyklopädist, Schulreformer und Volksbildner', *Erwachsenenbildung in Österreich* **40** (1989), H. 5-6, 33-36, © 1989, Österreichischer Bundesverlag, Wien. Translated with kind permission of BMUK, Abteilung Erwachsenenbildung, Wien, and the author by T. E. Uebel.

1 [Translator's note: For a bibliography of Anna Schapire-Neurath see Fleck 1982b.]
2 [Translator's note: For a bibliography of Marie Neurath see Stadler (ed.) 1982, p. 237.]
3 Besides the ongoing Neurath *Gesamtausgabe* (Neurath 1981, 1991), a research project financed by the Austrian Federal Ministry for Science and Research has begun to develop a computer program for the visualization of economic and social statistics that builds on the Vienna Method of Picture Statistics. In 1990 this research project was turned into a permanent post for visual pedagogy in the *Institut für Wissenschaft und Kunst*. And in March 1990 a film about Otto Neurath, produced by Austrian TV (ÖRF), was transmitted. For literature see also Neurath 1973, Neurath 1979, Neurath 1936 (repr. 1980), Stadler (ed.) 1982a, and Stadler 1989b (on G. Arntz).

JOHANN DVORAK

OTTO NEURATH AND ADULT EDUCATION: UNITY OF SCIENCE, MATERIALISM AND COMPREHENSIVE ENLIGHTENMENT*

1. *The Scientific World-Conception, Education and Everyday Life*

> ... because we cannot achieve anything, nor even talk about anything properly, if we have not learnt a proper understanding of that which is to be done or talked about. There is nothing in the mind, however, which has not been previously in the senses. Since the visible world does not possess anything which could not be seen, heard, smelled, tasted or touched, so that from this we can decide what and of what kind it is, it follows that the world does not contain anything which a human being endowed with sensuality and reason could not comprehend. If it is possible to start a light of comprehensive wisdom, it will spread its rays across the whole field of human spirit, waken contentedness in the hearts of men, and change their minds. For if they see their own destiny and the world clearly in this light and learn to apply the means which unfailingly lead to good goals, why should they not make use of them?
>
> Jan Amos Comenius

In the 1920's and the early 1930's there existed in Vienna a group of scientists from both within and outside of the University who sought to establish a connection between science, education and working life, between scientific investigations and their meaning for the everyday life of the wider population. Important representatives of the Vienna Circle—members of this discussion group around Moritz Schlick as well as the *Verein Ernst Mach* (e.g. Rudolf Carnap, Hans Hahn, Otto Neurath, Friedrich Waismann, Edgar Zilsel, and, independently, Ludwig Wittgenstein)—were concerned as much with the reform of scientific and philosophical activity as with school reform, work in adult education and the related reform of social life in general.[1]

Otto Neurath was principally involved in the program and organisation of the Vienna Circle. Common to the members of the Circle was the aim to develop by collective work (and in contact with similarly minded scholars abroad) a 'system' of rational scientific thought and action that was opposed to the irrationalism pervading religion and politics. Their instruments were scientific societies, journals, a series of monographs, and international congresses. Their endeavors were to "link and harmonise the achievements

T. E. Uebel (ed.), Rediscovering the Forgotten Vienna Circle, 265–274.

of individual investigators in their various fields of science" and also to familiarise the wider public with "the present position of the scientific world-conception" by means of "lectures and publications". The point was to "fashion intellectual tools for everyday life, for the daily life of the scholar, but also for the daily life of all those who in some way join in working at the conscious re-shaping of life"; it was intended that "intellectual tools be formed for modern empiricism, tools that are also needed in forming public and private life" (Carnap, Hahn, Neurath 1929, pp. 306, 305).

A primary aim of the representatives of the scientific world-conception was the 'unity of science'. From this goal derived

> the emphasis on collective efforts, and also the emphasis on what can be grasped intersubjectively; from this springs the the search for a neutral system of formulae, for a symbolism freed from the slag of the natural languages... Neatness and clarity are striven for, and dark distances and unfathomable depths rejected. In science there are no 'depths'; there is surface everywhere: all experience forms a complex network which cannot always be surveyed and can often be grasped only in parts. Everything is accessible to man; and man is the measure of all things. Here is an affinity with the Sophists, not with the Platonists; with the Epicureans, not with the Pythagoreans; with all those who stand for earthly being and the here and now. The scientific world-conception knows no unsolvable riddle. (Ibid., p. 306.)

It is remarkable to what degree the necessity of the collective work of the scientists and the unity of the sciences are stressed.

Already in May 1928, Carnap had written in the Preface to his book, *The Logical Structure of the World*:

> The thoughts which I have written down here are supported by a group of active or receptive collaborators.... The individual no longer undertakes to erect in one bold stroke an entire system of philosophy. Rather, each works at his special place within the one unified science. (Carnap 1928a, p. xvi.)

And Otto Neurath expressed the following view in his *Empirical Sociology*, published in 1931:

> The development of modern science, which in the end includes the whole life, consolidates the close connection of theoretical and practical workers... 'Philosophizing' is then replaced by work on 'unified science'. All endeavors toward clarity and self-reflection are therefore not independent, but dovetailed into a workmanlike concern that helps to build a tool for life. (Neurath 1931c, p. 329/30.)

Again, science is here conceived of as a collective activity, as the consciously planned and executed work of diverse specialists, as a contribution to the betterment of human existence. No longer was it to proceed in abstraction and in isolation from the everyday life of the wider population.

The program of unified science was not a capriciously invented and propagated idea, but the expression of a process already under way, of the industrialisation and socialisation of science. In the course of the 19th and the 20th century, scientific knowledge and its transformation into technologies and technical inventions had become increasingly integrated into the processes of industrial production; not the genial scholar in his study, but the co-worker of a (publically or privately financed) research institute or department of an industrial company had become (and still is) the normal exemplar of a scientist. Of course, this 'industrialisation' of science proceeded under the aspect of the private capitalist profit motive and thus by no means in the interest of as many people as possible. But in opposition to this, a number of the representatives of the scientific world-conception concerned themselves with the *non-capitalist socialisation of science*. And this was to succeed to the degree to which the scientific world-conception pervaded "the forms of public and private life" and helped "the creation of economic and social life according to rational principles". Scientific knowledge and insights accordingly could not remain the monopoly of a few, but were to be made available to as many people as possible: everybody was to cooperate "actively or receptively".

Thus the necessary close connection between scientific activity and work in education.[2] These ideas of a democratic science and a democratic education system could and can only be realised by radical social transformations. Particularly those representatives of the scientific world-conception who felt themselves connected to the workers' movement (like Carnap, Hahn, Neurath, Zilsel) were fully aware of this; for them the scientific world-conception constituted an important contribution to the realisation of a socialist society.

In 1928 Rudolf Carnap formulated this as follows:

> We do not deceive ourselves about the fact that movements in metaphysical philosophy and religion which are critical of such an orientation have again become very influential of late. Whence then our confidence that our call for clarity, for a science that is free from metaphysics, will be heard? It stems from the knowledge or, to put it somewhat more carefully, from the belief that these opposing powers belong to the past. We feel that there is an inner kinship between the attitude on which our philosophical work is founded and the intellectual attitude which presently manifests itself in entirely different walks of life; we feel this orientation in artistic movements, especially in architecture, and in movements which strive for meaningful forms of personal and collective life, of education, and of external organisation in general. We feel all around us the same basic orientation, the same

style of thinking and doing. It is an orientation which demands clarity everywhere, but which realizes that the fabric of life can never quite be comprehended. It makes us pay careful attention to detail and at the same time recognizes the great lines which run through the whole. It is an orientation which acknowledges the bonds that tie men together, but at the same time strives for free development of the individual. Our work is carried by the faith that this attitude will win the future. (Carnap 1928a, pp. xvii/xviii.)

2. *Materialism and Enlightenment*

> 'Education' is a word and possesses just as much content as he is able to think with it who uses this word. That which is commonly understood by it, that which the bourgeois society recognises as education, is most of all the ability to write correctly, speak correctly, eat correctly and dress correctly. One needs to know a certain amount of poets, composers and philosophers by name and also know when to shout 'Bravo' in the theater. That is the education of bourgeois society. One is educated also if one is for enlightenment and progress, but already is rather uneducated if one decides to understand something clear by this. In a word: there is a vast difference between what is accepted as education, and proletarian education. We do not ask of you any sort of correctness, we demand of you nothing but self-knowledge. To think about what you have become and what you should become, that I call education. And you reach a still higher level of education, if knowledge has awakened the will—if it emerges as your conscious goal, from your awareness of being products of society, that you are to become its masters, creators and stewards. The education of the working class consists in this that it consciously set itself a great task and is fully engaged by it, that it seeks to bring about with clear knowledge the creation of a social order which will bring totally different possibilities of education to the proletarians than our poor People's Education Society (*Volksbildungsverein*) has been able to do with its limited means.
>
> Victor Adler[3]

For Otto Neurath, people's education does not consist only in the dissemination of scientific knowledge amongst the uneducated masses. Educational work for him meant a comprehensive enlightenment, meant the creation of those intellectual tools which could contribute to the betterment of their life conditions. This point of view was part of the program of the Vienna Circle and was practically implemented at least partly at the Viennese institutes of adult higher education (*Volkshochschulen*). Representatives of the scientific world-conception, like Otto Neurath, Edgar Zilsel, Friedrich Waismann, and Viktor Kraft, taught there. The Viennese people's education movement seems to have been remarkably successful in the communication of scientific knowledge and its connection with the world-outlook and the personal development of the participants (particularly at the *Volksheim*

Ottakring). At the time, adult higher education reached all strata of society; in some years, blue-collar and office workers were even overrepresented in comparison to their share of the population (Kutalek, Fellinger 1969, p. 199).

The strict observance of neutrality of world-view and political orientation was one of the principles of the Viennese adult education institutes, but many workers were obviously of the opinion that the offered courses were of great importance for their life, and that they could enable them to create their own life-plan and picture of the world. The people's education, in principle neutral, then did not at all have an effect that was neutral or unpolitical. It seems to have communicated and transmitted knowledge and abilities which enabled the participants to continue their thought independently and, in cooperation with others, better understand and change the social conditions. In this respect, adult education provided an important contribution to the education of the working class, and the Viennese adult education movement of the First Republic can for this reason not be separated from the education of the workers.

Otto Neurath taught both at the adult higher education institutes and at the party education institute of the Social Democratic Workers Party, and also participated in essential ways in the education work of the free trade unions. An important contribution to adult education in general and worker education in particular was the work of his Social and Economic Museum in Vienna. His method of picture statistics served the "visual representation of social relations".[4] Numerous exhibitions were held which were devoted, amongst others, to the progress and successes of Viennese city politics in the field of health and education, housing and social security. Visits to the Social and Economic Museum were a regular part of the fare offered by the Education Institute of the Free Trade Unions, which was founded in 1926 (Wagner 1927, p. 695).

Otto Neurath's hope was expressed as follows:

> The number of social museums is likely to increase mightily in the near future. None of the large worker organisations, none of the city administrations active in adult education, will want to miss the opportunity to order social picture displays or models for their particular purposes. Through this, the social museums receive the task to become centers for initiatives of all sorts, and to arrange exhibitions, especially travelling exhibitions. By publicising their experiences gained in this, it will become possible to enliven the common understanding of such efforts and to effect a kind of school reform for adults in this field by the cooperation of all interested organisations. (Neurath 1926, p. 301.)

Beyond his contributions to the concept, and to the practical realisation, of adult education described so far, Neurath also wrote a number of theoretical works which to this day remain of fundamental importance for the theoretical foundations of adult education—that is, if it is viewed as aiming for comprehensive enlightenment.

Workers' education contributes to the development of the individual. It influences his actions, the pleasure he takes in people and things, his integration into the workers' organisations. In all this, enlightenment about humans and material objects, about connections of all sorts, plays a central role. Excitation and the description of glorious goals is not what is most important, rather, the more factual description of that which is. The struggling socialist worker must know what prospects are possessed by the workers' movement and what obstacles can be anticipated, because confidence is strengthened, hope is heightened and disappointments are minimised by predictions which are as exact as possible. (Neurath 1929b, p. 345.)

But the role of educational work must not be seen in isolation from the related reform of the totality of life conditions. Neurath's ideas are distinguished from others that found expression in the workers' movement in this respect: first of all, he investigates the real life conditions of the working class (and its causes) and then, based on this analysis, he sets the emphasis for the educational efforts. Any moralising attitude, any educational 'realism' is alien to him: education is for him not a higher good but a tool for daily life. According to Kautsky's and Lenin's conception of the place of the intellectual in the workers' movement, knowledgeable intellectuals have the role of disseminating the truth to ignorant proletarians. Neurath, however, begins in the middle of the processes of social transformation and sees that the workers could only be interested in education in relation to the immediate improvement of their material life conditions.

Already an essay from the year 1922 with the title '*Die Betriebsräteschulen als Schulen des Sozialismus*' ('The Schools for Trade Union Stewards as Schools for Socialism') is characterised by this attitude. There Neurath says:

What is the task of schools for trade union stewards? They are meant to train workers of average intelligence as trade union stewards. What is the task of trade union stewards? The task of trade union stewards is threefold. They must (1) contribute to the improvement of the condition of the workers within the traditional social order; (2) contribute to the transformation of the structure of companies and aim to increase the influence of the workers by their mediating efforts, so that they gain control of the companies; (3) contribute to the transformation of the entire economic order and to gaining control of the different branches of the economy which are to be unified in trusts. It is of great importance that none of these tasks overshadows the others, even more so that the third task is not neglected. We know from experience that the danger is great that the training of trade union stewards serves to provide good employees for the employers. By stimulating the workers to make 'their' company more productive, one easily encourages a tendency towards communities of employers and employees which are of little use in the transformation of the economic order. Just by participating more deeply in their companies, trade union stewards often are drawn into capitalistic thinking....

If then the trade union stewards on the one hand and the workers on the other are to be made into conscious creators of the processes of production, then one must teach the struggle against the traditional order as a counterweight. It will not suffice to describe in general the nature of socialisation and the nature of the socialist life order, but it must also be shown what can be done for socialism today and tomorrow when the larger questions are raised. The unification of the workers into federations that control entire branches of the economy must be taught systematically. The reform of the trade unions, the replacement of the principle of organisation by the trade of the workers, by the principle of organisation, by the type of company, must be discussed by experienced trade unionists, so that every worker will help to create this large organisation in the interest of socialism. Again and again we must point out that wherever a large conglomerate of companies comes into being, there the trade union stewards must create an appropriate loose counter-organisation which under certain circumstances may provide resistance to the capitalist organisation, but in any case could keep it under continual observation. (Neurath 1922b, p. 177f.)

In his *Lebensgestaltung und Klassenkampf* ('Personal Life and Class Struggle') of 1928 Neurath summarised many of his ideas about education (amongst other things). In his introduction he writes:

A new system of principles for living can be the basis of the community of proletarian fighters today, and later become the core of a peaceful community embracing more and more men The labor movement anticipates something of the future community of mankind. While aiming at this future and its happiness, the individual will fight and make sacrifices the more eagerly, however, the more the suffering *of today* is relieved by his action at the same time. (Neurath 1928a, p. 250.)

Already in the present, the cultural achievements of the working class—solidarity, cooperative action, the formation of mass organisations (trade unions, workers' parties)—present the principles and practices of a future better life, at least in outline. But it would be wrong to want to reduce this to a concern with the reform of merely individual lives.

It cannot be demanded that the proletariat must always act 'socialistically', even if it could be defined what a socialist way of acting is. It would weaken action and damage the socialist future if one tried to act today as if a socialist community existed... The longing to work and fight with 'purely socialist principles' is mostly found among young proletarians and among those intellectuals whose emotions made them join the labor movement. But the bulk of the fighters have no time to choose. The pressure of the day leaves no choice. But how deeply and firmly holds that part of socialist thinking and feeling which is conditioned by such pressure! If one is aware of such limitations, it makes sense to take trouble in the anticipation of socialist existence. The call of the proletarian youth for socialist community, the longing of mature workers for socialism which can be experienced today—that belongs to the transformation of our way of life and our humanity. The wish is to experience future humanity at present, and not only in festive mood at celebrations, but also in everyday life. Those who cut themselves off in sects to satisfy such longings more thoroughly in a close circle of equally minded persons often disturb the proletarian community and harm what they strive to achieve. A new community of living will arise, eliminating much hatred, suppression and heavy suffering; but the condition for this is that the whole social and economic order be transformed; community life is tied to the economic plan. He who furthers planned economy is more closely linked with the transformation of social life than is he who concentrates his efforts on special reforms without regarding their connection with the fight for the economic order. (Ibid., p. 250/1.)

The proletariat does not fight for the realisation of any ideals, "but above all in order to better its daily existence" (ibid., p. 297). Thus the daily way of life of many workers does not appear so optimistic and open towards new trends, but rather petty and conservative to the eyes of the bourgeois world improvers who due to their privileged social position have the time to consider questions of the reform of personal life and alternative lifestyles in detail. Otto Neurath has the following explanation.

The 'spirit of modern times' is more apparent in the proletariat when there are great common enterprises than where the individual proletarian arranges his personal life. City planning and housing may only serve as an example. The style of personal life does not make for intensive concern amongst proletarians. It could not be otherwise. In bourgeois circles one has much more time for questions of lifestyle. Questions of dress, of housing etc. are of late being discussed in several countries. But individuals mostly expressed their accidental ideals without great discernment, and without reflecting about how the new life of the individual is dependent upon the total transformation already in progress. Whoever wishes to take seriously housing, dress, reading, love life, sociability, and everyday just as much as holidays, in short he, who wishes to take personal life seriously, must desire the change of the whole social order. He must ask himself how these changes fit into the social changes. These questions, however, proletarians are only able to ask themselves very rarely, for the daily struggle permanently demands their strength for different things. What is at issue first of all is to sort out much more rough and earthly matters, to care for basic daily needs of proletarians, not to change their furniture.[5]

The culture of the working class is, in contrast to bourgeois culture, not an individualistic, but a collective culture. The individual is defenseless and powerless against the conditions of capitalist society; only if they unite in as large organisations as possible would the workers be in a position to pursue their interests. Whatever they produce, whether in factories or culturally, they produce collectively.[6]

Among workers the realization grows that the single person achieves hardly anything, the closed ranks of the workers more than expected. Thus common action, solidarity, becomes the first rule. This is what is valued, furthered and supported. Non-solidarity becomes the worst reproach. A strike-breaker's appeal to personal conviction may be valid within a bourgeois community, even if he is punished in the end; among socialist workers the answer will be: your personal conviction is not our concern, we demand solidarity above all. The basic attitude that solidarity is more important than acting on personal attitudes is common to several large organizations in history: it is typical for the workers' movement as well as for the Catholic Church which also submerged the individual in the whole. But there is an essential difference: the labor movement demands definite actions, picketing in strikes, stoppage of work, etc., but not a definite creed. It does not search the secret corners of the soul, it does not torture to find out whether a comrade is sinful in his soul. It does not believe that souls have to be saved. The Church is not content with action alone, it demands definite convictions. What characterizes the labor movement is that it does not demand confession, neither for atheism nor for Marxism nor for anything else. Whoever fulfils his union duties is accepted. (Neurath 1928a, p. 255/6.)

Through his contribution to the program of the Vienna Circle, his organisational activities towards its realisation, his theoretical writings on workers' and adult education, and in his practical work in education, Otto Neurath achieved the connection of science, education and everyday life—at least in outline. Only 'in outline', because little time remained for the practical application of his ideas: Austro-fascism and National Socialism destroyed not only the existing beginnings, but also for a long time even any perspective on a radical reform of society for the better and any hope for the creation of a socialist society by the workers' movement. This makes it all the more important to reconstruct the conceptions of adult education that arose during the 1920's and 1930's, and to reflect on how they could be relevant today.

Abteilung für Erwachsenenbildung,
Bundesminsterium für Unterricht, Kunst und Sport, Wien

NOTES

* First published as 'Otto Neurath und die Volksbildung - Einheit der Wissenschaft, Materialismus und umfassende Aufklärung', in *Arbeiterbildung in der Zwischenkriegszeit*, Hrsg. F. Stadler, © 1982 Österreichisches Gesellschafts- und Wirtschaftsmuseum, Wien, pp. 149-156. Translated with kind permission of Österreichisches Gesellschafts- und Wirtschaftsmuseum and the author by T. E. Uebel.

1 These aspects of the activity of the Vienna Circle have only been recently brought to the fore in the literature. Particularly to be noted are Bartley 1977, Stadler 1979a and 1979b; see also Dvorak 1981a, chs. 3 and 5.

2 Accordingly, Carnap, Hahn and Neurath stressed in the manifesto of 1929 also their affinity with "endeavors ... toward a reform of school and reformation" (ibid., p. 304/5) and saw themselves in the tradition of the liberal scholars of the Vienna of the 19th century who had acted in the spirit of the enlightenment: "Thanks to this spirit of enlightenment, Vienna has been leading in a scientifically oriented people's education. With the colloboration of Victor Adler and Friedrich Jodl, the society for popular education was founded and carried forth; the 'popular university course' and the 'people's college' were set up by the well-known historian Ludo Hartmann whose anti-metaphysical attitude and materialist conception of history expressed itself in all his actions. The same spirit also inspired the movement of the 'Free School' which was the forerunner of today's school reform." (Ibid., 301/02.) On the relation between the scientific world-conception and work in education see also Zilsel 1921.

3 Adler 1902, p. 21f., a lecture to the *Arbeiter-Bildungsverein* (Workers Education Society) *Gumpendorf* which was founded in 1867.

4 "Ethnological museums show how people live and are housed, but social museums show how many people share a certain way of life and how this way of life depends on income, environment etc... Social museums take the human being as a social being as their object; they are placed where the method of visual representation of social relations is developed." (Neurath 1927, p. 130.)

5 Ibid., (1981, p. 236f.; only the first sentence of the present quote is translated in the selection in 1973). Neurath continues with respect to art, culture and science: "The proletarian revolution is simply different from what the bourgeois revolution was. A new class arose then, in whose struggle thinkers, poets, and creative men of all kinds took part in great numbers. The bourgeoisie already had the new education and the new literature when it began its fight for power against the nobles, throne and altar. Today is the reverse. Not until after the proletariat's victory will education and art be brought to all who today must still go without. Not until then will there arise in a greater measure an art more adapted to the socialist order. The practice of art and science today lies mainly in the hands of opponents to the proletariat. Only rarely will bourgeois recalcitrants wage single-handed war against the past or break through to the proletarian front. And in the proletariat there are, for the present, few possibilities to work in a scientifically or artistically creative field. Traditional means and forms are predominantly in use. The proletariat aims above all at securing what until now was reserved to the bourgeoisie; it conquers existing positions in schooling, art and so on. The conscious fight against 'bourgeois' ways does not on the whole play a very great part as yet, even if it is attempted by individuals and groups more often than in the past. Most innovations flow from the needs of the day. It will be mainly an advance on a broad front. The ideas of individuals are rarely furthered by the proletarian front, unless they can become an expression of the total proletarian endeavor. Here is the soil for those who fashion life creatively in the spirit of the revolution, partly consciously, partly unconsciously. Socialist yearning is not enough, nor talent alone, nor mere Marxist thinking. The coming man demands a living connection with the social transformation." (Ibid., p. 258/9.)

6 The stress in the program of the Vienna Circle on the collective nature of scientific activity thus shows its high degree of affinity with the culture of the working class. On the culture of the working class see also Zilsel 1981b.

ELISABETH NEMETH

THE UNITY OF PLANNED ECONOMY AND THE UNITY OF SCIENCE*

The concepts 'unity', 'uniformity', 'unification' are all-pervasive in the writings of Otto Neurath and may serve as the leading themes in the discussion of what it was that Neurath wanted to achieve. Their repeated appearances in different contexts make evident the connectedness of his efforts and show up a line of development whose path cannot be reconstructed merely systematically or theoretically, but must be comprehended as the result of a *praxis* which found its direction in the confrontation of individual will and social exigencies.

The present lecture cannot, of course, provide a comprehensive reconstruction of this sort. Yet I shall attempt to characterise several aspects of Neurath's ideas on the topic of unity in such a way that the concretely practical intention which determined his thought is at least not barred from view by the abstract explication of concepts, but brought into view to some extent.

In Neurath's writings on economics and political economy prior to and during World War I we meet our topic primarily in the juxtaposition of monetary economy (connected with 'market economy') and economy-in-kind (connected with 'planned economy') (1917b, p. 24).[1] In both systems, certain economic movements or processes—simply put: production and consumption—are brought into relation with each other, are formed into a whole, however ordered. Corresponding to the two economic systems, there are two points of view, two methods of theoretically comprehending economic processes: monetary calculation and natural calculation.

Monetary calculation presupposes a 'unitary measure' in which every change must be represented in order to be comprehended as an economic process. This 'unitary measure' is money; every change is measured by it as a plus or minus, a gain or a loss.

It was monetary calculation that was used when Colquhoun described England's war economy during the Napoleonic wars by measuring the production of milk, meat, the import of cotton etc. not in quantities of these respective goods but in pounds sterling. It was monetary calculation that was used when it was recently shown that city administrations incur a loss in

T. E. Uebel (ed.), Rediscovering the Forgotten Vienna Circle, 275–283.

summertime, because its savings on gas for lighting its own offices was offset by the lower sales of gas to the inhabitants. It was a sign of the advance of natural calculation that many objected that summertime was after all characterised by the saving of coal and labour. If summertime depended on the free market economy, the gas companies would seek to prohibit it and propagate the view that it was no longer modern. (Ibid., p. 9f.)

By contrast, natural calculation places in parallel "an overview of the available quantities of productive means" and an overview of needs.

Once the *technical possibilities* are recognised, it is possible to determine how much the technical means can serve to satisfy the needs. The quantitative comprehension of technical possibilities, whose organisation and realisation is in question, may be juxtaposed as *natural calculation* to *monetary calculation*, which concerns itself primarily with prices and net profits. (Ibid., p. 9.)

Natural calculation does not recognise a unitary measure for economic processes. It proceeds from the quantitative comprehension of the means of production and asks for the technical and organisational possibility of need-satisfaction. The unity of the model of an economy-in-kind is from the start an 'organisational' unity; it places elements of different kinds into a whole whose order cannot be derived from the elements. For this reason it is important for the proper understanding of natural calculation "to discard all presuppositions concerning money, quantities of the same type of goods, etc., and to attempt to operate with complexes which do not *even contain elements of the same type*." (1911, p. 107.) Thus natural calculation needs a 'unitary organisation' (ibid., p. 69) which integrates the elements of different types into a unified order. Monetary calculation does not need such an organisation, because in it all relations are reduced to their equality in the value of the means of exchange and because the unified order is, as it were, automatically established through the unitary means of exchange: all movements are but successive acts of exchange between two persons.

Monetary calculation then is abstract in two respects. First, it comprehends wealth only in terms of money and its changes only as losses or gains. The real alterations of wealth, however, may differ from their representation in monetary terms (e.g. there may be an increase in the yield of a field, but a decrease in the amount of money gained by the sale of this increased yield (ibid., p. 90)). Monetary calculation leaves behind what really happens in production and consumption and hypostatises it: concepts like 'real income' are juxtaposed to 'monetary income'. Neurath calls them "derivatives of monetary calculation" (1917a, p. 124). But monetary calculation is abstract also in a second sense: it comprehends changes in wealth under only one principle, namely, in the singular form of individual acts of exchange between two persons. Changes in wealth thus consist in the iteration of the same 'unit', individual acts of exchange. These chains are in principle unlimited, but their direction—their 'logic'—is fixed by the loss or gain of the means of exchange. "Money is therefore the product of a deficient if more comprehensive organisation." (1911, p. 69.)

The 'uniform organisation' which is required for an economy-in-kind is concretely limited, does not possess a logic that follows from a single principle and so allows a mulitiplicity of different constellations of wealth. Because it needs such a uniform organisation, economy-in-kind is related—at least "currently" (1917b, p. 24)—to planned economy. This planning need not be from the start a central state administration of an entire country. In 1917, Neurath is able to imagine a 'transition economy' consisting of a system of different workers' organisations which have taken over different branches of the economy. Not even money would have to abolished in its entirety, for it could continue to be used for certain social purposes. Neurath's main concern is to prevent money from determining the economic process as an 'active factor', and that instead a 'comprehensive system of prices' is established which "is determined by uniform considerations and already shows in principle the character of an economy-in-kind" (ibid., p. 23).

'Uniform considerations' are to create economic life in its entirety. The result—in the cited case, a certain system of prices—cannot be any longer represented and justified by the principle of gain and loss.

> Many a thing will become clearer and more transparent in an economy of producers' organisations. Power will appear as power and will not act behind the veil of the monetary form of exchange. (Ibid., p. 19.)
>
> The mere fact that export brings money into the country will not suffice to make the state desire it. The export of important metal products etc. would happen only if that brought in important goods, not expensive lingerie or precious stones... Prices as automatically determined magnitudes have not been judged by the economy as a whole in a free market economy; by contrast, prices determined by producers' organisations under the control of the state are the direct result of power factors and will be considered the result of a communal effort. (Ibid., p. 25.)

To comprehend need-satisfaction as an achievement of the entire whole is to extract it from the automatism of the profit motive, is to make visible that need-satisfaction depends on power factors in a society and not on a principle which keeps the economy going 'above the heads' of the human beings involved. Just as this would allow people to work directly for the "improvement of the whole" (ibid., p. 32), and not merely indirectly via the money-directed mechanism of competition, so that it would be possible to calculate and create wealth itself in a new fashion without this mechanism. Natural calculation throws into relief the thought that "production exists only for consumption, and an increase in production can only be used if the time for consumption has been lengthened" (ibid., p. 14).

Already in 1911 Neurath had demanded that economic science must regain its object which was repressed by monetary calculation and was hypostatised as an abstract 'purpose' of science. For him, wealth "is but the the pleasure and pain which we find in individuals and groups of individuals." (1911, p. 53.) In his discussions of the approaching transition economy towards the end of World War I, Neurath defines wealth as the "income-in-kind" independent of monetary income, namely, as "the goods consumed by an individual in the widest sense, whether bought or not" (1917b, p. 16f.). This effort to comprehend what is being consumed as wealth, to make 'pleasure' the object of economic science and to understand by it not only a consequence of, and reaction to, production brings Neurath into an unexpected intellectual proximity to a contemporary, whose economic theory turned out to be still considerably more controversial than his own, to Georges Bataille:

> ... on the whole, any general judgement of social activity implies the principle that all individual effort, in order to be valid, must be reducible to the fundamental necessities of production and conservation. Pleasure, whether art, permissable debauchery, or play, is definitely reduced, in the intellectual representations *in circulation*, to a concession; in other words it is reduced to a diversion whose role is subsidary. The most appreciable share of life is given as the condition—sometimes even the regrettable condition—of productive social activity. (Bataille 1933, p. 117.)

I believe that Neurath's 'natural calculation' wanted to avoid the exclusion of this 'most appreciable share of life', indeed wanted to integrate it into theoretical considerations and into the economic organisation. Of course—here again he agrees with Bataille—such an economic theory would be "theory in a different sense than before. The material nature of all efforts would be considered far more so than in all previous theory." (1917b, p. 33.) The concern to comprehend 'the material nature of all efforts' changes the theory as a whole and changes the value of theory. And it seems to me that here lies a key to Neurath's understanding of 'empiricism'. For this is not a new empiricist 'system' of science, but an ever new demand to transcend the traditional understanding of theory and organisation whose abstract universality has made 'the material nature' of reality ungraspable and repressed it 'behind the scene'.

Of course, it will still take some time until it is generally recognised that an economic order is conceivable in which production and distribution are arranged not on the basis of a unitary measure, neither of common money nor of work money, nor anything else. The different branches of production would produce comparable amounts of bread, clothing, housing, and the income of individuals would consist of housing, clothing, food etc. The incomes in kind could be grouped into types which appear characterised by certain amounts and qualities of food, housing, clothing etc. We must insist on this possibility, because even radical reformers still wish to retain the tradition of individualism and atomism which seeks to separate everything into its individual parts, each of which would be quantifiable all by itself, so as to derive the whole by their addition, just as money was supposed to do. (Ibid., p. 31.)

The end of the war seemed to have brought within reach the possibility of realising such a new economc order. In 1917 already Neurath thought he could observe as a consequence of war economy a decrease of the fatalism

with which earlier on crises were tolerated that were tended to be viewed as inevitable side effects of production trends or their changes. People have become accustomed to the fact that everything is open to intervention and that severe measures could be taken overnight. (Ibid., p. 8.)

As is known, Neurath wanted to realise the model of a planned economy-in-kind in his role of economic commisar of the Munich soviet republic. The destruction of the soviet republic did not immediately shake his perspective of an imminent restructuring of the entire economic order. In the course of World War I Neurath drew closer to Marxism and, after his imprisonment due to his role in the soviet republic, engaged himself in the Viennese social democratic movement. He intervened in the so-called 'socialisation debate' by presenting his ideas of an 'economic plan', a 'planned economy', and of 'scientific utopianism'. What was decisive in all this was his attempt to retain the possibility of making decisions about an ecomomic order in its entirety. What had been experienced by many during the world war, namely, that the economic order was not ruled by laws of fate, that now seemed possible to find its historical conclusion through political developments. "Socialism stands at the door." (1920a, p. 224.) It needed Utopias, that is, pictures of a social order not yet historically given (ibid.), but realisable after the assumption of power by the proletariat (ibid., p. 227). The different models of economies-in-kind now emerged in a "comparative science of Utopias" as the possible aims for action in a future socialist society. Scientists and all those who possessed enough imagination and insight (ibid.) designed models that competed amongst themselves and amongst which the politician was to select one and realise it (ibid., p. 226).

Already in 1913 Neurath had considered in detail the necessity that in order to act, one had to decide between two theoretically equally valued possibilities, and had to decide without recourse to a theoretical justification. For the 'Lost Wanderers of Descartes' (1913b) it was necessary to chose *one* way if they wanted to leave the forest. The 'uniformity' of the direction chosen could not be derived from a theoretical insight into their situation, but could only spring from the will to reach a goal (to get out of the forest). Neurath's conviction that the uniqueness of a choice does not derive from logical reasons but from practical considerations, and that for every action we require a "uniform influence on the will" (1917b, p. 6)—and let it be an artificial one (an 'auxiliary motive' (1913b, p. 4))—gains concrete content in this time of hope for a soon-to-come revolution. The construction of a socialist society of the future becomes his practical goal, the 'socialist aim' gives direction for the selection of tasks to be undertaken. The administrative economy-in-kind of 'the future' becomes the concrete task of a centralist socialist planned economy (1923a, p. 153). Of course, still under socialism there exists the danger that the 'uniform unit' of money which determined production and distribution under capitalism (1923b, p. 283) is reintroduced in a different form. Thus Neurath writes against Helene Bauer:

> She thinks... that the selection of an economic plan must proceed on the basis of a calculation which determines by means of *units of labour*, the effort of human labour in production, the distribution of this labour amongst the different groups of humans, and the distribition of objects of consumption on the basis of this distribution of labour. There is a tension here. I believe that even in this limited field one cannot operate with natural units of labour and that only a conventional determination of units is possible, for there are no natural ones. (Ibid., p. 285.)

According to Neurath, whatever is counted as a unit must be made to depend on the agreement of human beings, and it is essentially artificial and can never be justified by reference to an underlying 'natural' principle. The 'theory of life situations' which was to inform the choice of an economic plan in a socialist society does not provide a uniform principle, but presents us with a complex structure in which "everything that determines human happiness and human suffering is considered as long as we possess data about it" (ibid., p. 284). What characterises Neurath's conception of a socialist society, then, is the conscious recognition of the conventional, artificial character of every order and the unreducible plurality of its elements.

We can observe Neurath's efforts for uniformity still before his activities in the Vienna Circle also on the level of pure theorising. In 1917 he demands the "unified construction of economic science" so that "in the investigation of the smallest detail the connection with the basic questions is preserved and that it would not be the case, as at present, that investigations for which no overarching concept could be found are unified under the name of 'economic science'." (1917a, p. 114.) When in the inter-war years the hope for a soon-to-come social revolution was waning, Neurath's concern with scientific thinking as such began to gain greater prominence in his work. As is known, it led to the program of the 'scientific world-conception', an essential aspect of which was the construction of 'unified science'. In what remains of my time I want to provide at least a rough sketch of how Neurath's efforts to introduce a new concept of uniformity into economic science found its continuation in 'unified science'. I must confine myself to only some aspects of Neurath's explicit remarks on the 'unity of science'. It would be more interesting to trace his search for an unmetaphysical uniformity in other contexts, for instance in his polemics against 'consciousness' (the modern principle of the unity of knowledge and the world), his opposition against individual protocol languages and against pseudo-rationalism, but these hints must suffice.

Unified science seeks to unite science as a whole (1938a, p. 3/4). "It creates an *ersatz* for the comprehensive perspective of metaphysics by its uniformisation of the language of science and its cross-connections between individual sciences." (1935c, p. 659.) According to Neurath, the comprehensive perspective of metaphysics is characterised by the construction of knowledge as a unique system, corresponding to a definite reality, and typically proceeding "from the 'simplicity' of certain initial elements—statements or concepts.." (1935a, p. 118.) Metaphysics wants to proceed from a *tabula rasa* and so seeks for its overview a point which yields a principle of comprehension that does not emerge from the given connections themselves. The metaphysical unity is, in the words of Nietzsche, a mode of the godhead.

By contrast, the uniformity of unified science is not a 'logical consequence' of the program of the scientific world-conception, but is a "historical fact in a sociological sense" (ibid., p. 115). That is to say, it is grounded in decisions, it is determined from the start by the accidents and uncertainties of human practices. The "scientific synthesis" (1938a, p. 1) which Neurath has in mind is not based on ultimate simple elements from which the whole could be built up consistently and systematically, but on "a great many scientific units" of varying magnitude (1937/8, p. 200) and differing provenance (1938a, p. 4). Not only the size and the number of elements are determined by the practice of science; the integration itself is the aim and result of a historical effort by the scientists. The "super science" (ibid., p. 20) of philosophy is deprived of the construction of the whole of science, it becomes the achievement of the collective of "scientifically oriented" (ibid., p.23) people. By contrast, under the abstract unity introduced by philosophy, a specialisation could develop whose isolationism can be viewed like the "fracturing wild economy" of capitalism (1921b, p. 289), and whose hierarchical organisation causes the limitation and prevention of many scientifically fruitful questions and investigations (1937/8, p. 200). The 'uniformisation of science' therefore does not wish to replace metaphysical principles of ordering with new ones, but seeks to make the establishment of order a matter of *praxis*, not theory.

For this reason, the construction of unified science can only proceed in an 'encyclopedic' fashion.

An encyclopedia (in contradistinction to an anticipated system or a system constructed *a priori*) can be regarded as a model of man's knowledge. For, since one cannot compare the historically given with 'the real science,' the most one can achieve in integration of scientific work seems to be an encyclopedia, constructed by scientists in co-operation. (1938a, p. 20)

Natural calculation, too, comprehended economic relations in a model which represented production and distribution as organisational, and so as practical questions. And just as the point of view of an economy-in-kind sought for methods that made representable 'the material nature of all efforts' instead of hypostatising it into an abstract universality, just so the encyclopedic point of view was to make transparent the historically given science in its 'material nature'—this means now: as a result of creative activity—and so make science available, place it within the reach of its practitioners.

The conception of science which Neurath intended seems to me to be empirical also in the sense indicated earlier: it is a matter of structuring the "scientific enterprise" (ibid., p. 26) in such a way that what is empirical becomes comprehensible and formulatable as a complex and many-sided edifice from the beginning. "The encyclopedia will express the situation of a living thing, not of a phantom: those who read the encyclopedia should feel that scientists are speaking of science as a being of flesh and blood." (Ibid., p. 25/6.)

This conception of scientific knowledge as a 'being of flesh and bood' contains another transgression of the traditional concept of knowledge, its transgression into the future that is characteristic for Neurath. The descriptive 'behavioristic of scientists' is only the other side—that of the history of science—of the concern which is directed towards a more daring goal than that of a rewriting of the history of science: to recognise the creation of scientifc knowledge as determined by historical conditions—i.e. made possible and limited as a social *praxis*—not only in hindsight, in a "retrospective synthesis" (1936b, p. 140), but to approach contemporary science with this insight, to comprehend it as historically and socially determined—and to practice it accordingly. What is at issue, so it seems to me, is this: how is it possible to do science without disregard of the fact that it constitutes a historically determined social practice? And that is no other question than: how is it possible to do science as *political praxis*?

The contemporary reception of Neurath tends to remove the point of this question. On the one hand, it reduces the attempt to reach a conception of science as a social practice to a discipline all its own—say, a 'behaviouristic of scientists'—while on the other, it misunderstands and discredits as unjustified and unjustifiable caprice the 'decisions' on which, according to Neurath, the acceptance of scientific sentences ultimately rests. But that, I think, is itself a consequence of our political and social situation. Apparently it has become very difficult or nearly impossible to allow a Utopian conception of social praxis, let alone to think it through.

Universität Wien

NOTES

* First published as 'Die Einheit der Planwirtschaft und die Einheit der Wissenschaft', in *Schlick und Neurath. Ein Symposion*, Hrsg. R. Haller, *Grazer Philosophische Studien* **16/7** (1982) 437-449, © 1982, Rodopi, Amsterdam. Translated with kind permission of Editions Rodopi and the author by T. E. Uebel.

1 Except where noted, all references will be to the writings of Otto Neurath.

ELISABETH NEMETH

OTTO NEURATH'S UTOPIAS - THE WILL TO HOPE*

> To aim for a life order as a goal means to produce a picture of it. Such pictures of historically not given life conditions we shall call 'Utopias.' The state of the future, as Ballod and Bebel tried to describe it, is a Utopia. Today, when socialism stands at the door, it is of greatest importance to know how the new life order looks, which one wants to bring into existence. (Neurath 1920a, p. 224.)[1]

When Neurath wrote these lines in the Social-Democratic monthly *Der Kampf* in 1920, he already had behind himself a concrete attempt to realise the Utopia of a new economic order: the Munich soviet republic had been destroyed! But not much had changed concerning the importance of the concept *Utopia*: the view that socialism stood at the door was still common in Austrian Social Democratic circles, and Neurath's Utopias were after all designs which were to be realised after the proletariat had gained power.

Neurath had used the concept 'Utopia' already before World War I (1913b, p. 11), but it only gained a central importance in the writings towards its end and shortly thereafter: "Utopias have become respectable in society", he writes in 1917 (1917b, p. 8). He thought that due to the experience of a world war which lasted several years and brought large-scale administrative economic measure for military purposes

> there obtains today a decrease of the tendency of people to tolerate reductions in production... just as of the fatalism with which earlier on crises tended to be tolerated that were tended to be viewed as inevitable side effects of production trends or their changes. People have become accustomed to the fact that everything is open to intervention and that severe measures could be taken overnight. (Ibid..)

The experience of the constructability and conscious direction of economic processes stimulated the creation of pictures of different possible forms of organisation. These pictures have two aspects: they are 'constructed scientifically', and they are 'practical', they can serve as goals for action which seeks to realise a new life order. As products of "purely scientific work" (1920a, p. 224), Neurath's Utopias are "thought-constructs" (1922a, p. 55), theoretically equivalent possibilities whose development "had nothing to do with action or affect" (1920a, p. 224). Accordingly, it is the task of science to develop "groups of Utopias" and to make transparent the

T. E. Uebel (ed.), Rediscovering the Forgotten Vienna Circle, 285–292.

differences between these models in a "comparative utopistics" (ibid.). Which of these models is to be preferred is impossible to say on grounds of theory alone: it is the politician who must select one of them (ibid., p. 226).

Practical life stands in contrast to the unlimited possibilities of theoretical modelling: the will can only be directed toward one goal (ibid.). But the uniqueness of willing does not diminish the ambiguity of knowledge; to the contrary: both correspond to each other and together characterise a situation which Neurath had already described in 1913 in a picture of Descartes.

This is the situation of the wanderers who are lost in a forest and must now decide upon one direction in order to reach open land. The actions of the "men of modern society" can be typically characterised in just this fashion (1913b, p. 5): for them, the progress of enlightenment has broken the power of instinct and tradition which in earlier times determined their actions. The obvious breakdown of the remaining traditional institutions in World War I made it possible to interpret the new situation in an original and euphoric fashion. The architect Josef Frank wrote: "The war placed us in the situation we have long since hoped for, namely, of being able to begin anew, and we are searching for a way." (J. Frank 1931, p. 1.)

Towards the end of the war, Neurath became a Marxist committed to the class struggle (cf. Glaser 1982); in the Munich soviet republic he saw the chance to realise the Utopia of an economic plan for the organisation of production and distribution. Still before the war, his studies of war economics had convinced him that the measures of an administrative economy were often better suited to the realisation of economic goals than the free competition of a market economy (Hegselmann 1979a, p. 21). Now he made common cause with the workers movement and stressed the demand of Marxist theory for the abolition of the capitalist market and the introduction of a planned economy (1920b, p. 137). The total socialisation of Bavaria was intended to effect the abolition of the market in a clearly delimited area.

The area at issue was clearly delimited in two respects: first, as an agrarian country, Bavaria was not dependent on imports for its food supplies; second, the economic plan was conceived to be independent of the political constitution and to be practicable under different forms of political organisation (ibid.). It is characteristic for Neurath's Utopias that they represent "comprehensive wholes" (1920a, p. 225), whose realisation—e.g. the socialisation of Bavaria in 5 to 10 years—could be effected in one sweep. (That war economics could become a model for his Utopias has a related reason: war determines human behaviour more uniquely than peace (1917b, p. 5), and so provides an economic structure which allows for a comprehensive view.)

Still after the destruction of the Munich soviet republic, Neurath retained the idea of the introduction of a planned economy as a comprehensive whole—against the tendencies in the Social Democratic movement to think of the socialised companies within the capitalist order as germ cells for the development of socialism (1922a, esp. pp. 56-59). Of course, there was for Neurath a historical tendency towards socialism, but this consisted in that the solutions for the technical problems of a planned economy had already been found in trusts or in armies (ibid., p. 59), and in the struggle for positions of political power by those who wanted to introduce a planned economy, the workers' parties. Yet the socialist economic order itself—socialisation—did not develop; it could only be introduced, effected by fiat, for human consciousness was to construct society by its means. The conscious planning of the social whole was to take the place of the unconscious functions of the market.

It is this will to make the social whole available to conscious intervention that distinguished socialism from capitalism, which allowed society to sink into anarchy and chaos under the banner of the abstract unity of market and money (1923a, p. 149), and also distingushed it from the tendencies which tended to transcend capitalism from within (1911, p. 69), but were not consciously unified in a new social organisation. To harness this will the workers' movement required Utopias, *pictures* that brought into plain view how society as a whole could function differently. It needed them immediately after World War I in order to be prepared for the imminent task of reorganisation, and it needed them all the more since their realisation once again receded into the distance: these Utopias prevented it from being forgotten in the course of the conquest of power positions in the capitalist market economy that the goal lay in the reorganisation of life-order as a whole.

These pictures are, as noted, 'thought-constructs': the limits which in earlier times had been drawn by instinct or tradition are not encountered in their development by the imagination and by the principles of scientific construction, nor in their realisation.

Just as we developed biotechnology—changed the sex of animals, had lizards grow feet on their backs, created animals whose left hand side differed from their right—so we shall develop sociotechnology. We shall reach a stage at which we can manipulate life orders to a remarkable degree, that is, to structure them from the base upward according to thought-constructs (Utopias). The word 'Utopia' must stop being a term of abuse. Every machine which an engineer invents remains a Utopia for so long as it has not become a reality. The construction with which we are concerned is the historical reality, the movement of human masses, and the organisational and other conditions have to be taken into account. (1922a, p. 55.)

The fright which bests us today when confronted by such formulations should not dominate our estimation of Neurath's ideas of social technology. Neurath understood 'centralist socialism' only as a necessary step in the rectification of the chaotic capitalist conditions.

There are many indications that the great centralist organisation, which we are now approaching and which appears as the only way to overcome the chaos and the suffering due to the anachronistic present order, will again be ultimately overcome. The age of the fracturing wild economy will first be followed by the age of centralist socialism which, although it aims for the happiness of the individual, nevertheless subjects him to the collective. It may come to this that humans will turn away from an age which views the individual as a wheel in a machine, even if for his own good, and will not let him develop fully as a whole personality. Then perhaps an order will be realised in which people active in many different ways will work in smaller circles, an order in which, as the *Communist Manifesto* stresses, the difference between city and country will be overcome and in which the unification of agrarian work and commerce will be possible. But for the moment we will have to face the great tasks of the immediate future, even if already we can hear chords which will come together as the symphony of that future which will follow it in turn. (1921b, p. 288f.)

The place of the ungovernable social machine of capitalism should be taken by the governable machine of centralist socialism (ibid.). For Neurath, the creation of the socialist society was to give direction to the advanced possibilities of social technology and bind its application to the historical effort whose direction was set by the 'socialist goal' (1920a, p. 224). Building socialism was the point of view from which to make a choice between the theoretically equally valuable pictures.

As I noted also, the loss of hope for a soon-to-come revolution in the course of the 1920's at first made Neurath insist more than ever on the need for Utopias. Simultaneously, however, and increasingly so as time went on, his orientation towards an imminent revolution loses in strength, and the competition between scientific designs once again is viewed as that which is typical of the situation in 'modern societies', in which there is no uniform principle available for the decision between theoretical possibilities. And with this, rationalism itself becomes a problem and a concern.

Here I must return again to Neurath's 1913 lecture to the Philosophical Society of the University of Vienna, 'The Lost Wanderers of Descartes'. This lecture shows that 'rationalism' constitutes for him a practical attitude which tries to come to terms with the fact that the knowledge of the enlightenment put an end to the traditional social institutions of justifying decisions, but one which is incapable of replacing these. Insight into the limits of rational knowledge leads 'true rationalism' to the recognition of the importance of instinct and tradition. "He may perhaps even regret that the period of community life, in which tradition and instinct were decisive, has ended..." (1913b, p. 10.) The opponent of the rationalist is the 'pseudo-rationalist': he is governed by the wish for an ultimate rational justification of action and his argumentative constructions of justifications for decisions pretend to express a definite knowledge which is never available.

"The pseudorationalists always want to act from insight and are therefore grateful to anybody who is able to suggest to them that they had acted from insight." (Ibid., p. 8.) This holds for acts of thought just as much as for other actions. For this reason, pseudo-rationalism has the same function in "the field of thinking" as in "the field of action in the narrower sense" (ibid., p. 4). It replaces a tradition which up to then was authoritative for the production of knowledge by seemingly rational principles. The traditional institution which justified decisions in the field of thinking and which, like other institutions of social action, was destroyed by the progress of enlightenment is metaphysics. "But once tradition and community feeling are weakened, there is no choice but that between rationalism, which undoubtedly leads to drawing lots, and pseudo-rationalism which falsifies thinking and feeling." (Ibid., p. 9.)

Neurath's combative anti-metaphysics of the later 1920's and 1930's is a continuation of his fight against pseudo-rationalism, as he had already formulated it in 1913. With the 'scientific world-conception' also, he fights against the justification of decisions made in the practice of science by recourse to a supra-scientific theoretical point of reference. But now the pseudo-rationalist constructions are only accidental forms of a new opponent, revived metaphysics (e.g. 1935b), for in contrast to the pre-war years, during which pseudo-rationalism appeared to Neurath to be the only alternative to rationalism, the inter-war years presented much more direct, regressive appeals to metaphysics, appeals which no longer bothered to pretend to be rational.

Neurath's efforts were directed towards a *practical social problem* already in 1913: how was it possible to reach social decisions in full recognition of developed rationality such that the accidental and ungrounded character of every action was not repressed, but also such that action itself was not prevented? In a similar fashion, the fight against the rebounding forces of metaphysics and theology in the 1920's and 1930's does not constitute a scientific discussion of opposing theoretical views, but constitutes an intervention into *social practices* and the role which rational knowledge plays in it. The new anti-metaphysics, however, is more political than Neurath's earlier rejection of pseudo-rationalism. In his view, the violent recourse to forms of thinking—once held to have been overcome—are but part of the violence by means of which a redundant social system intended to remain victorious over the progressive forces of the revolution. In this respect, the fight against metaphysics is a practical, political activity for Neurath. Just how the will for radical social change, which found formulation in his Utopianism, found expression in his efforts towards an unmetaphysical concept of science I shall now show at least in rough outlines.

Neurath's seriousness about the political meaning of anti-metaphysics is shown by the fact that he does not view the fight against individual metaphysical expressions as a purpose sufficient onto itself, but that he subordinates it to the fight against metaphysics as a socially efficient force.

> In some cases it is possible that a group of thinkers affected by a rest of metaphysics can contribute to the overcoming of metaphysics as such. The *concrete situation* must always be judged and analysed anew. The empirical struggle between metaphysics and anti-metaphysics does not show definite battle lines, like many other struggles in the history of mankind from which it cannot be separated.[2]

For Neurath, the dependence of scientific thinking on social conditions (1929a, p. 340) is not only of importance for an analytical and historical consideration of the phenomenon of science, but even more so for the evaluation and integration of scientific activity into the current social movements and their effect on the future. His concern to draw consequences from the social dependence of science leads Neurath to 'unified science'. This concept is, as he stresses again and again, not a "logical consequence" of his program, but a "historical fact in a sociological sense" (1935a, p. 115). For him, the 'unity of science' is the result of a historical effort of scientists to understand their actual scientific procedures; under this banner, Neurath intended for science to become a transparent and socially available whole. By contrast, the metaphysical concept of science produces and justifies a wild growth of specialisms on the one hand, and the theoretical formulation of irrationalisms on the other. For as long as it is pretended that knowledge constitutes a uniform system[3]—the claim of metaphysics—real existing scientific knowledge remains uncontrollable, for it bars from view the historical decisions and accidents which it contains.

Against this, Neurath places the Utopia of a transparent organisation of the production and dissemination of scientific knowledge. As in the case of the transition of the abstract unity of market and money in centralist socialism, what is at issue here is the replacement of the abstract principle of unity by the organisation of unification according to historical necessities. The dissolution of the unity of the subject—the principle of modern metaphysics—thus becomes a recurring theme of Neurath's (e.g. 1931c, p. 416). As understood by Neurath—in contrast to the systematic, metaphysical conception—science constructs different collections or wholes of sentences which are consistent in themselves. These stand in competition with each other, however, and every step in the development of one of these consistent theories calls up and topicalises the competion between the theories. For if a new sentence cannot be integrated without trouble, that is, if it contradicts a given system of sentences, then either that sentence must be rejected or the entire system must be revised (1932d, p. 95). In this respect, "the whole of science is basically always under discussion" (1935a, p. 118) and reveals its artificial character, that is, its dependence upon practical decisions. That which politicians had to do with utopias, namely, select one, that was now what scientists did by expanding the "network" of science (Carnap, Hahn, Neurath 1929, p. 306). Now *they* are the practical men and women by deciding how to build and rebuild the scientific machine (1932d, p. 98). Neurath's Utopia was that the scientists' collectively gained consciousness of the artificial and practical nature of their knowledge would lay bare the functioning of science in its entirety and so make it manageable and usable—just as after World War I the entire life order was to have been made manageable.

The failure of the workers revolutions and the rise of fascism did *not* lead Neurath to develop a theory of *prevented* praxis, but to uphold and conceive of the Utopia of the radical restructuring of social conditions with different means. In all of Neurath's Utopian inventions—amongst which his picture statistics would also have to be counted—we can find a common intention: his is the aim to make individuals conscious of the threatening and breathtaking limitlessnes of the scientific rationality operative in modern societies—and to empower them to use it.[4]

That the given developed conditions have to be taken as starting point and that it is never possible to begin at point zero is something which Neurath often remarked upon (e.g. 1935a, p. 118). The historicism in which we are today engulfed makes it initially easy for us to concur. But Neurath's question was how, given this insight, it was still possible to do anything of consequence. And if I am not mistaken, then the disquiet and the will to hope, which lie in this question, have become so alien to the 'scientifically oriented' people of our day that they may already be readmitted as exotic stimuli and remembrances of better times.

Universität Wien

NOTES

* First published as 'Otto Neuraths Utopien - Der Wille zur Hoffnung', in *Arbeiterbildung in der Zwischenkriegszeit*, Hrsg. F. Stadler, © 1982, Österreichisches Gesellschafts- und Wirtschaftsmuseum, Wien, pp. 94-99. Translated with kind permission of Österreichisches Gesellschafts- und Wirtschaftsmuseum and the author by T. E. Uebel.

1 Except where noted, all references will be to the writings of Otto Neurath.

2 Neurath 1931c, (1981, p. 425; not translated in Neurath 1973).

3 "'The' system is the great scientific lie." (1935a, p. 116.)

4 The connections suggested here are further explored in Nehmet 1981.

BIBLIOGRAPHY AND NAME INDEX

BIBLIOGRAPHY

V. Adler, 1902, 'Rede zur Festversammlung des Arbeiter-Bildungsvereins Gumpendorf', in *Aufsätze, Reden, Briefe*, Heft 11, Wien, 1922.

M. Aebi, 1947, *Kants Begründung der deutschen Philosophie*, Verlag für Recht und Gesellschaft, Zürich-Basel.

Anon., n. d., *Wiener Internationale Hochschulkurse 1922-1972*, Wien.

Anon. (ed.), 1965, *Österreich—Geistige Provinz?*, Forum, Wien-Hannover-Bern.

Arbeiterkammer (eds.), 1977, *Österreicher im Exil 1934-1945*, Wien.

R. G. Ardelt, 1972, *Zwischen Demokratie und Faschismus. Deutschnationales Gedankengut in Österreich 1919-1930*, Veröffentlichungen d. Historischen Instituts d. Universität Salzburg 5, Geyer, Wien-Salzburg.

G. Arntz, 1982, 'Otto Neurath, ich und die Bildstatistik', in Stadler (ed.) 1982a, pp. 31-34.

G. Arntz, 1988, *Zeit unterm Messer. Holz- und Linolschnitte 1920-1970*, Köln.

A. J. Ayer, 1936, *Language, Truth and Logic*, 2nd ed., Gollancz, London, 1946, repr. Dover, New York.

A. J. Ayer, 1956, 'The Vienna Circle', in *Revolution in Philosophy*, Macmillan, London/St. Martin's Press, New York, pp. 70-87.

A. J. Ayer (ed.), 1959, *Logical Positivism*, Free Press, New York.

G. P. Baker, 1979, 'Verehrung und Verkehrung', in Luckhardt (ed.) 1979, pp. 243-285.

W. Balzer, D. A. Pearce, H. J. Schmidt (eds.), 1984, *Reduction in Science. Structure, Examples, Philosophical Problems*, Reidel, Dordrecht.

S. F. Barker, P. Achinstein (eds.), 1969, *The Legacy of Logical Positivism*, Johns Hopkins University Press, Baltimore.

F. Barone, 1982, 'Protocol Sentences and Scientific Anarchism', in Haller (ed.) 1982e, pp. 327-346.

F. Barone, 1984, 'La polemique sur les énoncés protocolaire dans l'epistemologie du Cercle de Vienne', in Sebestik, Soulez (eds.) 1984, pp. 335-349.

W. W. Bartley, 1977, *Wittgenstein*, Quartet Books, London.

G. Bataille, 1933, 'Le concept de l'expenditure', *La Critique sociale* 7 (1933) 7-15, transl. A. Stoekl 'The Concept of Expenditure', in *Visions of Excess. Secected Writings 1927-1939* (ed. by A. Stoekl), University of Minnesota Press, Minneapolis, 1985.

H. Bauer, 1923, 'Geld, Sozialismus und Otto Neurath', *Der Kampf* 16 (1923) 195-202.

R. Bauer, 1966, *Der Idealismus und seine Gegner in Österreich* (*Euphorion* Beiheft 3), Winter, Heidelberg.

A. Beckermann, 1979, 'Logischer Positivismus und radikale Gesellschaftsreform', *Analyse & Kritik* 1 (1979) 30-46.

H. Berghel, A. Hübner, E. Köhler (eds.), 1979, *Wittgenstein, der Wiener Kreis und der Kritische Rationalismus. Akten des 3. Internationalen Wittgenstein-Symposiums 1978*, Schriften der Wittgenstein-Gesellschaft Bd. 3, Hölder-Pichler-Tempsky, Wien.

G. Bergmann, 1954, *The Metaphysics of Logical Positivism*, Longmans Green, New York-London.

H. Bergson, 1903, *Introduction a la Metaphysique*, Paris, transl. T. E. Hulme *An Introduction to Metaphysics*, London, 1913.

L. Bergström, 1982, 'Interpersonal Utility Comparisons', in Haller (ed.) 1982e, pp. 283-312.

E. Beth, 1938, 'Het Wetenschapsbegrip bij Nietzsche en bij de Neo-Positivisten', *De Gids* 5 (1938).

E. Beth, 1965, *Mathematical Thought*, Reidel, Dordrecht.

H. Blumenberg, 1987, *Die Sorge geht über den Fluß*, Suhrkamp, Frankfurt.

G. Boole, 1854, *An Investigation of the Laws of Thought*, London.

F. Brentano, 1895, *Die vier Phasen der Philosophie*, repr. Meiner, Hamburg, 1968.

E. Broda, n. d., *Österreichische Naturwissenschaftler im Exil*, n. p.

E. Broda, 1955, *Ludwig Boltzmann, Mensch-Physiker-Philosoph*, Deuticke, Wien, transl. L. Gay, E. Broda *Ludwig Boltzmann, Man-Physicist-Philosopher*, Ox Bow Press, Woodbridge, Conn.

E. Broda, 1967, 'Warum ist es in Österreich um die Naturwissenschaften so schlecht bestellt?', in H. Fischer (ed.) 1967, pp. 87-108.

E. Broda, 1978, 'Das Jahr 1938 und die Naturwissenschaft in Österreich', *Forschungen und Beiträge zur Wiener Stadtgeschichte* 2 (1978) 230-236.

H. Burkhardt, 1980, *Logik und Semiotik in der Philosophie von Leibniz*, Philosophia, München.

R. Carnap, 1928a, *Der logische Aufbau der Welt*, Bernary, Berlin, 2nd. ed. Meiner, Hamburg, 1961, transl. R. A. George *The Logical Structure of the World / Pseudoproblems in Philosophy*, University of California Press, Berkeley, 1967.

R. Carnap, 1928b, *Scheinprobleme in der Philosophie*, Bernary, Berlin, transl. R. A. George *The Logical Structure of the World / Pseudoproblems in Philosophy*, University of California Press, Berkeley, 1967.

R. Carnap, 1932a, 'Überwindung der Metaphysik durch logische Analyse der Sprache', *Erkenntnis* 2 (1931/2) 219-241, transl. A. Pap with remarks by author 'The Elimination of Metaphysics through Logical Analysis of Language' in Ayer (ed.) 1959, pp. 60-81.

R. Carnap, 1932b, 'Die physikalische Sprache als Unversalsprache der Wissenschaft', *Erkenntnis* 2 (1931/2) 432-465, transl. M. Black with author's introduction *The Unity of Science*, Kegan Paul, Trench Trubner & Co., London,1934.

R. Carnap, 1932c, 'Psychologie in physikalischer Sprache', *Erkenntnis* 3 (1931/2) 107-142, transl. with remarks by author 'Psychology in Physicalist Language' in Ayer (ed.) 1959a, pp. 165-198.

R. Carnap, 1932d, 'Erwiderung auf die vorstehenden Aufsätze von E. Zilsel und K. Duncker', *Erkenntnis* 3 (1931/2) 177-188.

R. Carnap, 1932e, 'Über Protokollsätze', *Erkenntnis* 3 (1931/2) 215-228, transl. R. Creath, R. Nollan 'On Protocol Sentences', *Nous* 21 (1987) 457–470.

R. Carnap, 1934, 'Theoretische Fragen und praktische Entscheidungen, *Natur und Geist* 2 (1934) 257-60, repr. in Schleichert (ed.) 1975, pp.173-6.

R. Carnap, 1934/7, *Logische Syntax der Sprache*, Schriften zur wissenschaftlichen Weltauffassung Bd. 8, Springer, Wien, 1934, 2nd ed. Wien, 1968, rev. ed. transl. A. Smeaton *The Logical Syntax of Language*, Kegan Paul, Trench Trubner & Co., London, 1937, repr. Littlefield, Adams & Co., Paterson, N. J., 1959.

R. Carnap, 1935, 'Ein Gültigkeitskriterium für die Sätze der klassischen Mathematik', *Monatshefte für Mathematik und Physik* 42 (1935) 163-190.

R. Carnap, 1936/7, 'Testability and Meaning', *Philosophy of Science* 3 (1936) 419-471, 4 (1937) 1-40, repr. with corrigenda and additions, Yale Graduate Philosophy Club, New Haven, 1954, and in *Classics of Analytical Philosophy*, (ed. by R. R. Ammermann), New York, 1965, pp. 130-195.

R. Carnap, 1942, *Introduction to Semantics*, repr. as vol.1 of *Studies in Semantics*, Harvard University Press, Cambridge, Mass., 1961.

R. Carnap, 1947, *Meaning and Necessity*, 2nd ed. with supplementary essays, University of Chicago Press, Chicago, 1956.

R. Carnap, 1950, 'Empiricism, Semantics and Ontology', *Revue Internationale de Philosophie* 4 (1950) 20-40, repr. in 2nd ed. of Carnap 1947 (1956), pp. 205-221.

R. Carnap, 1963a, 'Intellectual Autobiography', in Schilpp (ed.) 1963, pp. 3-84.

R. Carnap, 1963b, 'Comments and Replies', in Schilpp (ed.) 1963, pp. 859-1016.

R. Carnap, H. Hahn, O. Neurath, 1929,*Wissenschaftliche Weltauffassung - Der Wiener Kreis*, Veröffentlichungen des Vereins Ernst Mach H. 2, Artur Wolf, Wien, repr. in Schleichert (ed.) 1975, pp. 201-222 (without the annotated bibliography), and in Neurath 1981, pp. 299-336 (with the annotated bibliography), transl. in Neurath 1973, pp. 299-318 (without the annotated bibliography).

N. Cartwright, 1990, 'Otto Neurath. Marxism, Materialism and Pseudorationalism', Lecture, History of Science Society, October 1990, Seattle.

G. Cat, H. Chang, N. Cartwright, 1991, 'Otto Neurath: Unification as the Way to Socialism', in *Einheit der Wissenschaften* (ed. by J. Mittelstrass), Akademie der Wissenschaften, Berlin.

R. M. Chisholm 1982, 'Schlick on the Foundations of Knowing', in Haller (ed.) 1982e, pp. 149-158.

M. Clavelin, 1980, 'Le deux Positivismes de Cercle de Vienne', *Archives de Philosophie* 43 (1980) 33-56.

A. Coffa, 1977, 'Carnap's *Sprachanschauung* ca. 1932', *Proceedings of the Philosophy of Science Association 1976 Vol. 2* (ed. by P. Suppe, F. Asquith), East Lansing, Mich., pp. 205-241.

A. Coffa, 1991, *To The Vienna Station. Semantics, Epistemology and the A Priori from Kant to Carnap*, forthcoming.

R. S. Cohen, 1956, 'Alternative Interpretations of the History of Science', in Frank 1956b, pp. 218-232.

R. S. Cohen, 1963, 'Dialectical Materialism and Carnap's Logical Empiricism', in Schilpp (ed.) 1963, pp. 99-158.

R. S. Cohen, 1967, 'Neurath, Otto', in Edwards *et al*, (eds.) 1967, vol. 6, pp. 477-479.

R. S. Cohen, R. J. Seeger (eds.) 1970, *Ernst Mach: Physicist and Philosopher*, Boston Studies in the Philosophy of Science vol. 6, Reidel, Dordrecht.

R. S. Cohen, M. Wartofsky (eds.), 1965, *Boston Studies in the Philosophy of Science Vol. II (In Honor of Philipp Frank)*, Humanities Press, New York.

R. Creath, 1991, 'Contribution to Carnap Centenary', in Fine, Forbes, Wessell (eds.) 1991.

A. D'Acconti, 1986, 'Note Sull'Olismo Neurathiano', *Annali della Facolta di Lettere e Filosophia*, Universita di Bari, 29 (1986) 409-441.

H. Dachs, 1974, *Österreichische Geschichtswissenschaft und Anschluss*, Veröffentlichungen d. Historischen Instituts d. Universität Salzburg Bd. 9, Geyer, Salzburg.

H.-J. Dahms (ed.), 1985a, *Philosophie, Wissenschaft und Aufklärung*, De Gruyter, Berlin-New York.

H.-J. Dahms, 1985b, 'Versuch einer Charakterisierung des Wiener Kreises', in Dahms (ed.) 1985a, pp. 1-29.

H.-J. Dahms, 1985c, 'Vertreibung und Emigration des Wiener Kreises zwischen 1931 und 1940', in Dahms (ed.) 1985, pp. 307-365.

D. Davidson, 1982, 'Empirical Content', in Haller (ed.) 1982e, pp. 471ff.

W. Del Negro, 1923, 'Die Rolle der Fiktionen in der Erkenntnistheorie Friedrich Nietsches', *Baustein* 5 (1923)

A. DeMorgan, 1847, *Formal Logic*, London.

A. Diamant, 1960, *Austrian Catholics and the First Republic: Democracy, Capitalism and the Social Order 1918-1934*, Princeton University Press, Princeton, N. J..

W. Diederich, 1974, *Konventionalität in der Physik*, Duncker & Humboldt, Berlin.

B. Dreben, 1991, 'Contribution to Carnap Centenary', in A. Fine, M. Forbes, L. Wessels (eds.) 1991.

F. I. Dretske, 1981, *Knowledge and the Flow of Information*, M.I.T. Press, Cambridge, Mass..

P. Duhem, 1906, *La Theorie Physique: Son Object Sa Structure*, Marcel Rivière & Cie., Paris, transl. F. Adler, with foreword by E. Mach, *Ziel und Struktur der physikalischen Theorien*, Barth, Leipzig, 1908, repr. with an introduction and bibliography (ed. by L. Schäfer), Meiner, Hamburg, 1978, transl. P. Wiener *The Aim and Structure of Physical Theory*, Princeton University Press, Princeton, N.J., 1954, 2nd ed. Atheneum, New York, 1962.

J. Dvorak, 1981a, *Edgar Zilsel und die Einheit der Erkenntnis*, Veröffentlichungen des Ludwig-Boltzmann-Institutes für Geschichte der Gesellschaftswissenschaften Bd. 6, Löcker, Wien

J. Dvorak, 1981b, 'Kultur und Arbeiterkultur', *Zeitgeschichte* Heft 11/12 (1981) 466-472.

J. Dvorak, 1982a, 'Otto Neurath und die Volksbildung - Einheit der Wissenschaft, Materialismus und umfassende Aufklärung', in Stadler, (ed.) 1982a, pp. 149-156, transl. T. E. Uebel 'Otto Neurath and Adult Education - Unity of Science, Materialism and Comprehensive Enlightenment', in this volume, ch. 20.

J. Dvorak, 1982b, 'Wahrscheinlichkeit, Logik, Empirie: Neurath, Schlick und Zilsel', in Haller (ed.) 1982e, pp. 465-470.

J. Dvorak, 1985, 'Wissenschaftliche Weltauffassung, Volkshochschule, und Arbeiterbildung in der Zwischenkriegszeit', in Dahms (ed.) 1985a, pp. 129-143.

J. Dvorak (ed.), 1990, *Spuren und Zusammenhänge. Zur Kultur des Roten Wien 1918-1934*, Wien-München.

R. Ebneth, 1976, *Die österreichische Wochenschrift 'Der Christliche Ständestaat'. Deutsche Emigration in Österreich 1933-1938*, Veröffentlichungen der Kommission für Zeitgeschichte. Reihe B: Forschungen Bd. 19, Grünewald, Mainz.

P. Edwards *et al* (eds.),1967, *Encyclopedia of Philosophy*, 8 volumes, Macmillan & The Free Press, New York.

P. K. Esser, 1984, 'Review: Neurath Philosophical Writings 1913-1945', *Methodology and Science* 17 (1984) 126-136.

H. Feigl, 1937/8, 'Moritz Schlick', *Erkenntnis* 7 (1937/8) 393-419, transl. P. Heath 'Moritz Schlick, A Memoir', in Schlick 1979b, pp. xv-xxxviii.

H. Feigl, 1963, 'Physicalism, Unity of Science and the Foundations of Psychology', in Schilpp (ed.) 1963, pp. 227-268.

H. Feigl, 1969a, 'The Wiener Kreis in America', in *The Intellectual Migration* (ed. by D. Fleming and B. Baylin), Harvard University Press, Cambridge Mass., repr. in Feigl 1981, pp. 57-93.

H. Feigl 1981, *Inquiries and Provocations* (ed. by R. S. Cohen), Vienna Circle Collection vol.14, Reidel, Dordrecht.

B. Fenz, 1978, 'Zur Ideologie der Volksbürgerschaft', *Zeitgeschichte* 5 no. 4 (1978) 125-145.

L. Fermi, 1968, *Illustrious Immigrants. The Intellectual Migration from Europe 1930-1941*, University of Chicago Press, Chicago-London.

P. K. Feyerabend, 1966, 'Herbert Feigl: A Biographical Sketch', in *Mind, Matter, Method. Essays in Philosophy and Science in Honor of Herbert Feigl* (ed. by P. K. Feyerabend, G. Maxwell), University of Minnesota Press, Minneapolis.

P. K. Feyerabend, 1970, 'Against Method: Outline of an Anarchist Thjeory of Knowledge', in *Minnesota Studies in the Philosophy of Science IV* (ed. by M. Radner and S. Winokur), University of Minnesota Press, Minneapolis.

P. K. Feyerabend, 1978, 'From Incompetent Professionalism to Professional Incompetence', *Philosophy of Social Science* **8** (1978) 37ff., repr. in Feyerabend *Science in a free society*, New Left Books, London, 1978, repr. 1983, pp. 183-209, partly repr. as 'Philosophy of Science vs. Scientific Practice. Observations on Mach, his Followers and Opponents' in *Philosophical Papers vol. 2. Problems of Empiricism*, Cambridge University Press, Cambridge, 1981.

A. Fine, M. Forbes, L. Wessels (eds), 1991, *PSA 1990 vol. 2*, Philosophy of Science Association, East Lansing, Mich.

H. Fischer, 1965, 'Universität zwischen Tradition und Fortschritt', in Anon. 1965, pp. 204-231.

H. Fischer (ed.), 1967, *Versäumnisse und Chancen. Beiträge zur Hochschulfrage in Österreich*, Forum-Verlag, Wien-Hannover.

K. R. Fischer, 1982a, 'Analytische Philosophie und Marxismus. Der Fall Otto Neurath', in F. Stadler (ed.) 1982a, pp. 88-93.

K. R. Fischer, 1982b, 'Nietzsche und der Wiener Kreis', in Haller (ed.) 1982e, pp. 255-270.

M. Fischer-Kowalski, 1977, 'Universität und Gesellschaft in Österreich', in *Das politische System Österreichs* (ed. by H. Fischer), Europaverlag, Wien.

F. Fistetti, 1985, *Neurath contro Popper*, Dedalo, Bari.

K. Fleck, 1979, *Otto Neurath. Eine biographisch-systematische Untersuchung*, Phil. Diss., Graz.

L. Fleck [=K. Fleck], 1982a, 'Otto Neurath's Beitrag zur Theorie der Sozialwissenschaften', in F. Stadler (ed.) 1982a, pp. 100-103, transl. T. E. Uebel 'Otto Neurath's Contribution to the Theory of the Social Sciences', in this volume, ch. 16.

L. Fleck [=K. Fleck], 1982b, 'Über Anna Schapire-Neurath', in Stadler (ed.) 1982a, pp. 229-230.

J. Frank, 1931, *Architektur als Symbol*, Schroll, Wien, repr. Löcker, Wien, 1981.

P. Frank, 1907, 'Kausalgesetz und Erfahrung', *Annalen der Naturphilosophie* **6** (1907) 443-450, transl. 'Experience and the Law of Causality' in Frank 1949b, pp. 53-60.

P. Frank, 1917, 'Die Bedeutung der physikalischen Erkenntnistheorie Ernst Machs für das Geisteslebens unserer Zeit', *Die Naturwissenschaften* **5** (1917) 65ff., transl. 'The Importance of Ernst Mach's Philosophy of Science for our Times', in Frank 1949b, pp. 61-79, repr. in Cohen, Seeger (eds.) 1970, pp. 219-234..

P. Frank,1932, *Das Kausalgesetz und seine Grenzen*, Schriften zur Wissenschaftlichen Weltauffassung, Bd. 6, Springer, Wien, repr. Suhrkamp, Frankfurt a.M., 1986, transl. *The Law of Causality and its Limitations* (ed. by R. S. Cohen), Vienna Circle Collection, Kluwer, Dordrecht, 1991.

P. Frank, 1934, 'Hans Hahn', *Erkenntnis* **4** (1934) 315-316.

P. Frank, 1938, 'Ernst Mach—the Centenary of his Birth', Erkenntnis 7 (1938) 247ff., repr. as 'Ernst Mach and the Unity of Science' in Frank 1949b, pp. 79-89, repr. in Cohen, Seeger (eds.) 1970, pp. 235-244.

P. Frank, 1946, 'Science Teaching and the Humanities', *Etc.: A Review of General Semantics* **4** (1946) no. 3, repr. in Frank 1949b, pp. 260-285.

P. Frank, 1949a, 'Introduction. Historical Background', in Frank 1949b, pp.1-51.

P. Frank, 1949b, *Modern Science and its Philosophy*, Harvard University Press, Cambridge, Mass..

P. Frank, 1956a, 'The Variety of Reasons for the Acceptance of Scientific Theories', in Frank 1956b, pp. 3-17.

P. Frank, 1956b, *The Validation of Scientific Theories*, Beacon Press, Boston.

P. Frank, 1957, *Philosophy of Science. The Link between Science and Philosophy*, Prentice-Hall, Englewood Cliffs, N. J..

G. Frege, 1892, 'Über Sinn und Bedeutung', *Zeitschrift für Philosophie und philosophische Kritik* Neue Folge **100** (1892) 25-50, transl. H. Feigl 'On Sense and Nominatum' in *Readings in Philosophical Analysis* (ed. by H. Feigl, W. Sellars), Appleton-Century-Crofts, New York, 1949 also M. Black 'On Sense and Meaning' in *Translations from the Philosophical Writings of Gottlob Frege* (ed. by P. Geach, M. Black), Blackwell, Oxford/Rowman & Littlefield, N.J., 3rd ed. 1980, pp. 56-78.

G. Frege, 1893-1903, *Grundgesetze der Arithmetik*, 2 vols., Jena, repr. Olms, Hildesheim 1962, selections transl. M. Furth *The Basic Laws of Arithmetic* (ed. by M. Furth), University of California Press, Berkeley, 1964.

G. Freudenthal, 1989, 'Otto Neurath: From Authoritarian Liberalism to Empiricism', in *Knowledge and Politics* (ed. by M. Dascal, O. Gruengard), Westview Press, Boulder, pp. 207-240.

M. Friedman, 1983, 'Critical Notice: M. Schlick, *Collected Papers*', *Philosophy of Science* **50** (1983) 498-514.

M. Friedman, 1987, 'Carnap's *Aufbau* Reconsidered', *Nous* **21** (1987) 521-545.

M. Friedman, 1988, 'Logical Truth and Analyticity in Carnap's *Logical Syntax of Language*', in *Essays in the History and Philosophy of Mathematics* (ed. by W. Aspray, P. Kitcher), University of Minnesota Press, Minneapolis, pp. 82-94.

M. Friedman, 1991, 'Contribution to Carnap Centenary', in Fine, Forbes, Wessels (eds.) 1991.

A. Fuchs, 1949, *Geistige Strömungen in Österreich 1867-1918*, Globus-Verlag, Vienna.

E. T. Gadol (ed.), 1982, *Rationality and Science*, Springer, Wien.

P. Galison, 1990, 'Aufbau/Bauhaus: Logical Positivism and Architectural Modernism', *Critical Inquiry* **16** (1990) no. 4.

H. Geiringer-Mises, 1965, 'To Philipp Frank', in R. S. Cohen, M. Wartofsky (eds.) 1965, pp. xxi-xxiii.

C. Giannoni, 1967, 'Quine, Grünbaum and the Duhem-Thesis', *Nous* **1** (1967) 287 ff., repr. in Harding (ed.) 1976, pp. 162-175

E. Glaser, 1982, 'Otto Neurath und der Austromarxismus', in Stadler (ed.) 1982, pp. 126-132.

J. W. v. Goethe, 1977, *Naturwissenschaftliche Schriften*, in *Gesammelte Werke* Bd. 17 (ed. by. E. Trunz), Artemis, München-Zürich.

R. Goldscheid, 1926, 'Die Pflanzstätten der Wissenschaft als Brutstätten der Reaktion', *Arbeit und Wirtschaft* **5** no. 4 (1926).

H. Gomperz, 1953, 'Autobiographical Remarks', in *Philosophical Studies*, Christopher Publishing House, Boston, pp. 15-28.

H. Grümm, 1950, 'Der Philosoph Kraft', *Tagebuch* **5** no. 20 (30 September 1950).

A. Grünbaum, 1960, 'The Duhemian Argument', *Philosophy of Science* **27** (1960) 75-87, repr. in Harding (ed.) 1976, pp. 116-131.

H. Hahn, 1924a, 'Lehr- und Lernfreiheit an den Hochschulen', *Die Wage* **10** (May 1924) 312-18.

H. Hahn, 1924b, 'Lehr- und Lernfreiheit an den Hochschulen', *Der Kampf* **17** (1924) 169-175.

H. Hahn, 1926, 'Universität und Schulreform', *Schulreform* **5** (1926).
H. Hahn, 1930, *Überflüssige Wesenheiten (Occam's Rasiermesser)*, Veröffentlichungen des Vereins Ernst Mach vol. 2, Wolf, Wien, repr. in Schleichert (ed.) 1975, pp. 95-117, transl. H. Kaal 'Superfluous Entities (Occam's Razor)' in Hahn 1980, pp. 1-19.
H. Hahn, 1980, *Empiricism, Logic, Mathematics* (ed. by B. McGuiness), Vienna Circle Collection vol.13, Reidel, Dordrecht.
R. Haller, 1966, 'Der "Wiener Kreis" und die analytische Philosophie', in *Forschung und Fortschritt* (ed. by F. Sauer), Graz, pp. 33-46, repr. in Haller 1979b, pp. 79-98.
R. Haller, 1968, 'Wittgenstein und die österreichische Philosophie', in *Wissenschaft und Weltbild* **21** (1968) H. 2/3, 77-87, repr. in Haller 1979b, pp. 107-121, transl. B. Smith 'Wittgenstein and Austrian Philosophy' in J. C. Nyiri (ed.) 1981, pp. 91-112 and Haller 1988a, pp.1-26.
R. Haller, 1971, 'Eine Bemerkung zu Theorie und Deskription', in *Abstracts des IV. Int. Kongresses für Logik, Methodologie und Philosophie der Wissenschaften*, Bucharest, 1971.
R. Haller, 1974, 'Über das sogenannte Münchhausen-Trilemma', *Ratio* **16** (1974) 113-127.
R. Haller, 1975, 'Wittgenstein und die "Wiener Schule"', in *Dauer im Wandel. Aspekte österreichischer Kulturentwicklung* (ed. by W. Strolz & O. Schatz), Herder, Wien-Freiburg-Basel, pp. 137-162, repr. in Haller 1979b, pp. 163-187.
R. Haller 1977a, 'Bemerkungen zum Problem der kumulativen Wissens', in *Probleme des Erkenntnisfortschritts in den Wissenschaften* (ed. by K. Freisitzer, and R. Haller), VWGÖ, Wien.
R. Haller, 1977b, 'Österreichische Philosophie', in Marek et al (eds.) 1977, pp. 57-66.
R. Haller, 1979a, 'Über Otto Neurath', in Haller 1979b, pp.99-106, transl. T. E. Uebel 'On Otto Neurath', in this volume, ch. 2.
R. Haller, 1979b, *Studien zur Österreichischen Philosphie. Variationen über ein Thema*, Studien zur österreichischen Philosophie Bd. 1, Rodopi, Amsterdam.
R. Haller, 1979c, 'Geschichte und wissenschaftliches System bei Otto Neurath', in Berghel, Hübner, Köhler (eds.) 1979, pp. 302-307, repr. in Haller 1986b, pp. 125-134, transl. T. E. Uebel 'History and the System of Science in Otto Neurath' in this volume, ch. 3.
R. Haller, 1979d, 'Gibt es eine Österreichische Philosophie?', *Wissenschaft und Weltbild* **31** (1979) 173-181, repr. in Haller 1986b, pp. 31-44.
R. Haller, 1979e, 'Über Phasen der Wissenschaft', in Salamun (ed.) 1979.
R. Haller 1981, 'Poetic Imagination and Economy. E. Mach as Theorist of Science', in *Scientific Philosophy Today* (ed. by J. Agassi & R. S. Cohen), Reidel, Dordrecht.
R. Haller, 1982a, 'Das Neurath-Prinzip—Grundlagen und Folgerungen', in Stadler (ed.), 1982a, pp. 79-87, also in Dahms (ed.), 1985, pp. 205-220, repr. in Haller 1986b, pp. 108-124, transl. T. E. Uebel 'The Neurath-Principle. Its Grounds and Consequences' in this volume, ch. 9.
R. Haller, 1982b, 'New Light on the Vienna Circle', *The Monist* **65** (1982) 25-37.
R. Haller, 1982c, 'Zwei Arten der Erfahrungsbegründung', in Haller (ed.) 1982e, pp. 19-33, transl. T. E. Uebel 'Two Ways of Experiential Justification' in this volume ch. 15.
R. Haller (ed.), 1982e, *Schlick und Neurath - Ein Symposion, Grazer Philosophische Studien* **16/7** (1982), Rodopi, Amsterdam.
R. Haller, 1982f, 'Bedingungen, Vorrausetzungen, Verursachungen', in *Urteile und Ereignisse*, Albers, Freiburg-München, pp. 76ff.

R. Haller, 1983, 'Als ob—Zu Stephan Körner's "Idealisierungen"', *Grazer Philosophische Studien* **20** (1983) 117-127.
R. Haller, 1984a, 'Was Wittgenstein a Neopositivist?', in Sebestik, Soulez (eds.)1984, pp. 271-284, repr. in Haller 1988, pp. 27-43.
R. Haller, 1984b, *Tra soggettivismo e oggettivismo*, Napoli.
R. Haller, 1985a, 'Der erste Wiener Kreis', *Epistemology, Methodology and Philosophy of Science. Essays in Honor of C. G. Hempel on the occasion of his 80th Birthday, January 8, 1985* (ed. by W. Essler), *Erkenntnis* **22** (1985) 341-359, transl. T. E. Uebel 'The First Vienna Circle' in this volume, ch. 7.
R. Haller, 1985b, 'Problems of Knowledge in Moritz Schlick', in McGuinness ed. 1985, pp. 283-296.
R. Haller, 1986a, 'Zur Historiographie der österreichischen Philosophie', in Nyiri (ed.) 1986, pp. 41-53, transl. T. E. Uebel 'On the Historiography of Austrian Philosophy', in this volume, ch. 4.
R. Haller, 1986b, *Fragen zu Wittgenstein und Aufsätze zur Österreichischen Philosophie*, Studien zur Österreichischen Philosophie Bd. 10, Rodopi, Amsterdam.
R. Haller, 1988a, *Questions on Wittgenstein*, University of Nebraska Press.
R. Haller, 1988b, 'Justification and Praxeological Foundationalism', *Inquiry* **31** (1988) 335-345.
R. Haller, H. Rutte, 1977, 'Gespräch mit Heinrich Neider, Wien: Persönliche Erinnerungen an den Wiener Kreis', in Marek et al (eds.) 1977, pp. 21-42.
R. Haller, H. Rutte, 1981, 'Vorwort', in Neurath 1981, pp. ix-xvi.
O. Hanfling, 1981, *Logical Positivism*, Columbia University Press, New York.
E. Hanisch, 1977, *Die Ideologie des Politischen Katholizismus in Österreich 1918-1938*, Geyer, Wien/Salzburg.
N. R. Hanson, 1958, *Patterns of Discovery*, Cambridge University Press, Cambridge.
S. G. Harding (ed.), 1976, *Can Theories be Refuted? Essays on the Duhem-Quine Thesis*, Reidel, Dordrecht.
R. Harris, 1988, *Language, Saussure and Wittgenstein. How to Play Games with Words*, London-New York.
J. Harsanyi, 1977, *Bargaining Equilibrium in Games and Social Situations*, Cambridge University Press, Cambridge.
F. Heer, 1957, 'Humanitas Austriaca', in *Spectrum Austriae* (ed. by O. Schulmeister u. a.), Herder, Wien.
R. Hegselmann 1979a, 'Otto Neurath—Empiristischer Aufklärer und Sozialreformer', in Neurath 1979, pp. 7-78.
R. Hegselmann, 1979b, *Normativität und Rationalität*, Campus, Frankfurt a.M.-New York.
R. Hegselmann, 1979c, 'Grenzen der wissenschaftlichen Weltauffassung des Wiener Kreises', *Analyse & Kritik* **1** (1979) 47-50.
R. Hegselmann, 1984, 'Logischer Empirismus und Ethik', in M. Schlick 1930b as repr. 1984, pp. 7-46.
R. Hegselmann, 1985, 'Die Korrespondenz zwischen Otto Neurath und Rudolf Carnap aus den Jahren 1934-1945', in Dahms (ed.) 1985a, pp. 276-290.
R. Hegselmann, 1987, 'Introduction', in *Unified Science* (ed. by B. McGuinness), Vienna Circle Collection vol. 19, Reidel, Dordrecht.
M. Heidelberger, 1985, 'Zerspaltung und Einheit: vom logischen Aufbau der Welt zum Physikalismus', in Dahms (ed.) 1985a, pp. 144-189.
C. G. Hempel, 1935a, 'On the Logical Positivists' Theory of Truth', *Analysis* **2** (1935) 49-59.

C. G. Hempel, 1969, 'Logical Positivism and the Social Sciences', in Barker, Achinstein (eds.) 1969, pp. 163-194.
C. G. Hempel, 1979, 'Der Wiener Kreis: Eine persönliche Perspektive', in Berghel, Hübner, Köhler (eds.) 1979, pp. 21-26.
C. G. Hempel, 1981, 'Der Wiener Kreis und die Metamorphosen seines Empirismus', in Leser (ed.) 1981, pp. 205-215.
C. G. Hempel, 1982, 'Schlick und Neurath: Fundierung vs. Kohärenz in der wissenschaftlichen Erkenntnis', in Haller (ed.) 1982e, pp. 1-18.
D. Hilbert, P. Bernays, 1934, *Grundlagen der Mathematik*, Springer, Berlin, repr. Heidelberg 1975.
W. Hösl, G. Pirhofer, 1982, 'Otto Neurath und der Städtebau', in Stadler (ed.) 1982a, pp. 157-161.
R. Hoffmann, 1982, 'Proletarisches Siedeln. Otto Neurath's Engagement für die Wiener Siedlungsbewegung und den Gildensozialismus 1920 bis 1925', in Stadler (ed.) 1982a, pp. 140-148.
F. Hofmann-Grüneberg, 1986, 'Wahrheit im Universalslang', in *Die Aufgabe der Philosophie in der Gegenwart. Akten des 10. Internationalen Wittgenstein Symposiums* (ed. by W. Leinfellner, F. M. Wuketits), Schriften der Wittgenstein-Gesellschaft Bd. 12/1, Hölder-Pichler-Tempsky, Wien, pp. 217-220.
F. Hofmann-Grüneberg, 1988, *Radikal-empiristische Wahrheitstheorie. Eine Studie über Otto Neurath, den Wiener Kreis und das Wahrheitsproblem*, Hölder-Pichler-Tempsky, Wien.
J. H. Holland, 1989, 'Using Classifier Systems to Study Nonlinear Networks', in *Lectures in the Sciences of Complexity* (ed. by D. L. Stein), Redwood City, pp. 463-499.
J. H. Holland, K. J. Holyak, R. E. Nisbett, P. R. Thagard, 1989, *Induction. Processes of Inference, Learning and Discovery*, M. I. T. Press, Cambridge, Mass..
W. Hollitscher, 1977, *Für und wider die Menschlichkeit. Essays*, Verlag Marxistischer Blätter, Frankfurt a.M.-Wien.
M. Horkheimer, 1937, 'Der neueste Angriff auf die Metaphysik' *Zeitschrift für Sozialforschung* **6** (1937) 4-51, transl. M. J. O'Connell a. o. 'The Latest Attack on Metaphysics', in M. Horkheimer, *Critical Theory*, Herder & Herder, New York, 1972, pp. 132-187.
P. Jacob, 1984, 'The Neurath-Schlick Controversy', in Sebestik, Soulez (eds.) 1984, pp. 351-366.
S. Jacob, K.-H. Otto, 1990, 'Otto Neurath—Marxist Member of the Vienna Circle', *Auslegung* **16** (1990) 175-189.
A. Janik, 1989, 'The Politics of Conciliation', in *Style, Politics and the Future of Philosophy*, Boston Studies in the Philosophy of Science vol. 114, Kluwer, Dordrecht, pp. 116-132.
A. Janik, S. Toulmin, 1973, *Wittgenstein's Vienna*, Doubleday, New York.
L. Jedlicka, 1970, 'Das Jahr 1934', in *Österreich 1918-1938* (ed. by Institut für Österreichkunde), Hirt, Wien, pp. 73-94.
W. S. Jevons, 1879, *The Principles of Science*, London.
M. Jodl (ed.), 1920, *Friedrich Jodl. Sein Leben und Wirken*, Stuttgart-Berlin.
W. M. Johnston, 1971, 'Neo-Idealists from Austria. 1870-1938', in *Modern Austrian Literature* **4** (1971) 7-17.
W. M. Johnston, 1972, *The Austrian Mind*, University of California Press, Berkeley-Los Angeles-London.

B. v. Juhos, 1965, 'Gibt es in Österreich eine wissenschaftliche Philosophie?', in Anon. 1965, pp. 232-244.

B. v. Juhos,1967, 'Moritz Schlick', in P. Edwards *et al* (eds.), vol. 7 pp. 319-324.

H. M. Kallen, 1946, 'Postscript: Otto Neurath 1882-1945', *Philosophy and Phenomenological Research* **6** (1946) 533ff.

F. Kaufmann, 1936, *Methodenlehre der Sozialwissenschaften*, Springer, Wien.

P. Kampitz, 1984, *Zwischen Schein und Wirklichkeit. Eine kleine Geschichte der österreichischen Philosophie*, Wien.

E. C. Kemble, 1965, 'To Philipp Frank', in R. S. Cohen, M. Wartofsky (eds.) 1965, p. xix.

J. Kim, 1988, 'What is "Naturalized Epistemology"?', in *Philosophical Perspectives vol. 2: Epistemology* (ed. by J. Tomberlin), Ridgeview, pp. 381-405.

R. Kinross, 1979, *Otto Neurath's Contribution to Visual Communication*, M.Phil Diss., University of Reading.

R. Kinross, 1982, 'Die Aufgabe der Transformation', in Stadler (ed.) 1982a, pp. 189-197.

R. Kinross, 1984, 'The Work of Otto Neurath in Visual Communication', in Sebestik, Soulez (eds.) 1984, pp. 185-199.

S. C. Kleene, 1939, 'Review: Carnap,*The Logical Syntax of Language* ', *Journal of Symbolic Logic* **4** (1939) 82-87.

Ch. Klusacek, 1966, *Österreichs Wissenschaftler und Künstler unter dem NS-Regime*, Europa-Verlag, Wien-Frankfurt-Zürich.

E. Köhler, 1968, *Logic and Metaphysics. A Logical Investigation with Special Attention to Aristotle*, mimeographed (avail. from author), München.

E. Köhler, 1969, 'Parallele zwischen der Aristotelischen Metaphysik und der Carnapschen Wissenschaftstheorie', *Akten der XIV. Internationalen Philosophentagung Bd. V* .

E. Köhler, 1982, 'Zu einigen Schriften von Otto Neurath über Logik, Ethik und Physik', in Stadler (ed.) 1982a, pp. 107-111, transl. T. E. Uebel 'On Neurath's Writings on Logic, Ethics and Physics', in this volume, ch. 8.

E. Köhler, 1983, 'The Vienna Circle and Foundations without Absolutism', *in Proceedings of the 7th International Wittgenstein Symposium 1982* (ed. by P. Weingartner, J. Czernak), Schriften der Wittgenstein-Gesellschaft vol. 9, Hölder-Pichler-Tempsky, Wien.

E. Köhler, 1985, 'Die Metaphysik beim Wiener Kreis', in Dahms (ed.) 1985a, pp. 190-204, transl. T. E. Uebel 'Metaphysics in the Vienna Circle', in this volume, ch. 10.

W. R. Köhler, 1979a, 'Zerstört der Logische Empirismus die praktische Rationalität', *Analyse & Kritik* **1** (1979) 51-59.

W. R. Köhler, 1979b, 'Logischer Empirismus und normative Entscheidungstheorie', *Analyse & Kritik* **1** (1979) 192-199.

L. Kolakowski, 1969, *The Alienation of Reason. The History of Positivist Thought*, (transl. N. Guterman of *Filozotia pozytywistyczna*), Doubleday/Anchor Books, Garden City, N. J..

D. Koppelberg, 1987, *Die Aufhebung der analytischen Philosophie*, Suhrkamp, Frankfurt.

D. Koppelberg, 1990, 'How and Why to Naturalize Epistemology', *in Perspectives on Quine* (ed. by R. B. Barrett, R. F. Gibson), Blackwells, Oxford, pp. 200-211.

S. M. Kosslyn, 1980, *Image and Mind*, Harvard University Press, Cambridge, Mass..

J. Kraft, 1932, *Von Husserl zu Heidegger. Kritik der phänomenologischen Philosophie*, Hans Buske, Leipzig, 2nd. ed. Frankfurt a. M.-Hamburg, 1977.

V. Kraft, 1937, *Die Grundlagen der wissenschaftlichen Wertlehre*, Schriften zur wissenschaftlichen Weltauffassung Bd. 11, Springer, Wien, 2nd ed. 1951.

V. Kraft, 1950, *Der Wiener Kreis. Der Ursprung des Neopositivismus. Ein Kapitel der jüngsten Philosophiegeschichte*, Springer, Wien, 2nd exp. ed. 1968. Engl. transl. A. Pap *The Vienna Circle*, Philosophical Library, New York, 1953, 2nd ed. 1968.
V. Kraft, 1973, 'Gespräch mit Viktor Kraft', in *Conceptus* 7 no. 21-2 (1973) 9-25.
D. Krantz, D. Luce, P. Suppes, A. Tversky, 1971, *Foundations of the Theory of Measurement I*, New York.
J. K. Kreibig, 1914, 'Bernard Bolzano. Eine Skizze aus der Geschichte der Philosophie in Österreich', *Archiv für Geschichte der Philosophie* **27** (1914), Sonderdruck, pp.1-15.
L. Krestan (ed.), 1972, *Dokumentation zur Österreichischen Akademie der Wissenschaften, vol. 3.* Die Mitglieder und Institutionen der Akademie, Oesterreichische Akademie der Wissenschaften, Wien.
T. S. Kuhn, 1962, *The Structure of Scientific Revolutions*, International Encyclopedia of Unified Science, vol.2, no.2, University of Chicago Press, Chicago, 2nd ed. 1970.
N. Kutalek, H. Fellinger, 1969, *Zur Wiener Volksbildung*, Verlag Jugend und Volk, Wien-München.
F. v. Kutschera, 1982, *Grundlagen der Ethik*, De Gruyter, Berlin-New York.
I. Lakatos, 1970, 'Falsification and the Methodology of Scientific Research Programs', in *Criticism and the Growth of Knowledge* (ed. by I. Lakatos, A. Musgrave), Cambridge University Press, Cambridge.
L. Laudan, 1987, 'Progress or Rationality? The Prospects for Normative Naturalism', *American Philosophical Quarterly* **24** (1987) 19-31.
L. Laudan, 1990, 'Normative Naturalism', *Philosophy of Science* **57** (1990) 44-59.
H. Lauener, 1982, 'Neurath's Protocol Sentences and Schlick's "Konstatierungen" versus Quine's Observation Sentences', in Haller (ed.) 1982e, pp. 129-148.
H. Lauener, 1983, 'Die empirische Grundlage wissenschaftlicher Theorien', *Revue Internationale de Philosophie* 144-145 (1983) 113-144.
K. Lehrer, 1982, 'Schlick and Neurath: Meaning and Truth', in Haller (ed.) 1982e, pp. 49-62.
W. Leinfellner, 1982, 'Schlick's kognitive Erkenntnistheorie als Wissenschaftstheorie', in Haller (ed.) 1982e, pp. 103-128.
N. Leser, 1968, *Zwischen Reformismus und Bolschewismus. Der Austromarxismus als Theorie und Praxis*, Europa Verlag, Wien-Frankfurt a. M..
J. Lewis, 1991, 'Hidden Agendas: Knowledge and Verification', in Fine, Forbes, Wessels (eds.) 1991.
P. F. Linke, 1961, *Niedergangserscheinungen in der Philosophie der Gegenwart*, Reinhardt, München-Basel.
L. Löwenheim, 1915, 'Über Möglichkeiten im Relativkalkül', *Mathematische Annalen* **76** (1915) 447-470.
C. G. Luckhardt (ed.), 1979, *Wittgenstein. Sources and Perspectives*, Harvester Press, Hassocks.
P.C. Ludz, 'Thomas S. Kuhns Paradigmathese', in Salamun (ed.) 1979.
R. Lüscher, 1979, 'Die linke Ecke des Wiener Kreises', *Analyse & Kritik* 1 (1979) 182-191.
E. Mach, 1883, *Die Mechanik in ihrer Entwicklung*, Brockhaus, Leipzig, transl. T.J. McCormick, Ph. E. B. Jourdain with introduction by K. Menger *The Science of Mechanics*, 6th ed, Open Court, Lasalle, Ill., 1960.
E. Mach, 1896, 'Über den Einfluss zufälliger Umstände auf die Entwicklung von Erfindungen und Entdeckungen', in *Populär-Wissenschaftliche Vorlesungen*, Leipzig, transl. T. J. McCormack 'On the Part played by Accident in Invention and Discovery', in *Popular Scientific Lectures*, Open Court, Chicago, 3rd rev. ed. repr. 1986, pp.259-281.

E. Mach, 1905, *Erkenntnis und Irrtum*, Barth, Leipzig, 5th ed. 1926, repr. Wissenschaftliche Buchgesellschaft, Darmstadt, 1980, transl. T. J. MacCormack, P. Foulkes, *Knowledge and Error*, Vienna Circle Collection vol. 3, Reidel, Dordrecht, 1976.

E. Mach 1908, 'Vorwort', in German ed. of Duhem 1906 (1908, repr. 1978).

S. Maderegger, 1973, *Die Juden im österreichischen Ständestaat 1934-38*, Phil. Diss., Universität Salzburg.

E. Mally, 1926, *Grundgesetze des Sollens*, Graz, repr. in *Logische Schriften* (ed. by K. Wolf, P. Weingartner), Reidel, Dordrecht, 1971.

J. C. Marek et al, eds., 1977, *Österreichische Philosophen und ihr Einfluss auf die Gegenwart*, Sonderband *Conceptus* **11** no. 28-20 (1977).

B. F. McGuinness, 1979, 'Editor's Preface', in Waismann 1979, pp. 11-31.

B. F. McGuinness (ed.) 1985, *Moritz Schlick, Synthese* **64** (1985) no. 3.

K. Menger, 1930, 'Der Intuitionismus', *Blätter für Deutsche Philosophie* **4** (1930) 311-325, transl. R. Kowalski 'On Intuitionism' in Menger 1979, pp. 46-58.

K. Menger, 1933, 'Die neue Logik', in *Krise und Neuaufbau in den exakten Wissenschaften. Fünf Wiener Vorträge* (ed. by K. Menger), Deuticke, Wien, transl. with revisions by H. B. Gottlieb, J. K. Senior 'The New Logic', *Philosophy of Science* **4**, 1937, pp. 299-236, repr. and re-rev. by author in Menger 1979, pp. 18-45.

K. Menger, 1934, *Moral, Wille, Weltgestaltung. Grundlegung zur Logik der Sitten*, Springer, Wien, transl. with E. van der Schalie *Morality, Decision and Social Organization*, Vienna Circle Collection vol. 6, Reidel, Dordrecht, 1974.

K. Menger, 1939, 'A Logic of the Doubtful. On Optative and Imperative Logic', *Reports of a Mathematics Colloquium, University of Notre Dame* (Indiana) 1 (1939) 53-64, repr. in Menger 1979.

K. Menger, 1965, 'Der Wiener Kreis', in *Die Presse*, May 8/9 1965.

K. Menger, 1979, *Selected Papers in Logic and Foundations, Didactics, Economics*, Vienna Circle Collection vol. 10, Reidel, Dordrecht.

K. Menger, 1980, 'Forword', in Hahn 1980, pp. ix-xviii.

R. A. Metall, 1969, *Hans Kelsen. Leben und Werk*, Deuticke, Wien.

R. v. Mises, 1938, *Ernst Mach und die empiristische Wissenschaftsauffassung. Zu Ernst Machs 100. Geburtstag*, Einheitswissenschaft vol. 7, Stockum, 's-Gravehage, transl. I. Marcuse 'Ernst Mach and the Empiricist Conception of Science' (ed. by R. S. Cohen, H. v. Mises), in Cohen, Seeger (eds.) 1970, pp. 245-269.

E. Mohn, 1977, *Der Logische Positivismus. Theorien und politische Praxis seiner Vertreter*, Campus, Frankfurt a. M.-New York.

E. Mohn, 1985, 'Die politische Praxis Otto Neurath's während der Räterepublik in Bayern', in Dahms (ed.) 1985a, pp. 30-39.

O. Morgenstern, 1936, 'Logistik und Sozialwissenschaft', *Zeitschrift für Nationalökonomie* 7 (1936), repr. in *Logik der Sozialwissenschaften* (ed. by E. Topitsch), Köln, 1970.

E. Morscher, 1969, 'Der Begriff des Seienden in der traditionellen Ontologie und der Begriff der Allklasse in der modernen Logik', in *Heuresis. Festschrift für Andreas Rohracher* (ed. by Th. Michels), O. Müller, Salzburg.

E. Morscher, 1973, *Das logische An-sich bei Bernhard Bolzano*, Pustet, Salzburg-München.

K. H. Müller, 1982a, 'Vermischte Gedanken über Wissenschaft und Krise', in Stadler (ed.) 1982a, pp. 16-22.

K. H. Müller, 1982b, 'Planwelten. Die ökonomischen Beiträge von Otto Neurath', in Stadler (ed.) 1982a, pp. 104-106.

K. H. Müller, 1985, 'Die verspätete Aufklärung', in Dahms (ed.) 1985, pp. 291-306.

K. H. Müller, 1987, *Zur Computervisualisierung von Wirtschafts- und Sozialstatistiken*, Bd. 1, Wien.

K. H. Müller, 1989, *Das Forschungsprogramm des Otto Neurath*, Institut für Höhere Studien, Wien.

K. H. Müller, 1991a, *Symbole-Statistik-Computer-Design. Otto Neurath's Bildpädagigik im Computerzeitalter*, Hölder-Pichler-Tempsky, Wien.

K. H. Müller, 1991b, 'Grundzüge einer konstruktivistischen Handlungstheorie', in *Im Auge des Betrachters. Festschrift für Heinz von Foerster* (ed. by P. Krieg), München-Zürich.

K.H. Müller, 1991c 'Neurath's Theory of Pictorial-Statistical Representation', in this vilume, ch. 18.

H. L. Mulder, 1968, 'Wissenschaftliche Weltauffassung—der Wiener Kreis', *Journal of the History of Philosophy* **6** (1968) 386-90.

H. L. Mulder, R. C. Cohen, B. McGuiness (eds.), 1973 ff., *The Vienna Circle Collection*, Reidel/Kluwer, Dordrecht.

L. Nelson, 1921, 'Spuk. Einweihung in das Geheimnis der Wahrsagekunst Oswald Spenglers', in *Gesammelte Schriften* (ed. by P. Bernays), Bd. 3, pp. 349-552.

E. Nemeth, 1981, *Otto Neurath und der Wiener Kreis: Wissenschaftlichkeit als revolutionärer politischer Anspruch*, Campus, Frankfurt a.M.-New York.

E. Nemeth, 1982a, 'Die Einheit der Planwirtschaft und die Einheit der Wissenschaft', in Haller (ed.) 1982e, pp. 437-449, transl. T. E. Uebel 'The Unity of Planned Economy and the Unity of Science', in this volume, ch. 21.

E. Nemeth, 1982b, 'Otto Neurath's Utopien - Der Wille zur Hoffnung', in Stadler (ed.) 1982a, pp. 94-99, transl. T. E. Uebel 'Otto Neurath's Utopias: The Will to Hope' in this volume, ch. 22.

E. Nemeth, 1990a, 'Wissenschaftliche Weltauffassung zwischen Sozialrevolution und Sozialreform', *Conceptus* **24** no. 61 (1990) 73-90.

E. Nemeth, 1990b, 'Zur emanzipatorischen Perspektive in Otto Neuraths Begriff des wissenschaftlichen Wissens', Lecture, April 1990, Academy of Science, Bratislava.

G. Nenning, 1968, *Karl Grünberg und die Anfänge des Austromarxismus*, Graz.

J. v. Neumann & O. Morgenstern, 1944, *Theory of Games and Economic Behaviour*, Princeton University Press, Princeton, 3rd ed. 1953.

M. Neurath, 1973, 'Preface I', in Neurath 1973, p. xiii.

O. Neurath, 1904, 'Geldzins im Altertum', *Plutus* **1** (1904) 569-573.

O. Neurath, 1906a, 'Zur Anschauung der Antike über Handel, Gewerbe und Landwirtschaft', *Jahrbücher für Nationalökonomie und Statistik*, 3. Folge, **32** (1906) 577-606, **34** (1907) 145-205.

O. Neurath, 1906b, *Zur Anschauung der Antike über Handel, Gewerbe und Landwirtschaft* (Teil I), Fischer, Jena.

O. Neurath (ed.), 1906c, *Faust. Ein dramtisches Gedicht in drei Abschnitten von F. Marlow (Ludwig Hermann Wolfram)*, with an introduction by O. Neurath 'Ludwig Hermann Wolframs Leben', Ernst Frensdorff, Berlin.

O. Neurath 1909a, *Antike Wirtschaftsgeschichte*, Teubner, Leipzig, 2nd rev. ed. 1918, 3rd ed. 1926.

O. Neurath, 1909b, 'Ernst Schröder's Beweis des 12. Theorems: Für die identischen Operationen gilt das "Kommutationsgesetz"', *Archiv für Philosophie 2. Abteilung: Archiv für systematische Philosophie* Neue Folge **15** (1909) 104-106, repr. in Neurath 1981, pp. 1-3.

O. Neurath 1909c, 'Eindeutigkeit und Kommutativität des logischen Produkts *ab* ', *Archiv für Philosophie 2. Abteilung: Archiv für systematische Philosophie*, Neue Folge **15** (1909), pp. 342-344, repr. in Neurath 1981, pp. 17-18.

O. Neurath, 1910a, 'Zur Theorie der Sozialwissenschaften', *Jahrbuch für Gesetzgebung, Verwaltung und Volkswirtschaft im Deutschen Reich* **34**, pp. 37-67, repr. in Neurath 1981, pp. 23-46.

O. Neurath, 1910b, 'Definitionsgleichheit und symbolische Gleichheit', *Archiv für Philosophie 2. Abteilung: Archiv für systematische Philosophie* Neue Folge **16** (1910), pp. 142-144, repr. in Neurath 1981, pp. 19-21.

O. Neurath, 1910c, *Lehrbuch der Volkswirtschaftslehre*, Hölder, Wien.

O. Neurath, 1911, 'Nationalökonomie und Wertlehre, eine systematische Untersuchung', *Zeitschrift für Volkswirtschaft, Sozialpolitik und Verwaltung* **20**, pp. 52-114.

O. Neurath, 1912, 'Das Problem des Lustmaximums', in *Jahrbuch der Philosophischen Gesellschaft an der Universität Wien 1912*, J. A. Barth, Leipzig, 1913, repr. in Neurath 1981, pp. 47-56, transl. 'The Problem of the Pleasure Maximum' in Neurath 1973, pp. 113-122.

O. Neurath, 1913a, 'Probleme der Kriegswirtschaftslehre', *Zeitschrift für die gesammte Staatswissenschaft* **69** (1913) 438-501.

O. Neurath, 1913b, 'Die Verirrten des Cartesius und das Auxiliarmotiv (Zur Psychologie des Entschlusses)', *Jahrbuch der Philosophischen Gesellschaft an der Universität zu Wien 1913*, pp. 45-59, repr. in Neurath 1981, pp. 57-67, transl. 'The Lost Wanderers and the Auxiliary Motive (On the Psychology of Decision)' in Neurath 1983, pp. 1-12.

O. Neurath, 1915a, 'Prinzipielles zur Geschichte der Optik', *Archiv für die Geschichte der Naturwissenschaften und Technik* **5** (1915) 371-389, repr. in Neurath 1981, pp. 71-84, transl. 'On the Foundations of the History of Optics' in Neurath 1973, pp. 101-112.

O. Neurath, 1915b, 'Zwei Briefe von Otto Neurath an Ernst Mach', in *Wissenschaftliche Kommunikation. Die Korrespondenz E. Machs*, (ed. by J. Thiele), Henn, Kalstellaun, 1978. Also in *Grazer Philosophische Studien* **10** (1980), 3-5.

O. Neurath, 1916a, 'Zur Klassifikation von Hypothesensystemen', *Jahrbuch der Philosophischen Gesellschaft an der Universität Wien 1914 und 1915*, pp. 39-63, repr. in Neurath 1981, pp. 85-101, transl. 'On the Classification of Systems of Hypotheses' in Neurath 1983, pp. 13-31.

O. Neurath, 1917a, 'Das Begriffsgebäude der Wirtschaftslehre und seine Grundlagen', *Zeitschrift für die gesamte Staatswissenschaft* **73**, pp. 484–520, repr. in Neurath 1981, pp. 103-129.

O. Neurath, 1917b, *Die Wirtschaftsordnung der Zukunft und die Wirtschaftswissenschaften*, Verlag f. Fachliteratur, Wien-Berlin, repr. in Neurath 1919b, pp. 159-173.

O. Neurath, 1918, 'Josef Popper-Lynkeus, seine Bedeutung als Zeitgenosse', *Neues Frauenleben* **20** (1918) 33-38, repr. in Neurath 1981, pp 131-136.

O. Neurath, 1919a, 'Geleitwort', on Neurath 1919b, transl. 'Preface' in Neurath 1973, pp. 123-4.

O. Neurath, 1919b, *Durch die Kriegswirtschaft zur Naturalwirtschaft*, Callway, München, excerpts transl. 'Through War Economy to Economy in Kind' in Neurath 1973, pp. 123-154.

O. Neurath, 1920a, 'Wirtschaftsplan, Planwirtschaft, Länderverfassung und Völkerordnung', *Der Kampf* **13** (1920) 224-227.

O. Neurath, 1920b, 'Die wirtschaftlichen Räte im Programm der bayerischen Vollsozialisierung', *Der Kampf* 13 (1920) 136-141.

O. Neurath, 1920c, *Vollsozialisierung. Von der nächsten und übernächsten Zukunft*, Deutsche Gemeinwirtschaft H. 15, Diederichs, Jena.

O. Neurath, 1921a, *Anti-Spengler*, Callway, München, repr. in Neurath 1981, pp. 139-196, excerpts transl. 'Anti-Spengler' in Neurath 1973, pp. 158-213.

O. Neurath, 1921b, 'Berufsberatung und Gesamtwirtschaft', *Der Kampf* **14** (1921) 286-290.

O. Neurath, 1922a, 'Vollsozialisierung und gemeinwirtschaftliche Anstalten', *Der Kampf* **15** (1922) 54-60.

O. Neurath, 1922b, 'Die Betriebsräteschulen als Schulen des Sozialismus', *Der Betriebsrat* **2** (1922) No. 12 177-179.

O. Neurath 1923a, 'Geld und Sozialismus', *Der Kampf* **16** (1923) 145-157.

O. Neurath 1923b, 'Geld, Sozialismus, Marxismus', *Der Kampf* **16** (1923) 283-288.

O. Neurath, 1925, *Wirtschaftsplan und Naturalrechnung*, Laub'sche Buchhandlung, Berlin.

O. Neurath, 1926, 'Statistische Hyroglyphen', *Österreichische Gemeindezeitung* **3** (1926), repr. in Neurath 1979.

O. Neurath 1927, 'Bildliche Darstellung gesellschaftlicher Tatbestände', *Die Quelle* **77** (1927) no. 1.

O. Neurath, 1928a, *Lebensgestaltung und Klassenkampf*, Laub'sche Buchhandlung, Berlin, repr. in Neurath 1981, pp. 227-293, excerpts transl. 'Personal Life and Class Struggle' in Neurath 1973, pp. 249-298.

O. Neurath, 1928b, 'Rezension: R. Carnap, *Der Logische Aufbau der Welt* und *Scheinprobleme der Philosophie*', *Der Kampf* **21** (1928) 624-626, repr. in Neurath 1981, pp. 295-7.

O. Neurath, 1929a, 'Bertrand Russell, der Sozialist', *Der Kampf* **22** (1929) 234-238, repr. in Neurath 1981, pp. 337-343.

O. Neurath, 1929b, 'Wissenschaftliche Weltauffassung', *Arbeiter-Zeitung* 13. Oktober 1929 (Sonntagsbeilage) p. 17, repr. in Neurath 1981, pp. 345-347.

O. Neurath, 1930a, 'Wege der wissenschaftlichen Weltauffassung, *Erkenntnis* **1**, pp. 106-125, repr. in Neurath 1981, pp. 371-385, transl. as 'Ways of the Scientific World Conception' in Neurath 1983, pp. 32-47.

O. Neurath, 1930b, 'Verein Ernst Mach', *Erkenntnis* **1** (1930) 76.

O. Neurath, 1931a, 'Bildstatistik nach Wiener Methode', *Die Volksschule* 27 (1931) H. 12, 571-579.

O. Neurath 1931b, 'Physikalismus', *Scientia* (Nov.1931) 297-303, repr. in Neurath 1981, pp. 417-421, transl. 'Physicalism' in Neurath 1983, pp. 52-57.

O. Neurath, 1931c, *Empirische Soziologie. Der wissenschaftliche Gehalt der Geschichte undf Nationalökonomie*, Schriften zur wissenschaftlichen Weltauffassung Bd. 5, repr. in Neurath 1981, pp. 423-527, transl. (except preface) 'Empirical Soziology', in Neurath 1973, pp. 391-421.

O. Neurath, 1931d, 'Weltanschauung und Marxismus', *Der Kampf* **24** (1931) 447-451, repr. in Neurath 1981, pp. 407-412.

O. Neurath, 1931e, 'Marxismus eines Jesuiten', *Der Kampf* **24** (1931) 271-274.

O. Neurath, 1932a, 'Soziologie im Physikalismus', *Erkenntnis* **2** (1931/2) 393-431, repr. in Neurath 1981, pp. 533-562, transl. 'Sociology in the Framework of Physicalism' in Neurath 1983, pp. 58-90.

O. Neurath, 1932b, 'Protokollsätze', *Erkenntnis* **3** (1932/3) 204-214, repr. in Neurath 1981, pp. 577-586, transl. 'Protocol Statements' in Neurath 1983, pp. 91-99.

O. Neurath, 1932c, 'Sozialbehaviourismus', *Sociologicus* **8** (1932) 281-288, repr. in Neurath 1981, pp. 563-570.

O. Neurath, 1932d, 'Einheitswissebschaft und Empirismus der Gegenwart', *Erkenntnis* **2** (1931/2) 310-311.

O. Neurath, 1932e, 'Soziologie in physikalischer Sprache', *Erkenntnis* **2** (1931/2) 312.

O. Neurath, 1932f, 'Das Fremdpsychische in der Soziologie', *Erkenntnis* **3** (1932/3) 105-106.

O. Neurath, 1932g, 'Einheitswissenschaft und Psychologie', *Erkenntnis* **3** (1932/3) 233-234.

O. Neurath, 1933a, *Einheitswissenschaft und Psychologie*, Einheitswissenschaft Bd.1, Gerold, Wien, repr. in Neurath 1981, pp. 587-610, transl. H. Kraal 'Unified Science and Psychology' in B. McGuiness (ed.) 1987, pp. 1-23.

O. Neurath, 1933b, *Bildstatistik nach Wiener Methode*, Deutscher Verlag für Jugend und Volk, Wien-Leipzig.

O. Neurath, 1934, 'Radikaler Physikalismus und "wirkliche Welt"', *Erkenntnis* **4** (1934) 346-362, repr. in Neurath 1981, pp. 611-624, transl. as 'Radical Physicalism and "the Real World"' in Neurath 1983, pp. 100-114.

O. Neurath, 1935a, 'Einheit der Wissenschaft als Aufgabe', *Erkenntnis* **5** (1935) 16-22, repr. in Neurath 1981, pp. 625-630, transl. 'The Unity of Science as a Task' in Neurath 1983, pp. 115-120.

O. Neurath, 1935b, 'Pseudorationalismus der Falsifikation', *Erkenntnis* **5** (1935) 353-365, repr. in Neurath 1981, pp. 635-644, transl. 'Pseudorationalism of Falsification' in Neurath 1983, pp. 121-131

O. Neurath, 1935c, '1. Internationaler Kongress für Einheit der Wissenschaft in Paris 1935', *Erkenntnis* **5** (1935) 377-406, repr. in Neurath 1981, pp. 649-672.

O. Neurath, 1936a, *Le developpement du Cercle de Vienne et l'avenir de l'Empiricisme logique*, Hermann & Cie, Paris, transl. B. Treschmitzer, H. G. Zilian 'Die Entwicklung des Wiener Kreises und die Zukunft des Logischen Empirismus' in Neurath 1981, pp. 673-703.

O. Neurath, 1936b, 'Une encyclopédie internationale de la science unitaire', in *Actes du Congres Internationale de Philosophie Scientifique, Sorbonne, Paris 1935*, Facs. II 'Unite de la Science', Herman & Cie, Paris, pp.32-40, transl. 'An International Encyclopedia of Unified Science' in Neurath 1983, pp. 139-144.

O. Neurath, 1936c, 'Physikalismus und Erkenntnisforschung', *Theoria* **2** (1936) 97-105, 234-7, repr. in Neurath 1981, pp. 749-760, transl. 'Physicalism and the Investigation of Knowledge' in Neurath 1983, pp. 159-167.

O. Neurath, 1936d, *International Picture Language*, Psyche Miniatures No. 83, Kegan Paul, Trench, Trubner & Co., London, repr. with transl. M. *Neurath Internationale Bildersprache*, University of Reading, 1980 [page ref. to original ed.].

O. Neurath, 1936e, 'Soziologische Prognosen', *Erkenntnis* **6** (1936) 398-405, repr. Neurath 1981, pp. 771-776.

O. Neurath, 1936f, 'L'encyclopedie comme "modele"', *Revue de Synthese* **12** (1936) 187-201, transl. 'Encyclopedia as Model' in Neurath 1983, pp. 145-158.

O. Neurath, 1936g, 'Einheitswissenschaft', *Actes du VIII. Congres International de Philosophie a Prague, 2-7 Septembre 1934*, Orbis, Prague, 1936, pp. 139-141, repr. in Neurath 1981, pp. 761-763.

O. Neurath, 1937a, 'Unified Science and its Encyclopedia', *Philosophy of Science* **4** (1937) 265-277, repr. in Neurath 1983, pp. 172-182.

O. Neurath, 1937b, *BASIC by ISOTYPE*, Psyche Miniatures No. 86, Kegan Paul, Trench, Trubner & Co., London.

O. Neurath, 1937c, 'Inventory of the Standard of Living', *Zeitschrift für Sozialforschung* **6** (1937) 140-151.

O. Neurath 1937d, 'Prognosen und Terminologie in Physik, Biologie und Soziologie', in *Traveaux du IXe Congrès Internationale de Philosophie, IV. L'Unité de la Science: la méthode et les méthodes, Ie partie.* Actualités Scientifiques et Industrielles No. 533, Hermann & Cie, Paris, pp. 77-85, repr. in Neurath 1981, pp. 787-794.

O. Neurath, 1937/8, 'The Departmentalisation of Unified Science', *Journal of Unified Science (Erkenntnis)* **7** (1937/8) 240-246, repr. in Neurath 1983, pp. 200-205.

O. Neurath, 1938a, 'Unified Science as Encyclopedic Integration', in Neurath et al, 1938, pp. 1-27.

O. Neurath, 1938b, 'Die neue Enzyklopädie', in *Zur Enzyklopädie der Einheitswissenschaft* (by O. Neurath, E. Brunswik, C. L. Hull, G. Mannoury, J. H. Woodger), Einheitswissenschaft Bd. 6, Van Stockum & Zoon, Den Haag, 1938, pp. 6-16, repr. in Neurath 1981, pp. 863-871, transl. by H. Kaal as 'The New Encyclopedia', in *Unified Science* (ed. by B. F. McGuiness), Vienna Circle Collection vol. 19, Kluwer, Dordrecht, 1987, pp. 132-141.

O. Neurath, 1938c, 'Weten en Zijn', *Algemeen Nederlands Tijdschrift voor Wijsbegeerte en Psychologie* **31** (1938) no. 4, 230-239, transl. H. Mittelberger 'Wissen und Sein' in Neurath 1981, pp. 833-843.

O. Neurath, 1939, *Modern Man in the Making*, Knopf, New York.

O. Neurath, 1939/46, 'Wichtigkeit vorsichtiger Formulierung', *Synthese* **5** (1946) 93-94.

O. Neurath,1941a, 'Universal Jargon and Terminology', *Proceedings of the Aristotelian Society* n.s. **41** (1941) 127-148, repr. in Neurath 1983, pp. 213-229.

O. Neurath, 1941b, 'The Danger of Careless Terminology', *New Era* **22** (1941)145-150, transl. 'Die Gefahr sorgloser Terminologie' in Neurath 1981, pp. 919-924.

O. Neurath, 1942, 'ISOTYPE Institute and Adult Education', *Second Series of the World Association for Adult Education*, Bulletin 11, pp. 12-17.

O. Neurath, 1944, *Foundations of the Social Sciences*, International Encyclopedia of Unified Science, vol.2 no.1, Chicago: University of Chicago Press.

O. Neurath, 1946a, 'The Orchestration of the Sciences by the Encyclopedism of Logical Empiricism', *Philosophy and Phenomenological Research* **6** (1945/6) 496-508, repr. Neurath 1983, pp. 230-242.

O. Neurath 1946b, 'For the Discussion: Just Annotations, Not a Reply', *Philosophy and Phenomenological Research* **6** (1945/6) 526-528.

O. Neurath, 1946c, 'After Six Years', *Synthese* **5** (1946) 77-82.

O. Neurath, 1973, *Empiricism and Sociology* (ed. by M. Neurath, R.S. Cohen, transl. P. Foulkes, M. Neurath), Vienna Circle Collection vol.1, Reidel, Dordrecht.

O. Neurath, 1979, *Wissenschaftliche Weltauffassung, Sozialismus und Logischer Empirismus* (ed. by R. Hegselmann), Suhrkamp, Frankfurt a.M..

O. Neurath, 1981, *Gesammelte philosophische und methodologische Schriften* (ed. by R. Haller, H. Rutte), 2 vols., Hölder-Pichler-Tempsky, Wien.

O. Neurath, 1983, *Philosophical Writings 1913-1945* (eds. and transl. R. S. Cohen, M. Neurath), Vienna Circle Collection vol. 16, Reidel, Dordrecht.

O. Neurath, 1991, *Gesammelte bildpädagogische Schriften* (ed. by R. Haller, R. Kinross), Hölder-Pichler-Tempsky, Wien.

O. Neurath (unsigned), Gesellschafts- und Wirtschaftsmuseum in Wien, 1929, *Die bunte Welt*, Wolf, Wien.

O. Neurath (unsigned), Gesellschafts- und Wirtschaftsmuseum in Wien, 1930, *Gesellschaft und Wirtschaft. Bildstatistisches Mappenwerk*, Bibliographisches Institut, Leipzig.
O. Neurath, R. Carnap, H. Hahn, 1929 [see R. Carnap, H. Hahn, O. Neurath 1929]
O. Neurath, J. Frank, 1930, 'Hannes Meyer', *Der Klassenkampf* **4** (1930) 573-575.
O. Neurath, O. Hahn, 1909, 'Zum Dualismus in der Logik', *Archiv für Philosophie 2. Abteilung: Archiv für systematische Philosophie* Neue Folge **15** (1909), pp. 149-162, repr. in Neurath 1981, pp. 5-16.
O. Neurath, A. Schapire-Neurath (eds.), 1910, *Lesebuch der Volkswirtschaftslehre*, 2 vols., Klinkhardt, Leipzig, 1910, 2nd ed. Glöckner, Leipzig, 1913.
O. Neurath, W. Schumann, 1919, 'Sozialisierungsprojekt', *Dresdener Anzeiger* 13/5/1919, repr. in K. Bühler, *Die Sozialisierung*, 2nd ed. 1919, Lauppsche Buchhandlung, Tübingen.
O. Neurath et al, 1938, *Encyclopedia and Unified Science*, (International Encyclopedia of Unified Science, vol.1 no.1), Chicago: University of Chicago Press.
P. Neurath, 1982, 'Otto Neurath und die Soziologie', in Haller (ed.) 1982e, pp. 223-240.
P. Neurath, 1984, 'Recollections of the Beginnings of Pictorial Statistics and ISOTYPE', in Sebestik, Soulez (eds.) 1984, pp. 201-208.
P. Neurath, 1991, 'Sociological Thought with Otto Neurath', in this volume, ch. 17.
J. C. Nyiri (ed.), 1981, *Austrian Philosophy. Studies and Texts*, Philosophia, München.
J. C. Nyiri (ed.), 1986, *Von Bolzano zu Wittgenstein. Zur Tradition der österreichischen Philosophie*, Schriften der Wittgenstein-Gesellschaft vol. 12/2 , Hölder-Pichler-Tempsky, Wien.
T. Oberdan, 1979, 'Neurath's Complaint', in Berghel, Hübner, Köhler (eds.) 1979, pp. 241-244.
O. Oberhuber, P. Weibel (eds.), 1976, *Österreichs Avantgarde 1900-1938. Ein unbekannter Aspekt*, Wien.
H. Parthey, H. Vogel (eds.), 1969, *Joachim Jungius und Moritz Schlick*, Rostock, 1969.
J. Passmore, 1957, *A Hundred Years of Philosophy*, Duckworths, London, 2nd ed. 1968.
C. S. Peirce, 1883, *Studies in Logic by Members of the John Hopkins University*, Boston.
A. Pfoser, 1982, 'Glück in der Vernunft. Überlegungen zu Otto Neuraths Kulturtheorie und zur austromarxistischen Lebensform', in Stadler (ed.) 1982a, pp. 168-172.
W. Pircher, 1982, 'Bemerkungen zu "Statistik und Sozialismus"', in Stadler (ed.) 1982a, pp. 162-167.
J. H. Poincaré, 1902, *La Science et l'Hypothese*, Paris, transl. F. and L. Lindemann *Wissenschaft und Hypothese*, Teubner, Leipzig, 2nd ed. 1906, transl. W. J. Greenstreet *Science and Hypothesis*, Scott, London, 1905, repr. Dover, New York, 1952.
K. R. Popper, 1935, *Die Logik der Forschung*, Schriften zur wissenschaftlichen Weltauffasung, Bd. 9, Springer, Wien, transl. with addenda *The Logic of Discovery*, Hutchinson, London, 1959, 2nd ed., Harper Torchbooks, New York, 1968.
K. R. Popper, 1974, 'Intellectual Autobiography', in P.A. Schilpp, ed., *The Philosophy of Karl Popper*, (Library of Living Philosophers, vol.14), Open Court, Lasalle, Ill., repr. as *Unended Quest*, 1976, Open Court, Lasalle, Ill..
J. Proust, 1986, *Questions de Forme. Logique et proposition analytique de Kant à Carnap*, Fayard, Paris, transl. A. A. Brenner *Questions of Form. Logic and the Analytic Proposition from Kant to Carnap*, University of Minnesota Press, Minneapolis, 1989.

J. Proust, 1987, 'Formal Logic as Transcendental in Wittgenstein and Carnap', *Nous* **21** (1987) 501-520.

H. Putnam, 1982, 'Why Reason Can't Be Naturalized', repr. in *Realism and Reason. Philosophical Papers vol. 3*, Cambridge University Press, Cambridge, 1983, pp. 229-247.

W. V. O. Quine, 1951, 'Two Dogmas of Empiricism', *Philosophical Review* **60** (1951) 20-43, repr. in Quine, *From a Logical Point of View*, 1953, rev. ed. 1980, Harvard university Press, Cambridge, Mass., pp. 20-46.

W. V. O. Quine, 1960, *Word and Object*, MIT Press, Cambridge, Mass..

W. V. O. Quine, 1969, 'Epistemology Naturalized', in *Ontological Relativity and Other Essays*, Columbia University Press, New York, pp. 69-90.

G. Ramharter, 1973, *Geschichtswissenschaft und Patriotismus. Österreichische Historiker im Weltkrieg 1914-1818*, Verlag Geschichte und Politik, Wien-Oldenbourg, München.

R. Reininger (ed.), 1938, *50 Jahre Philosophische Gesellschaft an der Universität Wien 1888-1938*, Philosophische Gesellschaft, Wien.

A. Rey, 1905, *La Theorie de Physique chez les Physiciens Contemporains*, Paris, transl. R. Eisler, *Die Theorie der Physik bei den modernen Physikern*, Leipzig, 1908.

W. Rosar, 1971, *Deutsche Gemeinschaft. Seyss-Inquardt und der Anschluss*, Europaverlag, Wien-Frankfurt a. M..

M. Rosier, 1986, 'Otto Neurath et la critique du pseudo-rationalisme', *Dialogue* **25** (1986) 675-696.

R. Roszenich, 1974, 'Doch reformierbar? Sozialdemokratische Anmerkungen zur österreichischen Hochschulreform', *Neues Forum* (January/February 1974) 33-35.

E. Rüdegger, 1964, *Die philosophischen Studien an der Wiener Universität 1800-1848*, Phil. Diss., Wien.

E. Rungaldier, 1979, 'Das Problem der "Darstellung" bei Neurath und Carnap', in Berghel et al (eds.) 1979, pp. 245-247.

E. Rungaldier, 1984, *Carnap's Early Conventionalism: An Inquiry into the Background of the Vienna Circle*, Studien zur österreichischen Philosophie Bd. 7, Rodopi, Amsterdam.

B. Russell, 1903, *Principles of Mathematics*, Cambridge.

B. Russell, 1940, *An Inquiry into Meaning and Truth*, Allen & Unwin, London.

H. Rutte, 1973, 'Viktor Kraft. Eine philosophische Standortbestimmung', *Conceptus* **7** no. 21-22 (1973) 5-8. [S79] [S86]

H. Rutte, 1976, 'Moritz Schlick, der Positivismus und der Neupositivismus', *Zeitschrift für philosophische Forschung* **30** (1976) 246-286.

H. Rutte, 1977a, 'Positivistische Philosophie in Österreich' , in Marek et al (eds.) 1977, 43-56.

H. Rutte, 1977b, 'Moritz Schlick', *Neue Österreichische Biographie* 19 (1977) 120-128.

H. Rutte, 1979, 'Neurath contra Schlick. Zur Wahrheitsdiskussion im Wiener Kreis', in Berghel et al (eds.) 1979, pp.248-251, transl. T. E. Uebel 'Neurath contra Schlick. On the Discussion of Truth in the Vienna Circle', in this volume, ch. 13.

H. Rutte, 1982a, 'Der Philosoph Otto Neurath', in Stadler (ed.)1982a, pp. 70-78, transl. T. E. Uebel 'The Philosopher Otto Neurath', in this volume, ch. 6.

H. Rutte, 1982b, 'Über Neurath's Empirismus und seine Kritik am Empirismus', in Haller (ed.) 1982e, pp. 365-384, transl. T. E. Uebel 'On Neurath's Empiricism and his Critique of Empiricism', in this volume, ch. 14.

H. Rutte, 1985, 'Zum Problem "Wittgenstein, Ethik und Positivismus"', *Mitteilungen aus dem Brenner-Archiv* 1985 no. 4.

H. Rutte, 1986, 'Ethik und Werturteilsproblematik im Wiener Kreis', in Nyiri (ed.) 1986, pp. 162-172, transl. T. E. Uebel 'Ethics and the Problem of Value in the Vienna Circle', in this volume, ch. 11.

T. A. Ryckman, 1991a, 'Designation and Convention: a Chapter of Early Logical Empiricism', in Fine, Forbes, Wessels (eds.) 1991.

T. A. Ryckman, 1991b, 'Conditio Sine Qua Non? The Concept of *Zuordnung* in the Early Epistemologies of Cassirer and Schlick', forthcoming *Synthese*.

K. Salamun (ed.), 1979, *Sozialphilosophie als Aufklärung. Festschrift für E. Topitsch*, Mohr, Tübingen.

W. Sauer, 1982, *Österreichische Philosophie zwischen Aufklärung und Restauration. Beiträge zur Geschichte des Frühkantianismus in der Donaumonarchie*, Studien zur Österreichischen Philosophie Bd. 2, Würzburg-Amsterdam.

W. Sauer, 1985, 'Carnap's Aufbau in kantischer Sicht', *Grazer Philosophische Studien* 23 (1985) 19-35.

F. d. Saussure, 1915, *Introduction to General Linguistics* (transl. W. Baskin, ed. by C. Bally, A. Sechehaye), McGraw-Hill, New York-London, 1959, repr. 1966.

I. Scheffler, 1967, *Science and Subjectivity*, Bobbs-Merrill, Indianapolis-New York.

P. A. Schilpp (ed.), 1963, *The Philosophy of Rudolf Carnap*, Library of Living Philosophers Vol. 11, Open Court, LaSalle, Ill.

H. Schleichert (ed.), 1975, *Logischer Empirismus - der Wiener Kreis*, Fink, München.

M. Schlick, 1913, 'Gibt es intuitive Erkenntnis?', *Vierteljahresschrift für wissenschaftliche Philosophie und Soziologie* 37 (1913) 472-488, transl. P. Heath 'Is There Intuitive Knowledge?' in Schlick 1979a, pp. 141-152.

M. Schlick, 1918/1925, *Allgemeine Erkenntnislehre*, 1st ed. 1918, Springer, Berlin, 2nd rev. ed. Springer, Berlin, 1925, repr. Suhrkamp, Frankfurt a.M., 1979, transl. H. Feigl, A. Blumberg *General Theory of Knowledge*, Open Court, Lasalle, Ill., 1974.

M. Schlick, 1926, 'Erleben, Erkennen, Metaphysik', *Kant-Studien* 31 (1926) 146-158, repr. in Schlick 1938, pp. 1-18, transl. P. Heath 'Experience, Cognition and Metaphysics', in Schlick 1979b, pp. 99-111.

M. Schlick, 1927, 'Vom Sinn des Lebens', *Symposion* (1927) H. 6, transl. P. Heath 'On the Meaning of Life', in Schlick 1979b, pp. 112-129.

M. Schlick, 1930a, 'Die Wende in der Philosophie', *Erkenntnis* 1 (1930) 4-11, transl. D. Rynin 'The Turning Point in Philosophy', in Ayer (ed.) 1959b, pp. 53-59, repr. in Schlick 1979b, pp. 154-160.

M. Schlick, 1930b, *Fragen der Ethik*, Schriften zur wissenschaftlichen Weltauffassung, Bd. 4, Springer, Wien, repr. (ed. by R. Hegselmann) Suhrkamp, Frankfurt a.M., 1984, partial Engl. transl. D. Rynin *Problems of Ethics*, Prentice Hall, New York, 1939.

M. Schlick, 1931, 'Rezension: S. Weinberg, *Erkenntnistheorie* ', *Erkenntnis* 2 (1931/2) 466.

M. Schlick, 1934, 'Über das Fundament der Erkenntnis', *Erkenntnis* 4 (1934) 79-99, transl. D. Rynin 'The Foundation of Knowledge' in Ayer (ed.)1959a, pp. 209-227, repr. in Schlick 1979b, pp. 370-387.

M. Schlick, 1935a, 'Facts and Propositions', *Analysis* 2 (1935) 65-70, repr. in Schlick 1979b, pp. 400-404.

M. Schlick, 1935b, *Sur le Fondament du Conaissance*, Herman & Cie, Paris.

M. Schlick, 1935c, 'Introduction' in Schlick 1935b, , transl. P. Heath as 'Introduction' in Schlick 1979b, pp. 405-407.

M. Schlick, 1935d, 'Sur les "Constatations"', in Schlick 1935b, transl. P. Heath 'On "Affirmations"' in Schlick 1979b, pp. 407-413.

M. Schlick, 1937, , 'L'Ecole de Vienne et la Philosophie Traditionelle', *Travaux du IXeme Congres Internationale de Philosophie, IV . L'unite de la Science: la Methode et les Methodes*, Herman & Cie, Paris, pp. 99–107, repr. in Schlick 1938, pp. 389-398, transl. P. Heath 'The Vienna School and Traditional Philosophy' in Schlick 1979b, pp. 491-498.

M. Schlick, 1938, *Gesammelte Aufsätze 1926-1936*, Gerold, Wien, 2nd ed. 1968.

M. Schlick, 1952, *Natur und Kultur* (ed. from the Nachlass by J. Rauscher), Humboldt, Wien-Stuttgart, 1952.

M. Schlick, 1962, *Aphorismen* (ed. from the Nachlass by B. Hardy-Schlick), Wien, 1962.

M. Schlick, 1979a, *Philosophical Papers Vol.1 (1909-1922)* (ed. by H. L. Mulder and B. van de Velde-Schlick), Vienna Circle Collection vol.11, pt. 1, Reidel, Dordrecht.

M. Schlick, 1979b, *Philosophical Papers Vol.2 (1925-1936)* (ed. by H. L. Mulder and B. van de Velde-Schlick), Vienna Circle Collection vol.11, pt. 2, Reidel, Dordrecht.

H. Schnädelbach, 1971, *Erfahrung, Begründung, Reflexion. Versuch über den Positivismus*, Suhrkamp, Frankfurt a. M..

H. Scholz, 1930, 'Die Axiomatik der Alten', *Blätter für Deutsche Philosophie* **4** (1930), repr. in Scholz 1961.

H. Scholz, 1940, 'Was ist Philosophie', repr. in Scholz 1961.

H. Scholz, 1941, *Metaphysik als strenge Wissenschaft*, Staufenverlag, Köln.

H. Scholz, 1961, *Mathesis Universalis. Abhandlungen zur Philosophie als strenge Wissenschaft* (ed. by H. Hermes, F. Kambartel, J. Ritter), Schwabe, Basel-Stuttgart.

E. Schröder, 1890-95, *Vorlesungen zur Algebra der Logik*, 3 vols., Jena, repr. New York, 1966.

J. Schulte, 1989, *Wittgenstein. Eine Einführung*, Reclam, Stuttgart.

J. Sebestik, A. Soulez (eds.), 1984, *Journées Internationales sur le Cercle de Vienne: Doctrines et controverses*, in *Fundamenta Scientiae* **5** (1984) nos.2, 3/4.

C. Siegel, 1936, 'Unterrichtsform: Philosophie', in *Deutsch-Österreichische Literaturgeschichte. Ein Handbuch zur Geschichte der deutschen Dichtung in Österreich-Ungarn, Bd. 3, 1846-1890* (ed. by E. Castle), Wien, pp. 17-48.

M. Siegert, 1971, 'Warum Max Adler nicht Ordinarius wurde', *Neues Forum* (Nov./Dez. 1971).

M. Siegert, 1974, 'Numerus Juden raus. Professoren nehmen sich Freiheit der Wissenschaft', *Neues Forum* (Jan./Feb. 1974) 35-37.

K.-J. Siegfried, 1974, *Universalismus und Faschismus. Das Gesellschaftsbild Othmar Spanns*, Europaverlag, Wien.

Th. Skolem, 1970, *Selected Works in Logic*, Oslo.

H. Skolimowski, 1967, *Polish Analytic Philosophy*, Routledge & Paul, London/Humanities Press, New York.

E. W. Spaulding, 1968, *The Quiet Invaders: The Story of the Austrian Impact upon America*, Österreichischer Bundesverlag f. Unterricht Wiss. u. Kunst, Wien.

O. Spengler 1918, *Der Untergang des Abendlandes. Umrisse einer Morphologie der Weltgeschichte*, Beck, München, repr. dtv, München, 1973.

F. Stadler, 1979a, 'Aspekte des gesellschaftlichen Hintergrunds und Standorts des Wiener Kreises am Beispiel der Universität Wien', in Berghel et al (eds.) 1979, pp. 41-59, transl. T. E. Uebel 'Aspects of the Social Background and Position of the Vienna Circle at the University of Vienna', in this volume, ch. 5.

F. Stadler, 1979b, 'Bildstatistik in der Schule: Otto Neurath und sein Gesellschafts- und Wirtschaftsmuseum. Ein Beitrag zur Wiener Schulreform 1925-1934 in seiner heutigen Bedeutung', *Zeitgeschichte* **11/12** (1979) 401-421.

F. Stadler, 1981, 'Spätaufklärung und Sozialdemokratie in Wien 1918-1938', in *Aufbruch und Untergang. Soziologisches und Ideologisches zur Spätaufklärung in Österreich* (ed. by F. Kadrnoska), Europaverlag, Wien-München, pp. 441-473.

F. Stadler (ed.), 1982a, *Arbeiterbildung in der Zwischenkriegszeit*. Ausstellungskatalog mit Forschungsteil, Österreichisches Gesellschafts-und Wirtschaftsmuseum/Löcker, Wien.

F. Stadler, 1982b, *Vom Positivismus zur wissenschaftlichen Weltauffassung*, Veröffentlichungen des Ludwig-Boltzmann-Institutes für Geschichte der Gesellschaftswissenschaften Bd. 8/9, Löcker, Wien.

F. Stadler, 1982c, 'Otto Neurath - Moritz Schlick. Zum philosophischen und weltanschaulich-politischen Antagonismus im Wiener Kreis', in Haller (ed.) 1982e, pp. 451-463, transl. T. E. Uebel 'Otto Neurath-Moritz Schlick. On the Philosophical and Political Antagonisms in the Vienna Circle', in this volume, ch. 12.

F. Stadler, 1982d, 'Otto Neurath (1882-1945) - Zu Leben und Werk in seiner Zeit', in Stadler (ed.) 1982a, pp. 2-22.

F. Stadler, 1982e, 'Der Philosoph Moritz Schlick (1882-1936)', *Conceptus* **37** (1982), 75-80.

F. Stadler, 1984, 'Wiener Methode der Bildstatistik und politische Graphik des Konstruktivismus. Wien-Moskau 1931-1934', in *Österreich und die Sowjetunion 1918-1955. Beiträge zur Geschichte der Österreichisch-sowjetischen Beziehungen*, (ed. by Historikersektion der Österreichisch-Sowjetischen Gesellschaft), Wien, pp. 220-249.

F. Stadler, 1985, 'Popularisierungsbestrebungen im Wiener Kreis und "Verein Ernst Mach"', in H.-J. Dahms (ed.) 1985, pp. 101-128.

F. Stadler, 1984, 'Externe Faktoren in der österreichischen Philosophie', in Nyiri (ed.) 1986, pp. 147-160.

F. Stadler, 1989a, 'Otto Neurath (1882-1945) - Enzyklopedist, Schulreformer und Volksbildner', *Erwachsenenbildung in Österreich* **40** (1989) H. 5-6, 33-36, transl. T. E. Uebel 'Otto Neurath - Encyclopedist, Schoolreformer and People's Educator', in this volume, ch. 19.

F. Stadler, 1989b, '"Zeit unterm Messer". In memoriam Gerd Arntz', *Erwachsenenbildung in Österreich* **40** (1989) H. 5-6, 31-33.

G. Stadler-Türk, 1982, 'Lebensgestaltung und Persönlichkeit: Ein "Neuer Mensch" bei Otto Neurath?', in Stadler (ed.) 1982a, pp. 173-181.

G. Statera, 1967, *Logica, Linguaggio e Sociologica*, Taylor, Torino.

W. Stegmüller, 1960, 'Das Problem der Kausalität', in Topitsch (ed.) 1960.

W. Stegmüller, 1973, *Probleme und Resultate der Wissenschaftstheorie und analytischen Philosophie Bd. 2, 2. Halbband*, Springer, Berlin.

F. Suppe, 1977, 'The Search for a Philosophical Understanding of Theories', in *The Structure of Scientific Theories* (ed. by F. Suppe), 2nd. ed., University of Illinois Press, Urbana-Chicago-London, pp. 3-241.

A. Tarski, 1956, *Logic, Semantics, Metamathematics* (ed. by J. H. Woodger), Oxford University Press, Oxford, 2nd ed. (ed. by J. Corcoran) Hackett, Indianpolis, 1983.

H. Tennessen, 1982, 'Qualms about Otto Neurath's Cabby Language', in Haller (ed.) 1982e, pp. 385-398.

E. Topitsch, 1949, 'Kant in Österreich', in *Philosophie der Wirklichkeitsnähe. Festschrift zum 80. Geburtstag Robert Reiningers* (anon. ed.), Sexl, Wien, pp. 236-253.

E. Topitsch, 1958, *Vom Ursprung und Ende der Metaphysik*, Springer, Wien.
E. Topitsch, 1960a, 'Zum Geleit', in Topitsch (ed.) 1960b.
E. Topitsch (ed.), 1960b, *Probleme der Wissenschaftstheorie. Festschrift für Viktor Kraft*, Springer, Wien.
E. Topitsch, 1967a, 'Österreichs Philosophie—zwischen totalitär und konservativ, in Fischer (ed.) 1967, pp.29-52.
E. Topitsch, 1967b, *Die Sozialphilosophie Hegels als Heilslehre und als Herrschaftsideologie*, Luchterhand, Neuwied-Berlin, 2nd ed. Piper, München, 1981.
S. Toulmin, 1953, *The Philosophy of Science. An Introduction*, Hutchinson, London.
M. Twyman, 1975, 'The Significance of ISOTYPE', in *Graphic Communication through ISOTYPE* (ed. by J. A. Edwards and M. Twyman), University of Reading, pp. 7-17.
M. Twyman, 1982, 'ISOTYPE und die Universität Reading', in Stadler (ed.) 1982a, pp. 185-188.
T. E. Uebel, 1989, *The Vienna Circle's Debate about Protocol Sentences Revisited: Towards the Reconstruction of Otto Neurath's Epistemology*, Ph.D. Diss., Massachusetts Institute of Technology.
T. E. Uebel, 1991a, 'Zur Protokollsatzdebatte des Wiener Kreises', in *Jour Fixe der Vernunft. Der Wiener Kreis und die Folgen* (ed. by R. Haller, H. Kruntorad), Hölder-Pichler-Tempsky, Wien.
T. E. Uebel, 1991b, 'Otto Neurath's Idealistic Inheritance: The Social and Economic Thought of Wilhelm Neurath', forthcoming in *Synthese*.
T. E. Uebel, 1991c, 'Neurath's Program for Naturalistic Epistemology', forthcoming.
H. Wagner, 1977, 'Mehr als ein Jahrhundert seit dem Ende des deutschen Idealismus', *Hegel Studien* Beiheft 17, repr. in *Kritische Philosophie. Systematische und historische Abhandlungen* (ed. by K. Bärthlein, W. Flach), Königshausen & Neumann, Würzburg, 1980.
R. Wagner, 1927, 'Die Wiener Gewerkschaftsschule', *Arbeit und Wirtschaft* .(1927).
F. Waismann, 1938, 'Vorwort', in Schlick 1938, pp. vii-xv, transl. P. Heath 'Foreword' in Schlick 1979b, pp. xiii-xx.
F. Waismann, 1979, *Wittgenstein and the Vienna Circle. Conversations Recorded* (ed. by B. McGuinness, transl. J. Schulte & B. McGuinness), Blackwell, Oxford.
F. Waismann, 1983, 'Ethik und Wissenschaft', in F. Waismann, *Wille und Motiv. Zwei Abhandlungen über Ethik und Handlungstheorie* (ed. by J. Schulte), Stuttgart, 1983.
F. Wallner, 1982, 'Wittgenstein und Neurath. Ein Vergleich von Intention und Denkstil', in Haller (ed.) 1982e, pp. 419-424.
A. J. Walter, 1936, *Die Hochschulen im neuen Staate. Gedanken zu einer notwendigen Reform*, Deutscher Verlag Jugend und Volk, Wien-Leipzig.
A. Wandruszka, 1954, 'Österreichs politische Struktur. Die Entwicklung der Parteien und politischen Bewegungen', in *Geschichte der Republik Österreich*, (ed. by H. Benedikt), Oldenbourg, München, repr. 1977.
M. W. Wartofsky, 1973, 'Perception, Representation, and the Forms of Action: Towards an Historical Epistemology', repr. in *Models*, Boston Studies in the Philosophy of Science vol. 48, Reidel, Dordrecht, 1976, pp. 188-210.
M. W. Wartofsky, 1982, 'Positivism and Politics. The Vienna Circle as a Social Movement', in Haller (ed.) 1982e, pp. 79-102.

M. W. Wartofsky, 1987, 'Epistemology Historicized', in *Naturalistic Epistemology* (ed. by A. Shimony, D. Nails), Boston Studies in the Philosophy of Science vol. 100, Reidel, Dordrecht, pp. 357-374.

J. W. R. Watkins, 1974, 'Otto Neurath', *British Journal for the Philosophy of Science* **25** (1974) 343-352.

F. Weber, 'Otto Neurath und die Sozialisierung', in Stadler (ed.) 1982a, pp. 133-139.

P. Weingartner, 1983, 'The Ideal of the Mathematization of all Sciences and of "More Geometrico" in Descartes and Leibniz', in *Nature Mathematicized. Historical and Philosophical Studies in Classical Modern Natural Philosophy* (ed. by W. R. Shea), Reidel, Dordrecht.

E. Weinzierl, 1969, *Universität und Politik in Österreich*, Salzburger Universitätsreden 33, Pustet, Salzburg-München.

E. Weinzierl-Fischer, 1963, 'Österreichs Katholiken und der Nationalsozialismus', *Wort und Wahrheit* (1963).

J. E. Wheeler, 1969, 'Moritz Schlick', *Educational Theory* **19** (1969), 372-395.

W. Whewell, 1873, *History of the Inductive Sciences*, London.

G. Wielinger, 1977, 'Rechtstheorie in Österreich - Hans Kelsen und die Wiener rechtstheoretische Schule', in Marek et al (eds.) 1977, pp. 21-43.

A. R. Wieser, 1950, *Die Geschichte des Faches Philosophie an der Universität Wien 1848-1940*, Phil. Diss., Universität Wien.

E. & M. Winter, 1970, *Der Bolzanokreis 1824-1933 in Briefen von Anna Hoffmann, Michael Josef Fasl, Franz Schneider und Franz Prihonsky*, Österreichische Akademie der Wissenschaften, Philosophisch-Historische Klasse, Sitzungsberichte 266, Bd. 1, Abhandlung, Böhlau, Wien-Köln.

L. Wittgenstein, 1921, 'Logisch-Philosophische Abhandlung', *Annalen der Naturphilosophie* **14** (1921)185-262, transl. C. Ogden *Tractatus Logico-Philosophicus*, London, 1922, rev. ed. 1933, repr. 1983, Routledge & Kegan Paul, London.

L. Wittgenstein, 1953, *Philosophische Untersuchungen/Philosophical Investigations*, Blackwell, Oxford.

L. Wittgenstein, 1956, *Bemerkungen über die Grundlagen der Mathematik/Remarks on the Foundations of Mathematics* (ed. by G. H. v. Wright, R. Rhees, G. E. M. Anscombe), transl. G. E. M. Anscombe, Blackwell, Oxford, rev. ed. 1978.

L. Wittgenstein, 1965, 'A Lecture on Ethics', *The Philosophical Review* **74** (1965) 3-12.

L. Wittgenstein, 1969a, *Briefe an Ludwig von Ficker* (ed. by G. H. v. Wright, W. Methlagl), *Brenner Studien* Bd. 1, O. Müller, Salzburg, transl. B. Gillette 'Letters to Ludwig von Ficker' (ed. by A. Janik) in Luckhardt (ed.)1979, pp. 82-98.

L. Wittgenstein, 1969b, *On Certainty/Über Gewißheit* (ed. by G.E. M. Anscombe, G.H. v. Wright) transl. D. Paul, G. E. M. Anscombe, Harper, New York.

L. Wittgenstein, 1977, *Vermischte Bemerkungen*, Suhrkamp, Frankfurt a. M., transl. P. Winch *Culture and Value*, Blackwell, Oxford, 1980, University of Chicago Press, Chicago, 1984.

L. Wittgenstein, 1982, *Last Writings vol. 1 (Preliminary Studies for Part II of 'Philosophical Investigations')* (ed. by G. H. v. Wright, H. Nyman), transl. G. C. Luckhardt, M. A. E. Aue, University of Chicago Press, Chicago.

H. Zehentner, 1972, 'Gibt es eine wissenschaftliche Philosophie in Österreich?', *Zukunft* (January 1972) 49-51.

E. Zermelo, 1932, 'Einleitung', in G. Cantor, *Gesammelte Abhandlungen mathematischen und philosophischen Inhalts* (ed. by E. Zermelo), Springer, Berlin, repr. Olms, Hildesheim, 1962.

E. Zierngiebel, 1867, *Friedrich Heinrich Jacobi's Leben, Dichtung und Denken. Ein Beitrag zur Geschichte der deutschen Literatur und Philosophie*, Wien.

E. Zilsel, 1921, 'Der einführende Philosophieunterricht an den neuen Oberschulen', *Volkserziehung* (1921) 324-341.

E. Zilsel, 1924, 'Kant als Erzieher', *Schulreform* 3 (1924).

E. Zilsel, 1930, 'Soziologische Bemerkungen zur Philosophie der Gegenwart', *Der Kampf* 23 (1930) 410-424.

E. Zilsel, 1932, 'Bemerkungen zur Wissenschaftslogik', *Erkenntnis* 3 (1932/3) 143-161.

E. Zilsel, 1937, 'Moritz Schlick', *Die Naturwissenschaften* 25 no. 11 (1937) 161-167.

R. Zimmermann, 1855, *Bericht über ein unbekanntes rechtsphilosophisches Manuskript eines österreichischen Verfassers*, Prag.

H. Zoitl, 1976, *Kampf um Gleichberechtigung, Die sozialdemokratische Studentenbewegung in Wien 1914-1925*, Phil. Diss., Salzburg.

D. Zolo, 1986, *Scienza e Politica in Otto Neurath*, Feltrinelli, Milano, rev. ed. transl. D. McKie *Reflexive Epistemology. The Philosophical Legacy of Otto Neurath*, Boston Studies in the Philosophy of Science vol. 118, Kluwer, Dordrecht, 1989.

D. Zolo, 1990, 'Reflexive Epistemology and Social Complexity: The Philosophical Legacy of Otto Neurath', *Philosophy of the Social Sciences* 20 (1990) no. 2.

NAME INDEX

BOSTON STUDIES IN THE PHILOSOPHY OF SCIENCE

Editors:

ROBERT S. COHEN
(Boston University)

1. Marx W. Wartofsky (ed.), *Proceedings of the Boston Colloquium for the Philosophy of Science 1961–1962*. 1963.
2. Robert S. Cohen and Marx W. Wartofsky (eds.), *In Honor of Philipp Frank*. 1965.
3. Robert S. Cohen and Marx W. Wartofsky (eds.), *Proceedings of the Boston Colloquium for the Philosophy of Science 1964–1966. In Memory of Norwood Russell Hanson*. 1967.
4. Robert S. Cohen and Marx W. Wartofsky (eds.), *Proceedings of the Boston Colloquium for the Philosophy of Science 1966–1968*. 1969.
5. Robert S. Cohen and Marx W. Wartofsky (eds.), *Proceedings of the Boston Colloquium for the Philosophy of Science 1966–1968*. 1969.
6. Robert S. Cohen and Raymond J. Seeger (eds.), *Ernst Mach: Physicist and Philosopher*. 1970.
7. Milic Čapek, *Bergson and Modern Physics*. 1971.
8. Roger C. Buck and Robert S. Cohen (eds.), *PSA 1970. In Memory of Rudolf Carnap*. 1971.
9. A. A. Zinov'ev, *Foundations of the Logical Theory of Scientific Knowledge (Complex Logic)*. (Revised and enlarged English edition with an appendix by G. A. Smirnov, E. A. Sidorenka, A. M. Fedina, and L. A. Bobrova). 1973.
10. Ladislav Tondl, *Scientific Procedures*. 1973.
11. R. J. Seeger and Robert S. Cohen (eds.), *Philosophical Foundations of Science*. 1974.
12. Adolf Grünbaum, *Philosophical Problems of Space and Time*. (Second, enlarged edition). 1973.
13. Robert S. Cohen and Marx W. Wartofsky (eds.), *Logical and Epistemological Studies in Contemporary Physics*. 1973.
14. Robert S. Cohen and Marx W. Wartofsky (eds.), *Methodological and Historical Essays in the Natural and Social Sciences. Proceedings of the Boston Colloquium for the Philosophy of Science 1969–1972*. 1974.
15. Robert S. Cohen, J. J. Stachel, and Marx W. Wartofsky (eds.), *For Dirk Struik. Scientific, Historical and Political Essays in Honor of Dirk Struik*. 1974.
16. Norman Geschwind, *Selected Papers on Language and the Brain*. 1974
17. B. G. Kuznetsov, *Reason and Being: Studies in Classical Rationalism and Non-Classical Science*. 1987
18. Peter Mittelstaedt, *Philosophical Problems of Modern Physics*. 1976
19. Henry Mehlberg, *Time, Causality, and the Quantum Theory* (2 vols.). 1980.

20. Kenneth F. Schaffner and Robert S. Cohen (eds.), *Proceedings of the 1972 Biennial Meeting, Philosophy of Science Association.* 1974
21. R. S. Cohen and J. J. Stachel (eds.), *Selected Papers of Léon Rosenfeld.* 1978.
22. Milic Čapek (ed.), *The Concepts of Space and Time. Their Structure and Their Development.* 1976.
23. Marjorie Grene, *The Understanding of Nature, Essays in the Philosophy of Biology.* 1974.
24. Don Ihde, *Technics and Praxis. A Philosophy of Technology.* 1978.
25. Jaakko Hintikka and Unto Remes, *The Method of Analysis. Its Geometrical Origin and Its General Significance.* 1974.
26. John Emery Murdoch and Edith Dudley Sylla, *The Cultural Context of Medieval Learning.* 1975.
27. Marjorie Grene and Everett Mendelsohn (eds.), *Topics in the Philosophy of Biology.* 1976.
28. Joseph Agassi, *Science in Flux.* 1975.
29. Jerzy J. Wiatr (ed.), *Polish Essays in the Methodology of the Social Sciences.* 1979.
30. Peter Janich, *Protophysics of Time.* 1985.
31. Robert S. Cohen and Marx W. Wartofsky (eds.), *Language, Logic and Method.* 1983.
32. R. S. Cohen, C. A. Hooker, A. C. Michalos, and J. W. van Evra (eds.), *PSA 1974: Proceedings of the 1974 Biennial Meeting of the Philosophy of Science Association.* 1976.
33. Gerald Holton and William Blanpied (eds.), *Science and Its Public: The Changing Relationship.* 1976.
34. Mirko D. Grmek (ed.), *On Scientific Discovery.* 1980.
35. Stefan Amsterdamski, *Between Experience and Metaphysics. Philosophical Problems of the Evolution of Science.* 1975.
36. Mihailo Marković and Gajo Petrović (eds.), *Praxis, Yugoslav Essays in the Philosophy and Methodology of the Social Sciences.* 1979.
37. Hermann von Helmholtz, *Epistemological Writings. The Paul Hertz/Moritz Schlick Centenary Edition of 1921 with Notes and Commentary by the Editors.* (Newly translated by Malcolm F. Lowe. Edited, with an Introduction and Bibliography, by Robert S. Cohen and Yehuda Elkana). 1977.
38. R. M. Martin, *Pragmatics, Truth, and Language.* 1979.
39. R. S. Cohen, P. K. Feyerabend, and M. W. Wartofsky (eds.), *Essays in Memory of Imre Lakatos.* 1976.
40. B. M. Kedrov and V. Sadovsky. *Current Soviet Studies in the Philosophy of Science.* Forthcoming.
41. M. Raphael, *Theorie des Geistigen Schaffens auf Marxistischer Grundlage.* Forthcoming.
42. Humberto R. Maturana and Francisco J. Varela, *Autopoiesis and Cognition. The Realization of the Living.* 1980.
43. A. Kasher (ed.), *Language in Focus: Foundations, Methods and Systems. Essays Dedicated to Yehoshua Bar-Hillel.* 1976.
44. Trân Duc Thao, *Investigations into the Origin of Language and Consciousness.* (Translated by Daniel J. Herman and Robert L. Armstrong; edited by Carolyn

R. Fawcett and Robert S. Cohen). 1984.

45. A. Ishimoto (ed.), *Japanese Studies in the History and Philosophy of Science.*
46. Peter L. Kapitza, *Experiment, Theory, Practice.* 1980.
47. Maria L. Dalla Chiara (ed.), *Italian Studies in the Philosophy of Science.* 1980.
48. Marx W. Wartofsky, *Models: Representation and the Scientific Understanding.* 1979.
49. Trân Duc Thao, *Phenomenology and Dialectical Materialism.* 1985.
50. Yehuda Fried and Joseph Agassi, *Paranoia: A Study in Diagnosis.* 1976.
51. Kurt H. Wolff, *Surrender and Catch: Experience and Inquiry Today.* 1976.
52. Karel Kosík, *Dialectics of the Concrete.* 1976.
53. Nelson Goodman, *The Structure of Appearance.* (Third edition). 1977.
54. Herbert A. Simon, *Models of Discovery and Other Topics in the Methods of Science.* 1977.
55. Morris Lazerowitz, *The Language of Philosophy. Freud and Wittgenstein.* 1977.
56. Thomas Nickles (ed.), *Scientific Discovery, Logic, and Rationality.* 1980.
57. Joseph Margolis, *Persons and Minds. The Prospects of Nonreductive Materialism.* 1977.
58. G. Radnitzky and G. Andersson (eds.), *Progress and Rationality in Science,* 1978.
59. Gerard Radnitzky and Gunnar Andersson (eds.), *The Structure and Development of Science.* 1979.
60. Thomas Nickles (ed.), *Scientific Discovery: Case Studies.* 1980.
61. Maurice A. Finocchiaro, *Galileo and the Art of Reasoning.* 1980.
62. William A. Wallace, *Prelude to Galileo.* 1981.
63. Friedrich Rapp, *Analytical Philosophy of Technology.* 1981.
64. Robert S. Cohen and Marx W. Wartofsky (eds.), *Hegel and the Sciences.* 1984.
65. Joseph Agassi, *Science and Society.* 1981.
66. Ladislav Tondl, *Problems of Semantics.* 1981.
67. Joseph Agassi and Robert S. Cohen (eds.), *Scientific Philosophy Today.* 1982.
68. Władysław Krajewski (ed.), *Polish Essays in the Philosophy of the Natural Sciences.* 1982.
69. James H. Fetzer, *Scientific Knowledge.* 1981.
70. Stephen Grossberg, *Studies of Mind and Brain.* 1982.
71. Robert S. Cohen and Marx W. Wartofsky (eds.), *Epistemology, Methodology, and the Social Sciences.* 1983.
72. Karel Berka, *Measurement.* 1983.
73. G. L. Pandit, *The Structure and Growth of Scientific Knowledge.* 1983.
74. A. A. Zinov'ev, *Logical Physics.* 1983.
75. Gilles-Gaston Granger, *Formal Thought and the Sciences of Man.* 1983.
76. R. S. Cohen and L. Laudan (eds.), *Physics, Philosophy and Psychoanalysis.* 1983.
77. G. Böhme et al., *Finalization in Science,* ed. by W. Schäfer. 1983.
78. D. Shapere, *Reason and the Search for Knowledge.* 1983.
79. G. Andersson, *Rationality in Science and Politics.* 1984.
80. P. T. Durbin and F. Rapp, *Philosophy and Technology.* 1984.
81. M. Marković, *Dialectical Theory of Meaning.* 1984.

82. R. S. Cohen and M. W. Wartofsky, *Physical Sciences and History of Physics.* 1984.
83. E. Meyerson, *The Relativistic Deduction.* 1985.
84. R. S. Cohen and M. W. Wartofsky, *Methodology, Metaphysics and the History of Sciences.* 1984.
85. György Tamás, *The Logic of Categories.* 1985.
86. Sergio L. de C. Fernandes, *Foundations of Objective Knowledge.* 1985.
87. Robert S. Cohen and Thomas Schnelle (eds.), *Cognition and Fact.* 1985.
88. Gideon Freudenthal, *Atom and Individual in the Age of Newton.* 1985.
89. A. Donagan, A. N. Perovich, Jr., and M. V. Wedin (eds.), *Human Nature and Natural Knowledge.* 1985.
90. C. Mitcham and A. Huning (eds.), *Philosophy and Technology II.* 1986.
91. M. Grene and D. Nails (eds.), *Spinoza and the Sciences.* 1986.
92. S. P. Turner, *The Search for a Methodology of Social Science.* 1986.
93. I. C. Jarvie, *Thinking about Society: Theory and Practice.* 1986.
94. Edna Ullmann-Margalit (ed.), *The Kaleidoscope of Science.* 1986.
95. Edna Ullmann-Margalit (ed.), *The Prism of Science.* 1986.
96. G. Markus, *Language and Production.* 1986.
97. F. Amrine, F. J. Zucker, and H. Wheeler (eds.), *Goethe and the Sciences: A Reappraisal.* 1987.
98. Joseph C. Pitt and Marcella Pera (eds.), *Rational Changes in Science.* 1987.
99. O. Costa de Beauregard, *Time, the Physical Magnitude.* 1987.
100. Abner Shimony and Debra Nails (eds.), *Naturalistic Epistemology: A Symposium of Two Decades.* 1987.
101. Nathan Rotenstreich, *Time and Meaning in History.* 1987.
102. David B. Zilberman (ed.), *The Birth of Meaning in Hindu Thought.* 1987.
103. Thomas F. Glick (ed.), *The Comparative Reception of Relativity.* 1987.
104. Zellig Harris *et al.*, *The Form of Information in Science.* 1987
105. Frederick Burwick, *Approaches to Organic Form: Permutations in Science and Culture.* 1987.
106. M. Almási, *Philosophy of Appearances.* Forthcoming.
107. S. Hook, W. L. O'Neill, and R. O'Toole, *Philosophy, History and Social Action. Essays in Honor of Lewis Feuer.* 1988.
108. I. Hronszky, M. Fehér, and B. Dajka (eds.), *Scientific Knowledge Socialized. Selected Proceedings of the Fifth Joint International Conference on History and Philosophy of Science Organized by the IUHPS, Veszprém, 1984.* Forthcoming.
109. P. Tillers and E. D. Green (eds.), *Probability and Inference in the Law of Evidence. The Uses and Limits of Bayesianism.* 1988.
110. E. Ullmann-Margalit (ed.), *Science in Reflection. The Israel Colloquium: Studies in History, Philosophy, and Sociology of Science.* 1988.
111. K. Gavroglu, Y. Goudaroulis, and P. Nicolacopoulos (eds.), *Imre Lakatos and Theories of Scientific Change.* 1989.
112. Barry Glassner and Jonathan D. Moreno (eds.), *The Qualitative-Quantitative Distinction in the Social Sciences.* 1989.
113. K. Arens, *Structures of Knowing: Psychologies of the Nineteenth Century.* 1989.

114. A. Janik, *Style, Politics and the Future of Philosophy*. 1989.
115. F. Amrine (ed.), *Literature and Science as Modes of Expression*. 1989.
116. James Robert Brown and Jürgen Mittelstrass (eds.), *An Intimate Relation: Studies in the History and Philosophy of Science Presented to Robert E. Butts on his 60th Birthday*. 1989.
117. F. D'Agostino and I. C. Jarvie (eds.), *Freedom and Rationality: Essays in Honor of John Watkins*. 1989.
118. D. Zolo, *Reflective Epistemology: The Philosophical Legacy of Otto Neurath*. 1989.
119. Michael Kearn, Bernard S. Phillips and Robert S. Cohen (eds.), *George Simmel and Contemporary Sociology*. 1989.
120. Trevor H. Levere and William R. Shea (eds.), *Nature, Experiment, and the Sciences: Essays on Galileo and the History of Science in Honour of Stillman Drake*. 1989.
121. P. Nicolacopoulos (ed.), *Greek Studies in the Philosophy and History of Science*. 1990.
122. R. Cooke and D. Costantini (eds.), *Statistics in Science. The Foundations of Statistical Methods in Biology, Physics and Economics*. 1990
123. P. Duhem (ed.), *The Origins of Statics*. 1991.
124. K. Gavroglu and Y. Goudaroulis (eds.), Heike Kamerlingh Onnes, *Through Measurement to Knowledge. The Selected Papers of Heike Kamerlingh Onnes 1853–1926*. 1991.
125. M. Čapek, *The New Aspects of Time. Its Continuity and Novelties. Selected Papers in the Philosophy of Science*. 1991.
126. S. Unguru, *Physics, Cosmology and Astronomy, 1300–1700: Tension and Accommodation*. 1991.
127. Z. Bechler, *Newton's Physics of the Conceptual Structure of the Scientific Revolution*. 1991
128. E. Meyerson, *Explanation in the Sciences*. 1991
129. A.I. Tauber (ed.), *Organism and the Origins of Self*. 1991
130. F.J. Varela and J.-P. Dupuy (eds.), *Understanding Origins. Contemporary Views on the Origin of Life, Mind and Society*. 1991
131. G.L. Pandit, *Methodology Variance. Essays in Epistemological Ontology and the Methodology of Science*. 1991
132. G. Munévar (ed.), *Beyond Reason. Essays on the Philosophy of Paul Feyerabend*. 1991
133. T.E. Uebel (ed.), *Rediscovering the Forgotten Vienna Circle. Austrian Studies on Otto Neurath and the Vienna Circle*. 1991

MIX
Papier aus verantwortungsvollen Quellen
Paper from responsible sources
FSC® C105338

If you have any concerns about our products,
you can contact us on
ProductSafety@springernature.com

In case Publisher is established outside the EU,
the EU authorized representative is:
Springer Nature Customer Service Center GmbH
Europaplatz 3, 69115 Heidelberg, Germany

Printed by Libri Plureos GmbH
in Hamburg, Germany